VOLUME ONE HUNDRED AND THIRTY FOUR

Advances in
CANCER RESEARCH
Applications of Mass Spectrometry Imaging to Cancer

VOLUME ONE HUNDRED AND THIRTY FOUR

ADVANCES IN
CANCER RESEARCH

Applications of Mass Spectrometry Imaging to Cancer

Edited by

RICHARD R. DRAKE
Cell and Molecular Pharmacology and
Experimental Therapeutics,
Medical University of South Carolina,
Charleston, SC, United States

LIAM A. McDONNELL
Leiden University Medical Center,
Leiden, Netherlands
Fondazione Pisana per la Scienza ONLUS,
Pisa, Italy

An imprint of Elsevier
elsevier.com

Academic Press is an imprint of Elsevier
50 Hampshire Street, 5th Floor, Cambridge, MA 02139, United States
525 B Street, Suite 1800, San Diego, CA 92101-4495, United States
The Boulevard, Langford Lane, Kidlington, Oxford OX5 1GB, United Kingdom
125 London Wall, London, EC2Y 5AS, United Kingdom

First edition 2017

Copyright © 2017 Elsevier Inc. All rights reserved.

No part of this publication may be reproduced or transmitted in any form or by any means, electronic or mechanical, including photocopying, recording, or any information storage and retrieval system, without permission in writing from the publisher. Details on how to seek permission, further information about the Publisher's permissions policies and our arrangements with organizations such as the Copyright Clearance Center and the Copyright Licensing Agency, can be found at our website: www.elsevier.com/permissions.

This book and the individual contributions contained in it are protected under copyright by the Publisher (other than as may be noted herein).

Notices
Knowledge and best practice in this field are constantly changing. As new research and experience broaden our understanding, changes in research methods, professional practices, or medical treatment may become necessary.

Practitioners and researchers must always rely on their own experience and knowledge in evaluating and using any information, methods, compounds, or experiments described herein. In using such information or methods they should be mindful of their own safety and the safety of others, including parties for whom they have a professional responsibility.

To the fullest extent of the law, neither the Publisher nor the authors, contributors, or editors, assume any liability for any injury and/or damage to persons or property as a matter of products liability, negligence or otherwise, or from any use or operation of any methods, products, instructions, or ideas contained in the material herein.

ISBN: 978-0-12-805249-5
ISSN: 0065-230X

For information on all Academic Press publications visit our website at https://www.elsevier.com/

Publisher: Zoe Kruze
Acquisition Editor: Zoe Kruze
Editorial Project Manager: Fenton Coulthurst
Production Project Manager: Surya Narayanan Jayachandran
Senior Cover Designer: Greg Harris

Typeset by SPi Global, India

CONTENTS

Contributors ix
Preface xiii

1. The Importance of Histology and Pathology in Mass Spectrometry Imaging   1
K. Schwamborn

  1. Importance of Pathology   2
  2. Possible Errors Caused by Tissue Inherent Factors—Why Histology Is Important for Supervised Analysis   4
  3. Possible Errors Caused by Sample Inherent Factors—Small Pretherapeutic Biopsies   9
  4. Possible Errors Caused by Sample Preparations—Artifacts Are Not Your Friend   12
  5. Possible Errors Caused by Ill-Defined Sample Groups in the Training Set—Keep It Black and White/Shades of Gray Are Not Welcome   16
  6. Conclusion   19
  References   20

2. Applications of Mass Spectrometry Imaging to Cancer   27
G. Arentz, P. Mittal, C. Zhang, Y.-Y. Ho, M. Briggs, L. Winderbaum, M.K. Hoffmann, and P. Hoffmann

  1. Introduction   28
  2. Protein MSI in Cancer Research   33
  3. Lipid MSI in Cancer Research   45
  4. Glycan MSI in Cancer Research   47
  5. Drug Imaging in Cancer Research   50
  6. Data Analysis   53
  7. Concluding Remarks   57
  References   57

3. Assessing the Potential of Metal-Assisted Imaging Mass Spectrometry in Cancer Research   67
M. Dufresne, N.H. Patterson, N. Lauzon, and P. Chaurand

  1. Introduction   68
  2. Material   69

3. Silver-Assisted IMS	70
4. Gold-Assisted IMS	75
5. Applications to Cancer Research	78
6. Concluding Remarks	82
References	83

4. MALDI Mass Spectrometry Imaging of N-Linked Glycans in Cancer Tissues — 85

R.R. Drake, T.W. Powers, E.E. Jones, E. Bruner, A.S. Mehta, and P.M. Angel

1. Introduction	86
2. Glycosylation and Cancer	87
3. Methodology for N-Linked Glycan Detection by MALDI Imaging	92
4. N-Glycan Distribution Linked With Histopathology	101
5. Emerging Applications	106
6. Summary	111
References	112

5. In Situ Metabolomics in Cancer by Mass Spectrometry Imaging — 117

A. Buck, M. Aichler, K. Huber, and A. Walch

1. Metabolomics in Cancer	117
2. In Situ Metabolomics by MALDI Imaging	119
3. Fresh-Frozen- vs Formalin-Fixed Paraffin-Embedded Tissue Samples	120
4. Tissue-Based Disease Classification—Diagnostic Markers and Metabolic Signatures	124
5. Therapy Response Prediction and Prognosis	125
6. Intra- and Intertumoral Heterogeneity	126
7. Conclusion	127
References	128

6. Mass Spectrometry Imaging in Oncology Drug Discovery — 133

R.J.A. Goodwin, J. Bunch, and D.F. McGinnity

1. Introduction	134
2. How MSI Can Inform Our Understanding of Pharmacokinetic–Pharmacodynamic Relationships	135
3. Biodistribution	137
4. Tumor Metabolism: MSI Analysis for More Than Just Drug Distribution	138
5. Sample Preparation	139
6. Quantitation	141
7. Toxicity and Safety Assessment	142

8. Biomarkers for Efficacy	144
9. Drug Delivery	145
10. Tumor Microenvironment	147
11. Assessing Hypoxia	150
12. BBB Penetration	150
13. Beyond Small Molecules	153
14. Clinical Translation	156
15. Emerging Applications: Spheroids	157
16. Increased Spatial Resolution	159
17. Metrology for MS Imaging	160
18. Conclusion	162
References	163

7. MALDI IMS and Cancer Tissue Microarrays   173
R. Casadonte, R. Longuespée, J. Kriegsmann, and M. Kriegsmann

1. Introduction	174
2. TMA Technology	175
3. MALDI IMS Analysis of TMAs	178
4. Identification of Peptides	186
5. Application of MALDI IMS on FFPE TMAs	188
6. Perspectives and Concluding Remarks	192
Acknowledgments	193
References	193

8. Mass Spectrometry Imaging for the Investigation of Intratumor Heterogeneity   201
B. Balluff, M. Hanselmann, and R.M.A. Heeren

1. Tumor Heterogeneity	202
2. MSI to Study Tumor Heterogeneity	207
3. Multivariate Data Analysis Strategies in MSI	209
4. MSI Applications for the Investigation of ITH	213
5. Future Applications of MSI in ITH Research	223
6. Perspective	224
References	224

9. Ambient Mass Spectrometry in Cancer Research   231
Z. Takats, N. Strittmatter, and J.S. McKenzie

1. Desorption Electrospray Ionization	232
2. Intraoperative Mass Spectrometry	240

3.	REIMS Instrumentation	244
4.	DESI-MSI for Drug Imaging in Cancer Research	252
	References	253

10. Rapid Mass Spectrometry Imaging to Assess the Biochemical Profile of Pituitary Tissue for Potential Intraoperative Usage   257

K.T. Huang, S. Ludy, D. Calligaris, I.F. Dunn, E. Laws, S. Santagata, and N.Y.R. Agar

1.	Introduction	257
2.	Current Imaging and Visualization Techniques	259
3.	Mass Spectrometry in Clinical Usage	261
4.	Future Directions	275
	References	276

11. Mass Spectrometry Imaging in Cancer Research: Future Perspectives   283

L.A. McDonnell, P.M. Angel, S. Lou, and R.R. Drake

1.	MSI-Based Diagnostics	286
2.	Biological Insights	287
3.	Multimodal MSI	287
4.	Targeted MSI	288
5.	Summary	288
	References	289

Index   *291*

CONTRIBUTORS

N.Y.R. Agar
Brigham and Women's Hospital; Dana-Farber Cancer Institute, Harvard Medical School, Boston, MA, United States

M. Aichler
Research Unit Analytical Pathology, Helmholtz Zentrum München, Neuherberg, Germany

P.M. Angel
Medical University of South Carolina, Charleston, SC, United States

G. Arentz
Adelaide Proteomics Centre, School of Biological Sciences; Institute for Photonics and Advanced Sensing (IPAS), University of Adelaide, Adelaide, SA, Australia

B. Balluff
Maastricht University, Maastricht MultiModal Molecular Imaging institute (M4I), Maastricht, The Netherlands

M. Briggs
Adelaide Proteomics Centre, School of Biological Sciences; Institute for Photonics and Advanced Sensing (IPAS); ARC Centre for Nanoscale BioPhotonics (CNBP), University of Adelaide, Adelaide, SA, Australia

E. Bruner
Medical University of South Carolina, Charleston, SC, United States

A. Buck
Research Unit Analytical Pathology, Helmholtz Zentrum München, Neuherberg, Germany

J. Bunch
National Physical Laboratory, Teddington; School of Pharmacy, University of Nottingham, Nottingham, United Kingdom

D. Calligaris
Brigham and Women's Hospital, Harvard Medical School, Boston, MA, United States

R. Casadonte
Proteopath GmbH, Trier, Germany

P. Chaurand
Université de Montréal, Montreal, QC, Canada

R.R. Drake
Medical University of South Carolina, Charleston, SC, United States

M. Dufresne
Université de Montréal, Montreal, QC, Canada

I.F. Dunn
Brigham and Women's Hospital, Harvard Medical School, Boston, MA, United States

R.J.A. Goodwin
Mass Spectrometry Imaging Group, Pathology Sciences, Drug Safety & Metabolism, Innovative Medicines and Early Development, AstraZeneca, Cambridge, United Kingdom

M. Hanselmann
Heidelberg Collaboratory for Image Processing (HCI), Interdisciplinary Center for Scientific Computing (IWR), University of Heidelberg, Heidelberg, Germany

R.M.A. Heeren
Maastricht University, Maastricht MultiModal Molecular Imaging institute (M4I), Maastricht, The Netherlands

Y.-Y. Ho
Adelaide Proteomics Centre, School of Biological Sciences, University of Adelaide, Adelaide, SA, Australia

M.K. Hoffmann
Adelaide Proteomics Centre, School of Biological Sciences; Institute for Photonics and Advanced Sensing (IPAS), University of Adelaide, Adelaide, SA, Australia

P. Hoffmann
Adelaide Proteomics Centre, School of Biological Sciences; Institute for Photonics and Advanced Sensing (IPAS), University of Adelaide, Adelaide, SA, Australia

K.T. Huang
Brigham and Women's Hospital, Harvard Medical School, Boston, MA, United States

K. Huber
Research Unit Analytical Pathology, Helmholtz Zentrum München, Neuherberg, Germany

E.E. Jones
Medical University of South Carolina, Charleston, SC, United States

J. Kriegsmann
Proteopath GmbH; Institute of Molecular Pathology; Center for Histology, Cytology and Molecular Diagnostics, Trier, Germany

M. Kriegsmann
Institute of Pathology, University of Heidelberg, Heidelberg, Germany

N. Lauzon
Université de Montréal, Montreal, QC, Canada

E. Laws
Brigham and Women's Hospital, Harvard Medical School, Boston, MA, United States

R. Longuespée
Proteopath GmbH, Trier, Germany

S. Lou
Leiden University Medical Center, Leiden, The Netherlands

S. Ludy
Brigham and Women's Hospital, Harvard Medical School, Boston, MA, United States

L.A. McDonnell
Fondazione Pisana per la Scienza ONLUS, Pisa, Italy; Leiden University Medical Center, Leiden, The Netherlands

D.F. McGinnity
DMPK, Oncology IMED, Innovative Medicines and Early Development, AstraZeneca, Cambridge, United Kingdom

J.S. McKenzie
Imperial College London, London, United Kingdom

A.S. Mehta
Medical University of South Carolina, Charleston, SC, United States

P. Mittal
Adelaide Proteomics Centre, School of Biological Sciences; Institute for Photonics and Advanced Sensing (IPAS), University of Adelaide, Adelaide, SA, Australia

N.H. Patterson
Mass Spectrometry Research Center, Vanderbilt University School of Medicine, Nashville, TN, United States

T.W. Powers
Medical University of South Carolina, Charleston, SC, United States

S. Santagata
Brigham and Women's Hospital; Dana-Farber Cancer Institute, Harvard Medical School, Boston, MA, United States

K. Schwamborn
Institute of Pathology, Technische Universität München (TUM), Munich, Germany

N. Strittmatter
Drug Safety and Metabolism, AstraZeneca, Cambridge, United Kingdom

Z. Takats
Imperial College London, London, United Kingdom

A. Walch
Research Unit Analytical Pathology, Helmholtz Zentrum München, Neuherberg, Germany

L. Winderbaum
Adelaide Proteomics Centre, School of Biological Sciences, University of Adelaide, Adelaide, SA, Australia

C. Zhang
Adelaide Proteomics Centre, School of Biological Sciences; Institute for Photonics and Advanced Sensing (IPAS), University of Adelaide, Adelaide, SA, Australia

PREFACE

It was the pioneering work performed in the laboratory of Richard Caprioli that led to the current widespread application of imaging mass spectrometry to cancer research (Caprioli, Farmer, & Gile, 1997). Using spatially resolved matrix-assisted laser desorption/ionization (MALDI imaging mass spectrometry) they demonstrated how the distributions of hundreds of proteins in a tissue section could be simultaneously recorded, without prior knowledge and without labeling. Shortly after this first technical report it was demonstrated how the technique could reveal differential protein expression profiles throughout tumor samples, including proteins localized in the tumor's infiltrating zone (Stoeckli, Chaurand, Hallahan, & Caprioli, 2001).

Imaging mass spectrometry has grown rapidly, in terms of application and technology. New technologies now enable high speed, high spatial resolution, and high mass resolution analysis; new tissue preparation methods enable the label-free analysis of diverse molecular classes, including hormones, peptides, proteins, protein isoforms (Chapter 2), lipids (Chapter 3), glycans (Chapter 4), metabolites (Chapter 5), and drugs (as well as the drug's metabolites) (Chapter 6); new data registration methods enable the seamless integration of imaging mass spectrometry with histopathological analysis and in vivo imaging, in both two and three dimensions; and finally new data analysis routines have enabled all of these tools to be used to investigate the molecular changes that accompany cancer development. While earlier applications of imaging mass spectrometry focused on the identification of diagnostic biomarkers, the technique has developed rapidly and is now used to identify biomarkers of survival, metastatic status, hormone receptor status, microvascular invasion, and patient response to therapy; the technique is also used to investigate intratumor heterogeneity, from tumor cell density, tumor cell type, and even subpopulation phenotype perspectives (Chapter 8), and intraoperatively to rapidly assess tumor margins during surgery (Chapters 9 and 10).

In part the developments in imaging mass spectrometry reflect the demands of the application areas, as well as the active involvement of scientists from different disciplines. Pharmaceutical analysis (Chapter 6) is now established as one of the principal application areas of imaging mass spectrometry owing to its ability to help reduce the cost of developing and testing new drugs and drug formulations.

The active involvement of pathologists in the application of imaging mass spectrometry is of paramount importance (Chapters 1 and 7); it is the combination of histology and imaging mass spectrometry, enabling mass spectral profiles of specific histopathological features to be obtained, that is, the unique strength of the technique for cancer research. In this issue a series of reviews cover how the technique can be applied in cancer research, both from a diagnostic and biological perspective. The wide array of biomolecular targets central to cancer development and progression that can be detected in cancer tissues and emerging applications of imaging mass spectrometry technology are also evaluated.

RICHARD R. DRAKE
LIAM A. MCDONNELL

REFERENCES

Caprioli, R. M., Farmer, T. B., & Gile, J. (1997). Molecular imaging of biological samples: Localization of peptides and proteins using MALDI-TOF MS. *Analytical Chemistry*, *69*, 4751–4760.

Stoeckli, M., Chaurand, P., Hallahan, D. E., & Caprioli, R. M. (2001). Imaging mass spectrometry: A new technology for the analysis of protein expression in mammalian tissues. *Nature Medicine*, *7*, 493–496.

CHAPTER ONE

The Importance of Histology and Pathology in Mass Spectrometry Imaging

K. Schwamborn[1]

Institute of Pathology, Technische Universität München (TUM), Munich, Germany
[1]Corresponding author: e-mail address: kschwamborn@tum.de

Contents

1. Importance of Pathology . . . 2
2. Possible Errors Caused by Tissue Inherent Factors—Why Histology Is Important for Supervised Analysis . . . 4
3. Possible Errors Caused by Sample Inherent Factors—Small Pretherapeutic Biopsies . . . 9
4. Possible Errors Caused by Sample Preparations—Artifacts Are Not Your Friend . . . 12
5. Possible Errors Caused by Ill-Defined Sample Groups in the Training Set—Keep It Black and White/Shades of Gray Are Not Welcome . . . 16
6. Conclusion . . . 19
References . . . 20

Abstract

Mass spectrometry imaging (MSI) has become a valuable tool in cancer research. Even more, due to its capability to directly link molecular changes with histology, it holds the prospect to revolutionize tissue-based diagnostics. In order to learn to walk before running, however, information obtained through classical histology should not be neglected but rather used to its full capacity and integrated with mass spectrometry data to lead to a superior molecular histology synthesis. In order to achieve this, pathomorphological analyses have to be integrated into MSI analyses right from the beginning to avoid errors and pitfalls of MSI application possibly leading to incorrect or imprecise study outcomes. Such errors can be caused by different sample or tissue inherent factors or through factors in sample preparation. Future studies should, therefore, aim for a comprehensive incorporation of histology and pathology characteristics to ensure the generation of high-quality data in MSI to exploit its full capacity in tissue-based basic and translational research.

1. IMPORTANCE OF PATHOLOGY

The old saying that "a picture is worth a thousand words" is doubtlessly also true in science. Interpretation of histomorphological changes has substantially contributed to the understanding of physiological and pathological biological processes in different organs. Recently, however, the focus has shifted toward adding additional molecular components to such conventional imaging techniques. This so-called molecular histology is supposed to reveal the underlying biochemistry of tissues and organs, while simultaneously providing additional information on how therapeutics or toxins influence the function or misfunction of an organ (Vachet, 2015). However, correct interpretation of these molecular "pictures" can only be done in conjunction with simultaneous interpretation of the underlying conventional histomorphological "picture" by trained pathologists/morphologist. Having a profound knowledge and understanding of the biological basis of disease is the key for correct interpretation of tissue phenotypes. While, for example, biologist and other scientists can fulfil some of those needs, only a trained pathologist/morphologist is able to integrate the morphological, clinical, and molecular dimensions of a studied disease. A biologist can certainly learn specific morphology interpretation in selected (patho)physiological settings; however, for a full scale comprehensive interpretation usually a trained anatomist or pathologist is needed, who has trained on the interpretation of the whole spectrum of possible histomorphological tissue changes over several years/decades. Nevertheless, since only a few trained morphologists are dedicated to research and even less are working in the field of mass spectrometry imaging (MSI), the interface between basic research and diagnostic morphology is probably the weakest link in the whole interpretation chain in molecular imaging. It has become evident, however, that in order to advance the translation of biomarker discovery into diagnostic and therapeutic application, morphologists and especially pathologists have to play a major role in this endeavor (Salto-Tellez, James, & Hamilton, 2014).

Histology, the study of the microscopic structure of tissues, is an integral part of daily pathology routine. According to the Oxford English Dictionary, pathology is the science of the causes and effects of diseases (Stevenson, 2010); in particular, however, the term "pathology" also stands for a subspecialty of medicine that deals with the laboratory examination of samples of body tissues for diagnostic purposes (Stevenson, 2010). Pathology bridges science and medicine and underpins every aspect of patient care, from diagnostic testing, quality control, and treatment advice to the use of cutting-edge genetic technologies and the prevention of disease. The analysis of cytology, biopsy, or resection specimen often represents the final step in diagnostic processes and has remained the linchpin of cancer diagnosis for over 100 years (Lakhani & Ashworth, 2001). This knowledge provides a vital service; the pathology report is intended to guide the physician in finding the optimal treatment for a particular patient and builds the framework for future selection of new markers and new therapies for a given patient.

In the era of individualized medicine, standard clinical care is continuously becoming more complex, resulting in more complex and extensive diagnostic pathology workup. This becomes obvious by comparing guidelines for tissue preparation of lung cancer surgical specimens, where simple histochemistry stains such as hematoxylin and eosin and periodic acid–Schiff stain were sufficient in the 1960s. In contrast, today's pathology workup requires a plethora of additional information, including subtyping based on a number of immunohistochemical stains, semiquantitative assessment of growth pattern in the case of a pulmonary adenocarcinoma, and often additional molecular analyses. The increasing complexity in pathology workup is also reflected by a more than 60% increase in the number of tissue blocks per case and slides per tissue block over the last decade (e.g., immunohistochemistry and PCR) (Warth et al., 2016). Especially in the field of oncology, cell- and tissue-related molecular diagnostics results from techniques including immunohistochemistry, fluorescence in situ hybridization, and extraction-based molecular diagnostics such as PCR and sequencing today determine therapy decisions. And although extraction-based molecular diagnostics such as PCR and sequencing are molecular tests, they still strongly rely on an initial pathological examination of tissue samples to ensure sample adequacy, correct annotation of tumor areas for possible tumor cell enrichment, together with estimation of the tumor cell percentage. In particular, the percentage of tumor cells within a given area selected for extraction will ultimately determine the yield and possible "dilution" of tumor DNA by other cell types which is essential for the interpretation and accuracy of molecular test results. Consequently, inadequate sampling can result in a false-negative test result due to insufficient tumor DNA yield and ultimately in the wrong therapy (Thunnissen et al., 2012). Thus, correct a priori interpretation and annotation of tissue morphology by pathologists is key for the correct outcome of molecular testing.

Along the same line, pathological evaluation is also of utmost importance in MALDI imaging. This is obvious in all histology-directed approaches where only small predefined areas of a sample are analyzed (Cornett et al., 2006). However, histology coannotation is also of paramount importance in all other MSI approaches. Possible pitfalls caused by inexperience, explanations, and solutions will be given within the following paragraphs.

2. POSSIBLE ERRORS CAUSED BY TISSUE INHERENT FACTORS—WHY HISTOLOGY IS IMPORTANT FOR SUPERVISED ANALYSIS

With the advent and rise of molecular biology predictions on substituting histological assessment with molecular analyses were made. However, until now traditional microscopic morphological tissue analysis is still the gold standard and the conditio sine qua non for subsequent molecular testing. The rationale behind that fact is the need for high-quality microscopic tissue analysis prior to any other molecular test to establish adequacy of the sample (Szasz, Gyorffy, & Marko-Varga, 2016).

There is an equivalent need for high-quality histological examination in mass spectrometry imaging, in particular in studies with supervised data analysis. These types of studies typically aim to unravel differences between two or more different sample sets, e.g., normal and diseased tissue or two different subtypes of disease. Thus, in order to ensure, sample adequacy several factors have to be considered. First of all, it has to be established that the sample represents the right histological type for the intended analysis and is representative of the underlying disease. Second, the number of disease-defining cells within a given area has to be sufficient and above the detection/sensitivity limit of the analysis. Additionally, there should be no major interfering cell component (e.g., inflammatory cells or red blood cells) within the tissue area of interest and no strong morphological heterogeneity. To exclude samples with inadequate preanalytical conditions, features suggesting the sample has been compromised (e.g., autolysis) have to be recognized. These and many other aspects of histomorphological interpretation of tissue samples represent an essential first step to ensure the interpretability of the subsequent imaging mass spectrometry experiment. Only correct and precise annotation can lead to valid, robust, and meaningful data. In the following paragraphs, these aspects will be discussed in detail using the example of cancer.

When comparing different tumor types, it is obvious that the sample has to contain tumor cells. On the other hand, when comparing tumor and normal, one has to be careful not to choose an area of the tissue in close proximity to the tumor margin. Many tumors have a rim of reactive tissue containing different amounts of collagen matrix and inflammatory cells. Also, some tumors grow with a pushing and rather well-defined border, whereas others exhibit an infiltrative invasion front with small strands or nests of tumor cells or even single dispersed tumor cells (Friedl & Wolf, 2003). Thus, the area around the tumor margin neither represents the regular architecture of the tumor nor the normal tissue. On a molecular level, there are also differences between the center of a tumor and its margin or invasion front (Karamitopoulou et al., 2011). Additionally, several mass spectrometry studies have shown that even histologically normal appearing tissue might already show molecular signatures similar to the adjacent tumor (Caldwell, Gonzalez, Oppenheimer, Schwartz, & Caprioli, 2006; Herring, Oppenheimer, & Caprioli, 2007; Oppenheimer, Mi, Sanders, & Caprioli, 2010). Normal tissues are not composed of a single cell type. When primary cancer develops, it originates from a certain cell type within an organ. Thus, when comparing tumor and normal, the cell type of origin should be used for comparison (e.g., colon epithelium and colorectal adenocarcinoma), otherwise one would compare apples with oranges.

Histologically, cancers are very heterogeneous as they contain a diverse population of cells, including those harboring genetic mutations typically referred to as "tumor" or "cancer" cells as well as other cell types that are activated and/or recruited to the local microenvironment, e.g., fibroblasts, innate, and adaptive immune cells, and cells that line blood and lymphatic vessels (Tlsty & Coussens, 2006). Additionally, even within a relatively small area, cancer cells might exhibit different phenotypes and morphological features such as different growth pattern, mitotic rate, apoptosis, and necrosis (Fig. 1). This diversity results from different genetic and nongenetic determinants and has been noted since the earliest days of cancer (Marusyk, Almendro, & Polyak, 2012; Tabassum & Polyak, 2015). Phenotypic intratumor heterogeneity is generally considered a direct reflection of the underlying genetic diversity (Jones et al., 2008). However, a growing body of evidence suggests an evolutionary selection for nongenetic diversity-generating mechanisms in biological populations. To all appearances, genetically defined subclones and their evolution provide only the framework upon which other layers of variability produce further complexity (Caiado, Silva-Santos, & Norell, 2016). Four interdependent sources could play a role in nongenetic tumor cell heterogeneity by directly influencing gene expression: epigenetic modification (Berdasco & Esteller, 2010),

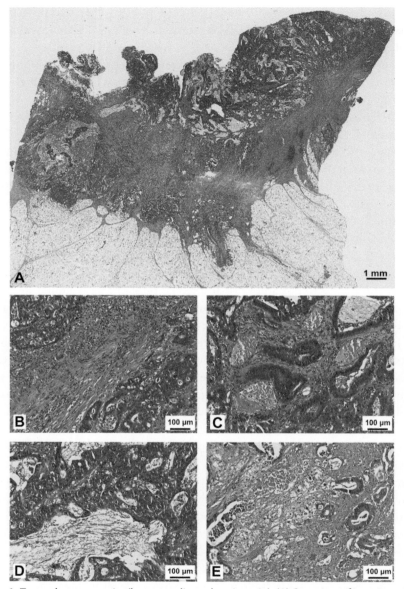

Fig. 1 Tumor heterogeneity (hematoxylin and eosin stain). (A) Overview of invasive adenocarcinoma of the colon with infiltration of subserosa. (B) Area with higher amount of stroma and inflammatory cells. (C) Area with a larger amount of blood vessels. (D) Area with a larger amount of extracellular mucin. (E) Area with extensive necrosis.

cellular differentiation hierarchies (Visvader & Lindeman, 2008), gene expression stochasticity (Almendro, Marusyk, & Polyak, 2013), and the influence of the microenvironment (Mueller & Fusenig, 2004).

Intratumor heterogeneity has also been investigated by MSI. In one study, 40 well-documented myxofibrosarcoma and myxoid liposarcoma cases were analyzed utilizing imaging mass spectrometry (Willems et al., 2010). Apart from classifying myxofibrosarcoma and myxoid liposarcomas according to tumor type and tumor grade by unsupervised clustering intratumor heterogeneity was investigated in myxofibrosarcoma samples. Microscopic evaluation of low- and intermediate-grade myxofibrosarcoma datasets revealed nodular tissue structure that seemed to be related to different peptide/protein spectra and reflected different nodules with high-grade (HG)- and low-grade (LG)-like character.

Furthermore, intratumor heterogeneity applies not only to tumor cells but also to components of the microenvironment (Fig. 1) (Tlsty & Coussens, 2006). Innate immune cells, such as lymphocytes, granulocytes, dendritic cells, macrophages, and mast cells, are found in different quantities dispersed throughout premalignant and malignant tissues. All of these cells are able to contribute to cancer development (Coffelt, Wellenstein, & de Visser, 2016; Dvorak, 1986; Elinav et al., 2013; Hanahan & Weinberg, 2011). The influence on inflammatory cells in regard to sample contamination is particularly important to keep in mind. As mentioned earlier, many types of cancer show a different amount of immune cells dispersed throughout the cancer region as well as at the boundary between cancer and adjacent tissue. While tumor cells are typically large, inflammatory cells are small, thus resulting in more cells per tissue area and more contribution to the mass spectrum. This can result in more m/z values from histones H4 peptides that are abundant in nucleated cells or in more m/z values from defensins which are found in neutrophils (Bauer et al., 2010). As a consequence, these signals can lead to suppression of other signals.

The second important cell type in the tumor microenvironment is the fibroblast. Although they were alleged to be passive participants in progression, growth, and spread of cancers, more recent studies suggest that they exert an active role as prominent modifiers of cancer progression (Kalluri & Zeisberg, 2006; Kuzet & Gaggioli, 2016).

Third, blood vessels are a key component of the tumor microenvironment contributing and essential for maintaining tissue homeostasis (Farnsworth, Lackmann, Achen, & Stacker, 2014). Since tumors require vasculature for the supply of oxygen and nutrients, angiogenesis to generate tumor blood vessels is considered essential for a tumor to grow beyond a few millimeters in size (Folkman & Hanahan, 1991). Histologically, tumor vessels are often abnormal, tortuous, deficient in pericyte coverage, and lack the normal hierarchical arrangement of arterioles, capillaries, and venules (Heath & Bicknell, 2009). Also, the blood vessel density within a sample can also have an impact on the mass spectrum. The major content of blood vessels are erythrocytes, which contain hemoglobin. Alike the defensins found in neutrophils, hemoglobin ionizes nicely and can lead to suppression of other m/z values (Amann et al., 2006). The importance of differences in vascularization in tumors has also a great impact in drug distribution studies. The distribution of three anticancer drugs (afatinib, erlotinib, and sorafenib) was studied by a combined approach of MALDI drug imaging and immunohistochemical vessel staining utilizing xenograft models (Huber et al., 2014). All anticancer drugs displayed a heterogeneous distribution throughout tumor xenografts. Coregistration and digital image analysis of MALDI drug imaging data and immunohistochemical vasculature staining utilizing the antibody CD31 revealed that tumor regions containing high drug levels were associated with a higher degree of vascularization than the regions without drug signals ($p < 0.05$). This strategy also uncovered differences in diffusion rates between the tyrosine kinase inhibitors afatinib and erlotinib by investigating the drug signal intensity as a function of distance to blood vessels. Finally, higher drug concentrations were observed in tumor regions containing small-sized vessels. This publication highlights the importance of correlating histological features with MALDI imaging data in order to obtain more insight into the underlying biology.

Many types of solid tumors display areas of necrosis that are surrounded by a rim of neoplastic tissue with very low oxygen concentration (Brown & Wilson, 2004). Heterogeneously distributed hypoxic and/or anoxic tissue areas within the tumor mass can be found in up to 50–60% of locally advanced solid tumors (Vaupel, Mayer, & Hockel, 2004). The presence of necrotic areas within tumors has been linked to adverse prognosis in several types of cancers including clear cell renal cell carcinoma and colorectal carcinoma (Klatte et al., 2009; Vayrynen et al., 2016). This negative effect is owed to the multiple contributions of hypoxia to tumor progression and resistance to therapy (Wilson & Hay, 2011). In response to hypoxia, cancer cells undergo a number of adaptive changes in order to survive including complex transcriptome changes of different signaling pathways such as pathways downstream of HIF-1α and mTOR (Harris, 2002; Wouters & Koritzinsky, 2008).

Several types of cancer (e.g., breast and prostate cancer) are even more challenging to analyze by MSI as they not only tend to diffusely infiltrate between the surrounding tissues but also display (sometimes extensively) different precursor lesions within the tumor bulk (Epstein, 2009; Lopez-Garcia, Geyer, Lacroix-Triki, Marchio, & Reis-Filho, 2010). Regarding breast cancer, ductal carcinoma in situ (DCIS) is an intraductal neoplastic proliferation of epithelial cells confined to the ductal–lobular system and a nonobligate precursor of invasive ductal carcinoma (Wiechmann & Kuerer, 2008). Although invasive ductal carcinoma and DCIS seem genetically closely related, some qualitative differences have been found (Cowell et al., 2013). For example, comparing matched DCIS and invasive ductal carcinoma, differences in HER2 expression levels and gene amplification were found (Latta, Tjan, Parkes, & O'Malley, 2002).

In summary, as tumors are generally heterogeneous in their overall composition due to many factors including regional differences in tumor cell content compared to cells belonging to the tumor microenvironment, inherent tumor cell heterogeneity, admixture with precursor lesions, composition of the microenvironment itself as well as differences in vascularization, and areas of necrosis trained morphologists/pathologists are needed to interpret all these tissue inherent factors to ensure adequacy of the sample/region used for classification. Since the percentage of analyzed tumor cells is one determinant of its contribution to the mass spectrum, evaluating the fraction of tumor cellularity is important in ensuring the accuracy of the test results.

3. POSSIBLE ERRORS CAUSED BY SAMPLE INHERENT FACTORS—SMALL PRETHERAPEUTIC BIOPSIES

In the era of modern and personalized medicine, decisions on cancer treatment are typically based on accurate tissue diagnosis from small biopsy samples. As a first step, such minuscule biopsy samples are processed, stained with hematoxylin and eosin, and analyzed under a light microscope. Following this initial evaluation of histopathology, additional immunohistochemical tests and in many cases also selected molecular tests are used to establish the diagnosis of a specific cancer subtype. It is well known, however, that small biopsy samples might not accurately reflect the tumor as a whole (Bedard, Hansen, Ratain, & Siu, 2013). For example, tumor biopsies may not represent the overall genetic heterogeneity of the sampled cancer, as each biopsy tends to carry some mutation(s) that are not present in the rest of the tumor (Caiado et al., 2016). Comprehensive characterization of multiple tumor specimens obtained from the same patient elucidate remarkable intratumor heterogeneity between geographical regions in the same tumor (spatial heterogeneity, see earlier), as well as between the primary tumor and a subsequent local or distant recurrence in the same patient (temporal heterogeneity) (Bedard et al., 2013). This inherent problem of small pretherapeutic biopsy samples is of particular importance in studies investigating prognosis and treatment response on the basis of biomarker profiles extracted from such tissue. Key aspects of genetic intratumor heterogeneity relate to its prognostic and predictive capacity (Caiado et al., 2016). Clinical studies of different types of cancer, including breast cancer, indicated that intratumor genetic heterogeneity is an independent risk factor for disease progression (Navin et al., 2010). Thus, if the sample that was used for the predictive analysis does not accurately reflect the underlying biology of the tumor, especially its aggressiveness or potential to metastasize, the prediction will not be precise. Of course, this problem is not restricted to MSI.

Many types of cancers are known to be very inhomogeneous or even develop multiple tumor nodules. One well-studied example is prostate cancer. Within the prostate, there is usually not one single tumor but instead there are multiple different tumors that also exhibit different degrees of differentiation and show genetic differences, suggesting independent clonal origins (Arora et al., 2004; Greene, Wheeler, Egawa, Dunn, & Scardino, 1991; Lindberg et al., 2013). Even within a single tumor nodule there are often different growth patterns (Fig. 2) which is reflected by the unique grading system for prostate cancer, the Gleason score (Epstein, Egevad, et al., 2016; Gleason & Mellinger, 1974). It resembles a sum of the two most prevalent growth patterns exhibited by the tumor. If a third growth pattern is present, Gleason grading guidelines vary between pretherapeutic needle core biopsies and resection specimen. In the needle core biopsies, the Gleason score is defined as the sum of the most prevalent and the worst growth pattern, however, in the resection specimen, the Gleason score is defined as the sum of the most and the second most growth pattern, whereas the third pattern can be added as an addendum (Moch, International Agency for Research on Cancer & World Health Organization, 2016). To complicate things even more, many studies have shown disparities between biopsy Gleason score and prostatectomy Gleason score (Dinh et al., 2015; Epstein, Feng, Trock, & Pierorazio, 2012). Thus, if studies are conducted to identify

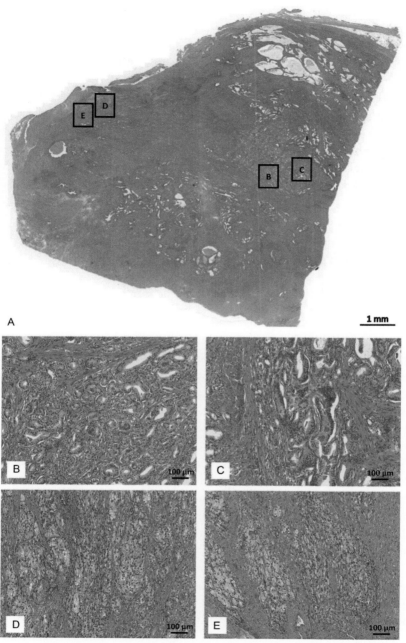

Fig. 2 Intratumor heterogeneity (hematoxylin and eosin stain). (A) Overview of prostate cross section with a large area of acinar (conventional) adenocarcinoma. (B–E) Different regions from within the tumor proper display different degrees of differentiation/different Gleason scores.

possible prognostic markers pretherapeutic biopsies might not adequately reflect the aggressiveness of prostate cancer.

The influence of different tumor regions on patients' prognosis has also been studied by MALDI imaging. Samples from 63 intestinal-type gastric cancer patients and 32 breast cancer patients were analyzed by MALDI MSI (Balluff et al., 2015). In order to obtain segmentation maps of molecularly distinct regions, mass spectra were grouped by a corroborated statistical clustering algorithm. Relating these regions, each characterized by different phenotypic tumor subpopulations, to the patients' clinical data revealed several subpopulations to be associated with a different overall survival or the presence of locoregional metastases.

4. POSSIBLE ERRORS CAUSED BY SAMPLE PREPARATIONS—ARTIFACTS ARE NOT YOUR FRIEND

When dealing with fresh frozen samples that have been stored in a freezer for a long period of time freezer burns can occur. This can result in the disruption of the normal tissue architecture. Freezer burns can also be caused by rapid freezing of tissues and may contribute to changes in the molecular profile of the tissue (Srinivasan, Sedmak, & Jewell, 2002). Compared to formalin-fixed and paraffin-embedded samples, fresh frozen sections in general do not allow for detailed evaluation of nuclear details such as nucleoli and nuclear inclusion (Fig. 3).

For formalin-fixed samples, optimal sample quality depends on the correct formalin fixation (Bass, Engel, Greytak, & Moore, 2014; Fox, Johnson, Whiting, & Roller, 1985). Many studies have shown that inappropriate formalin fixation hinders or even precludes proteomic analyses (Thompson et al., 2013). By formalin fixation, pathologists try to stabilize the microanatomy of tissue (Grizzle, 2009). In order to ensure optimal fixation, and thus achieve optimal tissue preservation and microanatomy, three aspects are important: tissue thickness, volume of formalin, and fixation time. Variation of tissue quality between different samples due to over- or underfixation is the cost to pay for ineffective optimization of these three factors. However, since clinical specimen largely vary in size (from small needle core biopsies up to whole organs or multiple organs) fixation procedures are very difficult to standardize. The tissue has to small enough to enable formalin penetration so that fixation takes place before autolysis starts. Thus, specimen have to be placed into a container that allows for the adequate amount of formalin

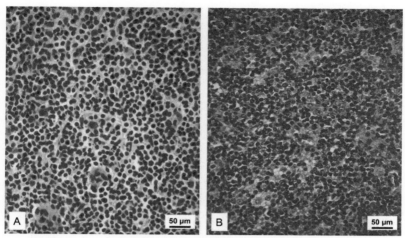

Fig. 3 Differences in histomorphology between fresh frozen and formalin-fixed and paraffin-embedded samples of a non-Hodgkin lymphoma (hematoxylin and eosin stain). (A) Fresh frozen section, nuclear details are not discernible. (B) Formalin-fixed and paraffin-embedded sample with clearly visible nuclear details.

(minimum formalin:tissue ratio 10:1) and larger specimens should be prepared (sectioned) to allow for a better formalin penetration (Maes et al., 2013). Regarding fixation time, there is a known paradox between the rate of penetration of formaldehyde and its rate of fixation. As a general rule of thumb, samples should be placed in neutral buffered formalin allowing for a fixation time of 24 h at a penetration of 1 mm/h (Howat & Wilson, 2014). Shorter fixation times interject the formalin fixation process. During the subsequent tissue dehydration steps, fixation is completed by coagulation fixation through alcohol. However, as a consequence variable admixtures of cross-linking and coagulation fixation can be observed within a tissue section (Werner, Chott, Fabiano, & Battifora, 2000). Whereas overfixation does not seem to have an adverse effect on morphology or immunohistochemistry (Arber, 2002; Engel & Moore, 2011; Goldstein, Ferkowicz, Odish, Mani, & Hastah, 2003), it does severely compromise proteomic analyses (Sprung et al., 2009; Tanca et al., 2011). On the other hand, if the sample is too large or the amount of formalin is too small fixation results will not be optimal. This leads to samples that are well fixed at the outer rim and poorly fixed in the center (Fig. 4) (Chesnick, Mason, O'Leary, & Fowler, 2010; Start, Layton, Cross, & Smith, 1992). Histologically, errors in sample fixation can be recognized by blurred nuclear details with indistinct nuclear chromatin, cytoplasmic clumping, poor differentiation of eosin color, and

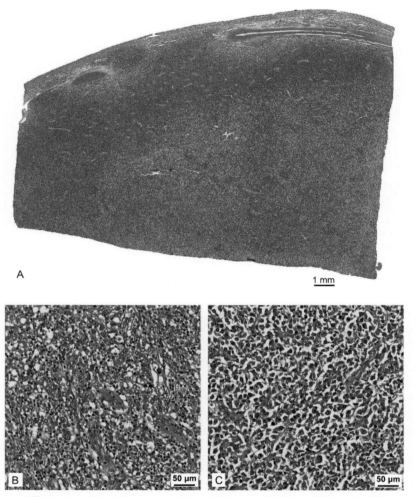

Fig. 4 Differences in formalin penetration and fixation of a testicular Non-Hodgkin lymphoma (hematoxylin and eosin stain). (A) Overview with visible differences between the outer edge (*top*) and the center (*bottom*) of the sample displaying a gradient in staining pattern. (B) Higher magnification of well-fixated outer edge of the sample with discernable nuclear and cytoplasmatic details. (C) Higher magnification of poorly fixed sample center with shriveled nuclei without nuclear details.

loss of detail in vascular structures, nerves, and glands (Chatterjee, 2014; Dapson, 2007). For immunohistochemistry, sections from such tissues can exhibit more intense staining of the center or of the periphery depending on the type of antibody used (Werner et al., 2000).

Tissue artifacts can also be caused by the way the tissue was removed from the patient. When surgeons use energy-based surgical dissection and coagulation devices tissue samples can experience different degrees of thermal damage (Taheri et al., 2014). The heat generated can cause marked alteration in both epithelium and connective tissues due to boiling of tissue fluids and precipitation/coagulation of proteins (Rastogi et al., 2013). As a consequence, cellular details are completely lost and cells exhibit a distorted appearance with nuclear elongation, nuclear palisading, and vacuolar degeneration, whereas connective tissue, fat, and muscle can acquire an opaque, amorphous appearance (Fig. 5) (Chatterjee, 2014; Logan & Goss, 2010). Coagulation artifacts are usually found at the surface of tissue samples and are clearly visible to the experienced pathologist. Depending on the device used, the extent of damage and the penetration depth is different

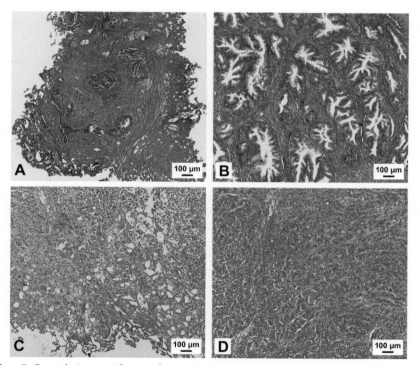

Fig. 5 Coagulation artifacts of transurethral resections specimen (hematoxylin and eosin stain); (A+B) Normal prostate. (C+D) Transitional carcinoma of the bladder. (A+C) Tissue with extensive coagulation artifacts leading to completely obscured histomorphology. (B+D) Adjacent areas from the same specimen without coagulation artifacts.

(Homayounfar et al., 2012; MacDonald, Bowers, Chin, & Burns, 2014). This is of particular importance when small samples are concerned as the thermal damage might compromise the whole sample. Crush or compression artifacts can be produced on the sample surface by using forceps with disproportionate force when handling unfixed samples (Zegarelli, 1978). Especially inflammatory tissue and tumor cells are prone to crush artifacts, where tissue morphology is rearranged and chromatin can be squeezed the out of nuclei (Bernstein, 1978; Chatterjee, 2014).

Some samples can contain large areas of hemorrhage. This can be inherent to the tissue or caused by the technique the sample was retrieved. Hemorrhage means spills of red blood cells that contain hemoglobin. This, as discussed previously, is a molecule that ionizes well in MALDI imaging studies of peptides and proteins and can thus be the major base peak in the spectrum and overshadow other analytes (Amann et al., 2006).

All these possible artifacts can lead to differences in the mass spectra, turning already inherently complex tissue analyzation into an even more challenging endeavor. Artifacts are also very important to bear in mind in analytical strategies that incorporate unsupervised analysis. If the analysis only relies on maximum changes between mass spectra within a tissue sample, these might be caused by artifacts rather than differences attributable to the underlying biology. Thus, even for unsupervised analysis, correlation with histology should be pursued to ensure biological significance of results.

5. POSSIBLE ERRORS CAUSED BY ILL-DEFINED SAMPLE GROUPS IN THE TRAINING SET—KEEP IT BLACK AND WHITE/SHADES OF GRAY ARE NOT WELCOME

In order to generate a robust classifier in MALDI imaging studies, classes within the training set have to be well defined. There is no one single type of cancer and also within an organ different types of cancer can occur. Thus, first of all, the type of cancer should be precisely defined. For example, regarding prostate cancer, according to the latest WHO (World Health Organization) classification there are nine different types of carcinoma, including the most common type acinar adenocarcinoma (Moch, Cubilla, Humphrey, Reuter and Ulbright, 2016) but also a bunch of other, biologically entirely different other types. Second, cancers exhibit different degrees of differentiation and are in general graded by means of a three-tiered grading system (G1–3) (Sobin, Gospodarowicz, Wittekind, & International Union Against Cancer, 2010).

Typically, both the tumor architecture and cytological features are evaluated and the most poorly differentiated part of the tumor establishes the tumor grade (Damjanov, Fan, & SpringerLink (Online service), 2007). LG cancers (G1) resemble the cell of origin, whereas HG cancers (G3) are less differentiated and the cell of origin can sometimes if at all only be discerned using immunohistochemistry. Some cancer subtypes (e.g., serous ovarian carcinoma) have two-tiered grading systems (low and high grade) and others such as prostate cancer have a unique grading system as discussed previously (Epstein, Zelefsky, et al., 2016; McCluggage et al., 2015). As a rule of thumb, HG cancers are more aggressive and have a worse prognosis, whereas LG cancers are more indolent (Damjanov et al., 2007; Kryvenko & Epstein, 2016). Third, cancers are staged according to the TNM classification of tumors (Sobin et al., 2010). The "T" stands for tumor stage (T1–T4), meaning the extent or size of the tumor within a particular organ or extension beyond the organ; the "N" and "M" stands for spread to lymph nodes or distant organs, respectively. The higher the T-stage the larger is the tumor, wherein T4 tumors are usually already involving neighboring tissues or organs. Additional components of the so-called tumor formula are the presence of perineural invasion (Pn1), lymph, or blood vessel infiltration (L1 and V1, respectively), and the status of resection: R0, tumor locally completely removed; R1, tumor is microscopically visible at the resection margin; and R2, tumor is macroscopically visible at the resection margin. Regarding the study design, all these cancer-specific details with respect to subtype, grading, and staging are important. Numerous studies have shown that many of these factors have a major impact on survival (Elenkov et al., 2016; Epstein, Zelefsky, et al., 2016; Lim, Chien, & Earle, 2003; Soerjomataram, Louwman, Ribot, Roukema, & Coebergh, 2008). Additionally, there are also differences between different tumor grade and stage on the proteomics and metabolomics level (Gan, Chen, & Li, 2014; Simeone et al., 2014; Wettersten et al., 2015). This has been shown by MALDI imaging for non-invasive papillary urothelial carcinomas (Oezdemir et al., 2012). Thus, only well-defined groups within the training set can lead to meaningful and robust classifiers.

Pathology is not a game of black and white, there are many shades of gray in between. Meaning, there is no one cancer and one normal but there are many different types of tumor and normal and many different states in between. Some of the contributing factors have been highlighted in the paragraphs earlier such as cancer precursor lesions and inflammation. Another degree of variation might be caused by slightly differing evaluation criteria between pathologists, ultimately resulting in a variance of the diagnoses of the very same sample. This dilemma has been well studied and described in the literature (Ozkan et al., 2016; Rakha, Ahmed, & Ellis, 2016; Suzuki et al., 2006) and many possible solutions have been suggested. However, the system is not easy to change and for the purpose of mass spectrometry imaging studies one has to bear two things in mind. First, with regard to choosing the right samples for a training set, only samples that fall into the black and white category are ideal. This means that only samples that can be clearly diagnosed by a morphologist/pathologist without any doubt should be included into the training set. It will be easier to look at the far ends of the spectrum between healthy and cancer or different cancer types than looking at the gray areas in between. Once a robust model is built, one might want to look at samples that are difficult to diagnose to fill in the gray areas. Second, samples should ideally be revised by a single morphologist/pathologist to ensure for accuracy and uniformity in diagnosis, grading, and staging. Relying on old pathology reports might lead to deviations. This can also be caused by changes in tumor subtyping, grading, and staging over the years. As pathology is an ongoing field of research and development where new tumor entities are discovered and grading or staging systems are refined according to new research findings archived samples might have been classified according to an outdated and obsolete scheme. For example, the latest TNM classification of malignant tumors (Sobin et al., 2010) did incorporate major changes in staging of cancers in six different organs including stomach, skin, lung and prostate, and several other minor changes. But there are not only changes in staging of different tumors over the years but also changes in grading. For example, in 2004 the WHO classification system replaced the three-tiered grading system (G1–3) of noninvasive papillary urothelial carcinomas by a two-tiered grading system (LG and HG) (Eble, World Health Organization, & International Academy of Pathology, 2004). Thus, the former G2 tumors have to be assigned into one or the other of these two groups. This change was motivated by research findings indicating that HG tumors are genetically instable, whereas LG tumors are genetically stable (Lopez-Beltran & Montironi, 2004). Also, the definition of the former grades 1–3 was vague, without specific histological criteria and resulted in most cases falling into the intermediate (G2) category (Miyamoto et al., 2010). Similarly, in serous ovarian adenocarcinoma, the continuous grading in G1–G3 has been eliminated and replaced by a subclassification into LG and HG carcinomas (Kurman, International Agency for Research on Cancer, & World Health Organization, 2014). Within the same classification, the subcategorization of mucinous borderline tumors into an intestinal and an endocervical type has been dispensed with. The formerly endocervical mucinous borderline tumors now fall in the newly created seromucinous tumor group, whereas the formerly intestinal type is now the new mucinous borderline tumor (Kurman et al., 2014; Meinhold-Heerlein et al., 2016). In addition, new tumor subtypes are being introduced as diagnosis becomes more refined. For instance, the new WHO classification of kidney cancers incorporated five new subtypes of renal cell carcinoma (Moch, Cubilla, et al., 2016). Taking all these dynamic changes occurring at many levels all the time into account, archived samples might need to be reviewed again in order to keep up with most recent classification, grading, and staging.

6. CONCLUSION

Since MSI can measure hundreds of molecules at the cellular level with direct correlation to histological feature, it has the potential to revolutionize pathology. It could overcome limitations of other approaches in the identification and routine diagnostic measurement of new marker molecules/profiles. However, in order to achieve this goal, close collaboration between physicians, especially pathologists, molecular biologists, and technology developers is mandatory. MSI generates molecular maps of tissue sections that can elucidate the underlying biochemistry or provide information on how therapeutics or toxins influence the function or misfunction of an organ. However, it can only fully achieve this goal when combined with state of the art tissue morphology evaluation. As discussed, many tissues or sample inherent factors can influence the outcome of the analysis. Tissue heterogeneity on many levels between different sample areas will result in differences between spectra as will a large variety of different technical artifacts that may occur when tissue is processed. Thus, correct interpretation of molecular maps or pictures obtained by MSI can only be attained by profound knowledge and understanding of the tissue architecture and by integrating morphological, clinical, and molecular dimensions.

REFERENCES

Almendro, V., Marusyk, A., & Polyak, K. (2013). Cellular heterogeneity and molecular evolution in cancer. *Annual Review of Pathology: Mechanisms of Disease, 8*(8), 277–302. http://dx.doi.org/10.1146/annurev-pathol-020712-163923.

Amann, J. M., Chaurand, P., Gonzalez, A., Mobley, J. A., Massion, P. P., Carbone, D. P., & Caprioli, R. M. (2006). Selective profiling of proteins in lung cancer cells from fine-needle aspirates by matrix-assisted laser desorption ionization time-of-flight mass spectrometry. *Clinical Cancer Research, 12*(17), 5142–5150. http://dx.doi.org/10.1158/1078-0432.CCR-06-0264.

Arber, D. A. (2002). Effect of prolonged formalin fixation on the immunohistochemical reactivity of breast markers. *Applied Immunohistochemistry and Molecular Morphology, 10*(2), 183–186.

Arora, R., Koch, M. O., Eble, J. N., Ulbright, T. M., Li, L., & Cheng, L. (2004). Heterogeneity of Gleason grade in multifocal adenocarcinoma of the prostate. *Cancer, 100*(11), 2362–2366. http://dx.doi.org/10.1002/cncr.20243.

Balluff, B., Frese, C. K., Maier, S. K., Schone, C., Kuster, B., Schmitt, M., ... McDonnell, L. A. (2015). De novo discovery of phenotypic intratumour heterogeneity using imaging mass spectrometry. *The Journal of Pathology, 235*(1), 3–13. http://dx.doi.org/10.1002/path.4436.

Bass, B. P., Engel, K. B., Greytak, S. R., & Moore, H. M. (2014). A review of preanalytical factors affecting molecular, protein, and morphological analysis of formalin-fixed, paraffin-embedded (FFPE) tissue: How well do you know your FFPE specimen? *Archives of Pathology and Laboratory Medicine, 138*(11), 1520–1530. http://dx.doi.org/10.5858/arpa.2013-0691-RA.

Bauer, J. A., Chakravarthy, A. B., Rosenbluth, J. M., Mi, D., Seeley, E. H., De Matos Granja-Ingram, N., ... Pietenpol, J. A. (2010). Identification of markers of taxane sensitivity using proteomic and genomic analyses of breast tumors from patients receiving neoadjuvant paclitaxel and radiation. *Clinical Cancer Research, 16*(2), 681–690. http://dx.doi.org/10.1158/1078-0432.CCR-09-1091.

Bedard, P. L., Hansen, A. R., Ratain, M. J., & Siu, L. L. (2013). Tumour heterogeneity in the clinic. *Nature, 501*(7467), 355–364. http://dx.doi.org/10.1038/nature12627.

Berdasco, M., & Esteller, M. (2010). Aberrant epigenetic landscape in cancer: How cellular identity goes awry. *Developmental Cell, 19*(5), 698–711. http://dx.doi.org/10.1016/j.devcel.2010.10.005.

Bernstein, M. L. (1978). Biopsy technique: The pathological considerations. *The Journal of the American Dental Association, 96*(3), 438–443.

Brown, J. M., & Wilson, W. R. (2004). Exploiting tumour hypoxia in cancer treatment. *Nature Reviews. Cancer, 4*(6), 437–447. http://dx.doi.org/10.1038/nrc1367.

Caiado, F., Silva-Santos, B., & Norell, H. (2016). Intra-tumour heterogeneity—Going beyond genetics. *The FEBS Journal, 283*(12), 2245–2258. http://dx.doi.org/10.1111/febs.13705.

Caldwell, R. L., Gonzalez, A., Oppenheimer, S. R., Schwartz, H. S., & Caprioli, R. M. (2006). Molecular assessment of the tumor protein microenvironment using imaging mass spectrometry. *Cancer Genomics & Proteomics, 3*, 279–288.

Chatterjee, S. (2014). Artefacts in histopathology. *Journal of Oral and Maxillofacial Pathology, 18*(Suppl. 1), S111–S116. http://dx.doi.org/10.4103/0973-029X.141346.

Chesnick, I. E., Mason, J. T., O'Leary, T. J., & Fowler, C. B. (2010). Elevated pressure improves the rate of formalin penetration while preserving tissue morphology. *Journal of Cancer, 1*, 178–183.

Coffelt, S. B., Wellenstein, M. D., & de Visser, K. E. (2016). Neutrophils in cancer: Neutral no more. *Nature Reviews. Cancer, 16*(7), 431–446. http://dx.doi.org/10.1038/nrc.2016.52.

Cornett, D. S., Mobley, J. A., Dias, E. C., Andersson, M., Arteaga, C. L., Sanders, M. E., & Caprioli, R. M. (2006). A novel histology-directed strategy for MALDI-MS tissue profiling that improves throughput and cellular specificity in human breast cancer. *Molecular and Cellular Proteomics, 5*(10), 1975–1983. http://dx.doi.org/10.1074/mcp.M600119-MCP200.

Cowell, C. F., Weigelt, B., Sakr, R. A., Ng, C. K., Hicks, J., King, T. A., & Reis-Filho, J. S. (2013). Progression from ductal carcinoma in situ to invasive breast cancer: Revisited. *Molecular Oncology, 7*(5), 859–869. http://dx.doi.org/10.1016/j.molonc.2013.07.005.

Damjanov, I., Fan, F., & SpringerLink (Online service). (2007). *Cancer grading manual*. Retrieved from http://dx.doi.org/10.1007/978-0-387-33751-7. Table of Contents/Abstracts.

Dapson, R. W. (2007). Macromolecular changes caused by formalin fixation and antigen retrieval. *Biotechnic and Histochemistry, 82*(3), 133–140. http://dx.doi.org/10.1080/10520290701567916.

Dinh, K. T., Mahal, B. A., Ziehr, D. R., Muralidhar, V., Chen, Y. W., Viswanathan, V. B., … Nguyen, P. L. (2015). Incidence and predictors of upgrading and up staging among 10,000 contemporary patients with low risk prostate cancer. *The Journal of Urology, 194*(2), 343–349. http://dx.doi.org/10.1016/j.juro.2015.02.015.

Dvorak, H. F. (1986). Tumors: Wounds that do not heal. Similarities between tumor stroma generation and wound healing. *The New England Journal of Medicine, 315*(26), 1650–1659. http://dx.doi.org/10.1056/NEJM198612253152606.

Eble, J. N., World Health Organization, & International Academy of Pathology. (2004). *Pathology and genetics of tumours of the urinary system and male genital organs*. Lyon: IARC Press.

Elenkov, A. A., Timev, A., Dimitrov, P., Vasilev, V., Krastanov, A., Georgiev, M., … Panchev, P. (2016). Clinicopathological prognostic factors for upper tract urothelial carcinoma. *Central European Journal of Urology, 69*(1), 57–62. http://dx.doi.org/10.5173/ceju.2016.713.

Elinav, E., Nowarski, R., Thaiss, C. A., Hu, B., Jin, C., & Flavell, R. A. (2013). Inflammation-induced cancer: Crosstalk between tumours, immune cells and microorganisms. *Nature Reviews. Cancer, 13*(11), 759–771. http://dx.doi.org/10.1038/nrc3611.

Engel, K. B., & Moore, H. M. (2011). Effects of preanalytical variables on the detection of proteins by immunohistochemistry in formalin-fixed, paraffin-embedded tissue. *Archives of Pathology and Laboratory Medicine, 135*(5), 537–543. http://dx.doi.org/10.1043/2010-0702-RAIR.1.

Epstein, J. I. (2009). Precursor lesions to prostatic adenocarcinoma. *Virchows Archiv, 454*(1), 1–16. http://dx.doi.org/10.1007/s00428-008-0707-5.

Epstein, J. I., Egevad, L., Amin, M. B., Delahunt, B., Srigley, J. R., Humphrey, P. A., & Grading, C. (2016). The 2014 International Society of Urological Pathology (ISUP) consensus conference on Gleason grading of prostatic carcinoma: Definition of grading patterns and proposal for a new grading system. *The American Journal of Surgical Pathology, 40*(2), 244–252. http://dx.doi.org/10.1097/PAS.0000000000000530.

Epstein, J. I., Feng, Z., Trock, B. J., & Pierorazio, P. M. (2012). Upgrading and downgrading of prostate cancer from biopsy to radical prostatectomy: Incidence and predictive factors using the modified Gleason grading system and factoring in tertiary grades. *European Urology, 61*(5), 1019–1024. http://dx.doi.org/10.1016/j.eururo.2012.01.050.

Epstein, J. I., Zelefsky, M. J., Sjoberg, D. D., Nelson, J. B., Egevad, L., Magi-Galluzzi, C., … Klein, E. A. (2016). A contemporary prostate cancer grading system: A validated alternative to the Gleason score. *European Urology, 69*(3), 428–435. http://dx.doi.org/10.1016/j.eururo.2015.06.046.

Farnsworth, R. H., Lackmann, M., Achen, M. G., & Stacker, S. A. (2014). Vascular remodeling in cancer. *Oncogene, 33*(27), 3496–3505. http://dx.doi.org/10.1038/onc.2013.304.

Folkman, J., & Hanahan, D. (1991). Switch to the angiogenic phenotype during tumorigenesis. *Princess Takamatsu Symposia*, *22*, 339–347.

Fox, C. H., Johnson, F. B., Whiting, J., & Roller, P. P. (1985). Formaldehyde fixation. *The Journal of Histochemistry and Cytochemistry*, *33*(8), 845–853.

Friedl, P., & Wolf, K. (2003). Tumour-cell invasion and migration: Diversity and escape mechanisms. *Nature Reviews Cancer*, *3*(5), 362–374. http://dx.doi.org/10.1038/nrc1075.

Gan, Y., Chen, D., & Li, X. (2014). Proteomic analysis reveals novel proteins associated with progression and differentiation of colorectal carcinoma. *Journal of Cancer Research and Therapeutics*, *10*(1), 89–96. http://dx.doi.org/10.4103/0973-1482.131396.

Gleason, D. F., & Mellinger, G. T. (1974). Prediction of prognosis for prostatic adenocarcinoma by combined histological grading and clinical staging. *The Journal of Urology*, *111*(1), 58–64.

Goldstein, N. S., Ferkowicz, M., Odish, E., Mani, A., & Hastah, F. (2003). Minimum formalin fixation time for consistent estrogen receptor immunohistochemical staining of invasive breast carcinoma. *American Journal of Clinical Pathology*, *120*(1), 86–92. http://dx.doi.org/10.1309/QPHD-RB00-QXGM-UQ9N.

Greene, D. R., Wheeler, T. M., Egawa, S., Dunn, J. K., & Scardino, P. T. (1991). A comparison of the morphological features of cancer arising in the transition zone and in the peripheral zone of the prostate. *The Journal of Urology*, *146*(4), 1069–1076.

Grizzle, W. E. (2009). Special symposium: Fixation and tissue processing models. *Biotechnic and Histochemistry*, *84*(5), 185–193. http://dx.doi.org/10.3109/10520290903039052.

Hanahan, D., & Weinberg, R. A. (2011). Hallmarks of cancer: The next generation. *Cell*, *144*(5), 646–674. http://dx.doi.org/10.1016/j.cell.2011.02.013.

Harris, A. L. (2002). Hypoxia—A key regulatory factor in tumour growth. *Nature Reviews. Cancer*, *2*(1), 38–47. http://dx.doi.org/10.1038/nrc704.

Heath, V. L., & Bicknell, R. (2009). Anticancer strategies involving the vasculature. *Nature Reviews. Clinical Oncology*, *6*(7), 395–404. http://dx.doi.org/10.1038/nrclinonc.2009.52.

Herring, K. D., Oppenheimer, S. R., & Caprioli, R. M. (2007). Direct tissue analysis by matrix-assisted laser desorption ionization mass spectrometry: Application to kidney biology. *Seminars in Nephrology*, *27*(6), 597–608. http://dx.doi.org/10.1016/j.semnephrol.2007.09.002.

Homayounfar, K., Meis, J., Jung, K., Klosterhalfen, B., Sprenger, T., Conradi, L. C., ... Becker, H. (2012). Ultrasonic scalpel causes greater depth of soft tissue necrosis compared to monopolar electrocautery at standard power level settings in a pig model. *BMC Surgery*, *12*, 3. http://dx.doi.org/10.1186/1471-2482-12-3.

Howat, W. J., & Wilson, B. A. (2014). Tissue fixation and the effect of molecular fixatives on downstream staining procedures. *Methods*, *70*(1), 12–19. http://dx.doi.org/10.1016/j.ymeth.2014.01.022.

Huber, K., Feuchtinger, A., Borgmann, D. M., Li, Z., Aichler, M., Hauck, S. M., ... Walch, A. (2014). Novel approach of MALDI drug imaging, immunohistochemistry, and digital image analysis for drug distribution studies in tissues. *Analytical Chemistry*, *86*(21), 10568–10575. http://dx.doi.org/10.1021/ac502177y.

Jones, S., Chen, W. D., Parmigiani, G., Diehl, F., Beerenwinkel, N., Antal, T., ... Markowitz, S. D. (2008). Comparative lesion sequencing provides insights into tumor evolution. *Proceedings of the National Academy of Sciences of the United States of America*, *105*(11), 4283–4288. http://dx.doi.org/10.1073/pnas.0712345105.

Kalluri, R., & Zeisberg, M. (2006). Fibroblasts in cancer. *Nature Reviews. Cancer*, *6*(5), 392–401. http://dx.doi.org/10.1038/nrc1877.

Karamitopoulou, E., Zlobec, I., Panayiotides, I., Patsouris, E. S., Peros, G., Rallis, G., ... Lugli, A. (2011). Systematic analysis of proteins from different signaling pathways in the tumor center and the invasive front of colorectal cancer. *Human Pathology*, *42*(12), 1888–1896. http://dx.doi.org/10.1016/j.humpath.2010.06.020.

Klatte, T., Said, J. W., de Martino, M., Larochelle, J., Shuch, B., Rao, J. Y., ... Pantuck, A. J. (2009). Presence of tumor necrosis is not a significant predictor of survival in clear cell renal cell carcinoma: Higher prognostic accuracy of extent based rather than presence/absence classification. *The Journal of Urology, 181*(4), 1558–1564. http://dx.doi.org/10.1016/j.juro.2008.11.098. discussion 1563–1554.

Kryvenko, O. N., & Epstein, J. I. (2016). Changes in prostate cancer grading: Including a new patient-centric grading system. *Prostate, 76*(5), 427–433. http://dx.doi.org/10.1002/pros.23142.

Kurman, R. J., International Agency for Research on Cancer, & World Health Organization. (2014). *WHO classification of tumours of female reproductive organs* (4th ed.). Lyon: International Agency for Research on Cancer.

Kuzet, S. E., & Gaggioli, C. (2016). Fibroblast activation in cancer: When seed fertilizes soil. *Cell and Tissue Research, 365*, 607–619. http://dx.doi.org/10.1007/s00441-016-2467-x.

Lakhani, S. R., & Ashworth, A. (2001). Microarray and histopathological analysis of tumours: The future and the past? *Nature Reviews. Cancer, 1*(2), 151–157. http://dx.doi.org/10.1038/35101087.

Latta, E. K., Tjan, S., Parkes, R. K., & O'Malley, F. P. (2002). The role of HER2/neu overexpression/amplification in the progression of ductal carcinoma in situ to invasive carcinoma of the breast. *Modern Pathology, 15*(12), 1318–1325. http://dx.doi.org/10.1097/01.MP.0000038462.62634.B1.

Lim, J. E., Chien, M. W., & Earle, C. C. (2003). Prognostic factors following curative resection for pancreatic adenocarcinoma: A population-based, linked database analysis of 396 patients. *Annals of Surgery, 237*(1), 74–85. http://dx.doi.org/10.1097/01.SLA.0000041266.10047.38.

Lindberg, J., Klevebring, D., Liu, W., Neiman, M., Xu, J., Wiklund, P., ... Gronberg, H. (2013). Exome sequencing of prostate cancer supports the hypothesis of independent tumour origins. *European Urology, 63*(2), 347–353. http://dx.doi.org/10.1016/j.eururo.2012.03.050.

Logan, R. M., & Goss, A. N. (2010). Biopsy of the oral mucosa and use of histopathology services. *Australian Dental Journal, 55*(Suppl. 1), 9–13. http://dx.doi.org/10.1111/j.1834-7819.2010.01194.x.

Lopez-Beltran, A., & Montironi, R. (2004). Non-invasive urothelial neoplasms: According to the most recent WHO classification. *European Urology, 46*(2), 170–176. http://dx.doi.org/10.1016/j.eururo.2004.03.017.

Lopez-Garcia, M. A., Geyer, F. C., Lacroix-Triki, M., Marchio, C., & Reis-Filho, J. S. (2010). Breast cancer precursors revisited: Molecular features and progression pathways. *Histopathology, 57*(2), 171–192. http://dx.doi.org/10.1111/j.1365-2559.2010.03568.x.

MacDonald, J. D., Bowers, C. A., Chin, S. S., & Burns, G. (2014). Comparison of the effects of surgical dissection devices on the rabbit liver. *Surgery Today, 44*(6), 1116–1122. http://dx.doi.org/10.1007/s00595-013-0712-4.

Maes, E., Broeckx, V., Mertens, I., Sagaert, X., Prenen, H., Landuyt, B., & Schoofs, L. (2013). Analysis of the formalin-fixed paraffin-embedded tissue proteome: Pitfalls, challenges, and future prospectives. *Amino Acids, 45*(2), 205–218. http://dx.doi.org/10.1007/s00726-013-1494-0.

Marusyk, A., Almendro, V., & Polyak, K. (2012). Intra-tumour heterogeneity: A looking glass for cancer? *Nature Reviews. Cancer, 12*(5), 323–334. http://dx.doi.org/10.1038/nrc3261.

McCluggage, W. G., Judge, M. J., Clarke, B. A., Davidson, B., Gilks, C. B., Hollema, H., ... International Collaboration on Cancer, R. (2015). Data set for reporting of ovary, fallopian tube and primary peritoneal carcinoma: Recommendations from the International Collaboration on Cancer Reporting (ICCR). *Modern Pathology, 28*(8), 1101–1122. http://dx.doi.org/10.1038/modpathol.2015.77.

Meinhold-Heerlein, I., Fotopoulou, C., Harter, P., Kurzeder, C., Mustea, A., Wimberger, P., ... Sehouli, J. (2016). The new WHO classification of ovarian, fallopian tube, and primary peritoneal cancer and its clinical implications. *Archives of Gynecology and Obstetrics*, *293*(4), 695–700. http://dx.doi.org/10.1007/s00404-016-4035-8.

Miyamoto, H., Miller, J. S., Fajardo, D. A., Lee, T. K., Netto, G. J., & Epstein, J. I. (2010). Non-invasive papillary urothelial neoplasms: The 2004 WHO/ISUP classification system. *Pathology International*, *60*(1), 1–8. http://dx.doi.org/10.1111/j.1440-1827.2009.02477.x.

Moch, H., Cubilla, A. L., Humphrey, P. A., Reuter, V. E., & Ulbright, T. M. (2016). The 2016 WHO classification of tumours of the urinary system and male genital organs-part A: Renal, penile, and testicular tumours. *European Urology*, *70*(1), 93–105. http://dx.doi.org/10.1016/j.eururo.2016.02.029.

Moch, H., International Agency for Research on Cancer, & World Health Organization. (2016). *WHO classification of tumours of the urinary system and male genital organs* (4th ed.). Lyon: International Agency for Research on Cancer.

Mueller, M. M., & Fusenig, N. E. (2004). Friends or foes—Bipolar effects of the tumour stroma in cancer. *Nature Reviews. Cancer*, *4*(11), 839–849. http://dx.doi.org/10.1038/nrc1477.

Navin, N., Krasnitz, A., Rodgers, L., Cook, K., Meth, J., Kendall, J., ... Wigler, M. (2010). Inferring tumor progression from genomic heterogeneity. *Genome Research*, *20*(1), 68–80. http://dx.doi.org/10.1101/gr.099622.109.

Oezdemir, R. F., Gaisa, N. T., Lindemann-Docter, K., Gostek, S., Weiskirchen, R., Ahrens, M., ... Henkel, C. (2012). Proteomic tissue profiling for the improvement of grading of noninvasive papillary urothelial neoplasia. *Clinical Biochemistry*, *45*(1-2), 7–11. http://dx.doi.org/10.1016/j.clinbiochem.2011.09.013.

Oppenheimer, S. R., Mi, D., Sanders, M. E., & Caprioli, R. M. (2010). Molecular analysis of tumor margins by MALDI mass spectrometry in renal carcinoma. *Journal of Proteome Research*, *9*(5), 2182–2190. http://dx.doi.org/10.1021/pr900936z.

Ozkan, T. A., Eruyar, A. T., Cebeci, O. O., Memik, O., Ozcan, L., & Kuskonmaz, I. (2016). Interobserver variability in Gleason histological grading of prostate cancer. *Scandinavian Journal of Urology*, *50*, 420–424. http://dx.doi.org/10.1080/21681805.2016.1206619.

Rakha, E. A., Ahmed, M. A., & Ellis, I. O. (2016). Papillary carcinoma of the breast: Diagnostic agreement and management implications. *Histopathology*, *69*, 862–870. http://dx.doi.org/10.1111/his.13009.

Rastogi, V., Puri, N., Arora, S., Kaur, G., Yadav, L., & Sharma, R. (2013). Artefacts: A diagnostic dilemma—A review. *Journal of Clinical and Diagnostic Research*, *7*(10), 2408–2413. http://dx.doi.org/10.7860/JCDR/2013/6170.3541.

Salto-Tellez, M., James, J. A., & Hamilton, P. W. (2014). Molecular pathology—The value of an integrative approach. *Molecular Oncology*, *8*(7), 1163–1168. http://dx.doi.org/10.1016/j.molonc.2014.07.021.

Simeone, P., Trerotola, M., Urbanella, A., Lattanzio, R., Ciavardelli, D., Di Giuseppe, F., ... Alberti, S. (2014). A unique four-hub protein cluster associates to glioblastoma progression. *PloS One*. *9*(7), e103030. http://dx.doi.org/10.1371/journal.pone.0103030.

Sobin, L. H., Gospodarowicz, M. K., Wittekind, C., & International Union Against Cancer. (2010). *TNM classification of malignant tumours* (7th ed.). Chichester, West Sussex, UK; Hoboken, NJ: Wiley-Blackwell.

Soerjomataram, I., Louwman, M. W., Ribot, J. G., Roukema, J. A., & Coebergh, J. W. (2008). An overview of prognostic factors for long-term survivors of breast cancer. *Breast Cancer Research and Treatment*, *107*(3), 309–330. http://dx.doi.org/10.1007/s10549-007-9556-1.

Sprung, R. W., Jr., Brock, J. W., Tanksley, J. P., Li, M., Washington, M. K., Slebos, R. J., & Liebler, D. C. (2009). Equivalence of protein inventories obtained from formalin-fixed paraffin-embedded and frozen tissue in multidimensional liquid chromatography-tandem mass spectrometry shotgun proteomic analysis. *Molecular and Cellular Proteomics*, *8*(8), 1988–1998. http://dx.doi.org/10.1074/mcp.M800518-MCP200.

Srinivasan, M., Sedmak, D., & Jewell, S. (2002). Effect of fixatives and tissue processing on the content and integrity of nucleic acids. *The American Journal of Pathology*, *161*(6), 1961–1971. http://dx.doi.org/10.1016/S0002-9440(10)64472-0.

Start, R. D., Layton, C. M., Cross, S. S., & Smith, J. H. (1992). Reassessment of the rate of fixative diffusion. *Journal of Clinical Pathology*, *45*(12), 1120–1121.

Stevenson, A. (2010). *Oxford dictionary of English* (3rd ed.). New York, NY: Oxford University Press.

Suzuki, N., Price, A. B., Talbot, I. C., Wakasa, K., Arakawa, T., Ishiguro, S., ... Saunders, B. P. (2006). Flat colorectal neoplasms and the impact of the revised Vienna Classification on their reporting: A case-control study in UK and Japanese patients. *Scandinavian Journal of Gastroenterology*, *41*(7), 812–819. http://dx.doi.org/10.1080/00365520600610345.

Szasz, A. M., Gyorffy, B., & Marko-Varga, G. (2016). Cancer heterogeneity determined by functional proteomics. *Seminars in Cell and Developmental Biology*, pii: S1084-9521(16), 30270-1. http://dx.doi.org/10.1016/j.semcdb.2016.08.026. [Epub ahead of print].

Tabassum, D. P., & Polyak, K. (2015). Tumorigenesis: It takes a village. *Nature Reviews. Cancer*, *15*(8), 473–483. http://dx.doi.org/10.1038/nrc3971.

Taheri, A., Mansoori, P., Sandoval, L. F., Feldman, S. R., Pearce, D., & Williford, P. M. (2014). Electrosurgery: Part II. Technology, applications, and safety of electrosurgical devices. *Journal of the American Academy of Dermatology*, *70*(4), 607.e601–607.e612. http://dx.doi.org/10.1016/j.jaad.2013.09.055. quiz 619–620.

Tanca, A., Pagnozzi, D., Falchi, G., Biosa, G., Rocca, S., Foddai, G., ... Addis, M. F. (2011). Impact of fixation time on GeLC-MS/MS proteomic profiling of formalin-fixed, paraffin-embedded tissues. *Journal of Proteomics*, *74*(7), 1015–1021, http://dx.doi.org/10.1016/j.jprot.2011.03.015.

Thompson, S. M., Craven, R. A., Nirmalan, N. J., Harnden, P., Selby, P. J., & Banks, R. E. (2013). Impact of pre-analytical factors on the proteomic analysis of formalin-fixed paraffin-embedded tissue. *Proteomics. Clinical Applications*, *7*(3-4), 241–251. http://dx.doi.org/10.1002/prca.201200086.

Thunnissen, E., Kerr, K. M., Herth, F. J., Lantuejoul, S., Papotti, M., Rintoul, R. C., ... Laenger, F. (2012). The challenge of NSCLC diagnosis and predictive analysis on small samples. Practical approach of a working group. *Lung Cancer*, *76*(1), 1–18. http://dx.doi.org/10.1016/j.lungcan.2011.10.017.

Tlsty, T. D., & Coussens, L. M. (2006). Tumor stroma and regulation of cancer development. *Annual Review of Pathology: Mechanisms of Disease*, *1*(1), 119–150. http://dx.doi.org/10.1146/annurev.pathol.1.110304.100224.

Vachet, R. W. (2015). Molecular histology: More than a picture. *Nature Nanotechnology*, *10*(2), 103–104. http://dx.doi.org/10.1038/nnano.2015.4.

Vaupel, P., Mayer, A., & Hockel, M. (2004). Tumor hypoxia and malignant progression. *Methods in Enzymology*, *381*, 335–354. http://dx.doi.org/10.1016/S0076-6879(04)81023-1.

Vayrynen, S. A., Vayrynen, J. P., Klintrup, K., Makela, J., Karttunen, T. J., Tuomisto, A., & Makinen, M. J. (2016). Clinical impact and network of determinants of tumour necrosis in colorectal cancer. *British Journal of Cancer*, *114*(12), 1334–1342. http://dx.doi.org/10.1038/bjc.2016.128.

Visvader, J. E., & Lindeman, G. J. (2008). Cancer stem cells in solid tumours: Accumulating evidence and unresolved questions. *Nature Reviews. Cancer*, *8*(10), 755–768. http://dx.doi.org/10.1038/nrc2499.

Warth, A., Stenzinger, A., Andrulis, M., Schlake, W., Kempny, G., Schirmacher, P., & Weichert, W. (2016). Individualized medicine and demographic change as determining workload factors in pathology: Quo vadis? *Virchows Archiv*, *468*(1), 101–108. http://dx.doi.org/10.1007/s00428-015-1869-6.

Werner, M., Chott, A., Fabiano, A., & Battifora, H. (2000). Effect of formalin tissue fixation and processing on immunohistochemistry. *The American Journal of Surgical Pathology*, *24*(7), 1016–1019.

Wettersten, H. I., Hakimi, A. A., Morin, D., Bianchi, C., Johnstone, M. E., Donohoe, D. R., ... Weiss, R. H. (2015). Grade-dependent metabolic reprogramming in kidney cancer revealed by combined proteomics and metabolomics analysis. *Cancer Research*, *75*(12), 2541–2552. http://dx.doi.org/10.1158/0008-5472.CAN-14-1703.

Wiechmann, L., & Kuerer, H. M. (2008). The molecular journey from ductal carcinoma in situ to invasive breast cancer. *Cancer*, *112*(10), 2130–2142. http://dx.doi.org/10.1002/cncr.23430.

Willems, S. M., van Remoortere, A., van Zeijl, R., Deelder, A. M., McDonnell, L. A., & Hogendoorn, P. C. (2010). Imaging mass spectrometry of myxoid sarcomas identifies proteins and lipids specific to tumour type and grade, and reveals biochemical intra-tumour heterogeneity. *The Journal of Pathology*, *222*(4), 400–409. http://dx.doi.org/10.1002/path.2771.

Wilson, W. R., & Hay, M. P. (2011). Targeting hypoxia in cancer therapy. *Nature Reviews. Cancer*, *11*(6), 393–410. http://dx.doi.org/10.1038/nrc3064.

Wouters, B. G., & Koritzinsky, M. (2008). Hypoxia signalling through mTOR and the unfolded protein response in cancer. *Nature Reviews. Cancer*, *8*(11), 851–864. http://dx.doi.org/10.1038/nrc2501.

Zegarelli, D. J. (1978). Common problems in biopsy procedure. *Journal of Oral Surgery*, *36*(8), 644–647.

CHAPTER TWO

Applications of Mass Spectrometry Imaging to Cancer

G. Arentz*,†, P. Mittal*,†, C. Zhang*,†, Y.-Y. Ho*, M. Briggs*,†,‡,
L. Winderbaum*, M.K. Hoffmann*,†, P. Hoffmann*,†,1

*Adelaide Proteomics Centre, School of Biological Sciences, University of Adelaide, Adelaide, SA, Australia
†Institute for Photonics and Advanced Sensing (IPAS), University of Adelaide, Adelaide, SA, Australia
‡ARC Centre for Nanoscale BioPhotonics (CNBP), University of Adelaide, Adelaide, SA, Australia
[1]Corresponding author: e-mail address: peter.hoffmann@adelaide.edu.au

Contents

1. Introduction	28
1.1 The Advantages of MSI	29
1.2 The Basic Principles of MSI	30
2. Protein MSI in Cancer Research	33
2.1 Distinguishing Tissue Types by Peptide MSI	36
2.2 Determining Tumor Margins by Peptide MSI	37
2.3 Prediction of Metastasis by Peptide MSI	37
2.4 Analysing Chemoresponse by Peptide MSI	39
2.5 Identification of Diagnostic and Prognostic Markers by Peptide MSI	39
2.6 Characterisation of Intra- and Intertumor Variability by Peptide MSI	41
2.7 Practical Considerations for Proteolytic Peptide MSI: Sample Preparation	42
2.8 Practical Considerations for Peptide MSI: Spatial Resolution	43
2.9 Practical Considerations for Peptide MSI: Mass Analysers	44
2.10 Practical Considerations for Peptide MSI: Identification	44
3. Lipid MSI in Cancer Research	45
3.1 Profiling Lipids in Cancer by DESI-MSI	45
3.2 Profiling Lipids in Cancer by SIMS-MSI	46
4. Glycan MSI in Cancer Research	47
5. Drug Imaging in Cancer Research	50
5.1 MALDI-MSI on Tissue Sections	51
5.2 MALDI-MSI on Whole Body Sections	53
5.3 MALDI-MSI on 3D Tissue Cultures	53
6. Data Analysis	53
6.1 Spatial Information	53
6.2 Preprocessing: Peak Detection	55
6.3 Classification of FFPE-TMAs and the Importance of Dimension Reduction	56
7. Concluding Remarks	57
7.1 The Future of MSI: Molecular Pathology	57
References	57

Abstract

Pathologists play an essential role in the diagnosis and prognosis of benign and cancerous tumors. Clinicians provide tissue samples, for example, from a biopsy, which are then processed and thin sections are placed onto glass slides, followed by staining of the tissue with visible dyes. Upon processing and microscopic examination, a pathology report is provided, which relies on the pathologist's interpretation of the phenotypical presentation of the tissue. Targeted analysis of single proteins provide further insight and together with clinical data these results influence clinical decision making. Recent developments in mass spectrometry facilitate the collection of molecular information about such tissue specimens. These relatively new techniques generate label-free mass spectra across tissue sections providing nonbiased, nontargeted molecular information. At each pixel with spatial coordinates (x/y) a mass spectrum is acquired. The acquired mass spectrums can be visualized as intensity maps displaying the distribution of single m/z values of interest. Based on the sample preparation, proteins, peptides, lipids, small molecules, or glycans can be analyzed. The generated intensity maps/images allow new insights into tumor tissues. The technique has the ability to detect and characterize tumor cells and their environment in a spatial context and combined with histological staining, can be used to aid pathologists and clinicians in the diagnosis and management of cancer. Moreover, such data may help classify patients to aid therapy decisions and predict outcomes. The novel complementary mass spectrometry-based methods described in this chapter will contribute to the transformation of pathology services around the world.

ABBREVIATIONS

CHCA α-cycano-4-hydroxycinnamic acid
DESI desorption electrospray ionization
DHB 2,5-dihydroxybenzoic acid
FFPE formalin-fixed paraffin embedded
FT-ICR Fourier transformed ion cyclotron resonance
IHC immunohistochemistry
LNM lymph node metastasis
MALDI matrix-assisted laser desorption/ionization
MSI mass spectrometry imaging
PCA principle component analysis
SIMS secondary ion mass spectrometry
TOF time-of-flight

1. INTRODUCTION

In order to optimize the treatment outcomes for cancer patients and even provide a cure, several issues regarding the clinical management of disease need to be addressed. Based on pathological and clinical parameters patients are classified to receive the best possible therapy. The accuracy of this process would be better if a more complete understanding of the molecular events involved in the development and progression of cancer was available. Currently, mass spectrometry imaging (MSI) is being extensively applied to the in situ molecular analysis of cancerous cells and tissues with the aim of identifying tumor margins, classifying primary tumor tissues with regards to chemoresponse and metastatic status, identifying diagnostic and prognostic markers, and the analysis of drug response rates and resistance. Elucidation of the spatial location and abundance of peptides, proteins, lipids, glycans, and drug metabolites in relation to cancerous tissues and cells has the potential to significantly improve the diagnosis, staging, and treatment of disease.

1.1 The Advantages of MSI

Interpatient heterogeneity significantly impacts the success of diagnostic and prognostic tests, because it influences the way patients respond to the same treatment. Moreover, proteomic analysis is impacted by intratumor heterogeneity, as tumors are complex structures comprised of a number of different cell types, such as epithelial or endothelial, stromal, vascular, and inflammatory cells. This is particularly important when considering concepts such as innate chemoresistance, metastasis, and stem-cell-like cancer cells, which are believed to consist of a small subset of cells within the bulk of the tumor tissue. Analysis methods that require homogenisation of a sample lose important information about the spatial location of the molecules being studied. Moreover, the information of a potential difference in protein abundance of a small number of cells might not be reflected in the average abundance of the tissue. Certain analysis methods, such as immunohistochemistry (IHC) or fluorescence microscopy, allow for the quantification of compounds within a spatial context. However, these methods are targeted, requiring prior knowledge about the target and the sample in question, and only a small number of molecules can be analyzed simultaneously.

The rapidly evolving technique of MSI allows for the in situ analysis of biological samples, combining classical mass spectrometry (MS) with histological tissue analysis. MSI has the capacity to determine the relative intensity and spatial distribution of several hundreds of compounds from cells and tissue while retaining important spatial information (Schwamborn &

Caprioli, 2010). In the context of cancer research this affords the technique the ability to characterize the molecular features of a sample while maintaining morphology and requiring no prior knowledge of the molecular expression profile. The type of molecules detected, or ions in the context of MS, is dependent on the sample preparation, instrumentation, and acquisition protocols used.

1.2 The Basic Principles of MSI

MS analysis includes the processes of analyte ionization, mass analysis, and detection (de Hoffmann & Stroobant, 2007). During ionization analyte molecules become electrically charged and liberated into the gas phase. Once ionized the mass of the charged molecules is determined by analysis in the mass spectrometer; the analyte ions are separated based on their mass-to-charge ratio (m/z) and detected, commonly by a microchannel plate detector. In terms of MSI, there are two basic forms: microprobe mode imaging and microscope mode imaging (Bodzon-Kulakowska & Suder, 2016; Klerk, Maarten Altelaar, Froesch, McDonnell, & Heeren, 2009; Luxembourg, Mize, McDonnell, & Heeren, 2004). In microprobe mode MSI, where information is collected sequentially from each pixel (pixel sizes typically 5–100 µm), each mass spectrum contains information about the m/z and relative intensity of the detected ions at each location. Following data acquisition, the collected information is reconstructed into a molecular image. In microscope mode MSI, spectral information is collected over a relatively large sample area (100–300 µm) simultaneously; in this form of MSI, the ions produced over the analyzed area are measured using a position-sensitive two-dimensional detection system. The information collected by both microprobe and microscope mode can be displayed as ion maps which show the distribution of the detected m/z values and their intensities, thereby characterizing cancer cells or tissues within their environment in a spatial context.

Following MSI the sample morphology is frequently retained and can be histochemically analyzed postdata acquisition (Crecelius et al., 2005; Schwamborn et al., 2007), making it feasible to directly compare MSI results with histological staining. These methods are complementary and have a high potential for aiding pathologists and clinicians in their diagnosis and management of disease (Rodrigo et al., 2014). As indicated earlier, MSI usually requires very little material and, for example, a single tissue section from an endoscopic biopsy is enough to perform a successful MSI experiment.

For MSI, the ionization methods matrix-assisted laser desorption/ionization (MALDI), secondary ion mass spectrometry (SIMS), and desorption electrospray ionization (DESI) are popular, and are frequently coupled to time-of-flight (TOF) or Fourier transform type mass analyzers. A very brief description of the commonly applied techniques and their capabilities is provided in the following sections.

1.2.1 MALDI-TOF

MALDI-TOF is commonly used for the analysis of proteins, peptides, lipids, metabolites, and glycans (Cho et al., 2015; Signor & Boeri Erba, 2013). The application of MALDI-TOF to imaging MS was first introduced in 1997 (Caprioli, Farmer, & Gile, 1997) and the technique has gained significant popularity over the last two decades. For analysis by MALDI-TOF-MSI, a sample is placed on a conductive surface such as a stainless steel plate or an indium tin oxide-coated glass slide. Commonly fresh frozen or formalin-fixed paraffin-embedded (FFPE) tissue sections 4–10 μm thick are analyzed by MALDI-TOF-MSI, with FFPE tissues requiring additional preparation steps of deparaffinization and antigen retrieval (Ronci et al., 2008). Once the tissue is placed on the slide, depending on the type of analysis being performed, a digestion step may be carried out. For proteolytic peptide analysis, the most common enzyme utilized is trypsin, for glycan analysis a peptide-N-glycosidase F (PNGase F) digestion may be employed to release N-linked glycan structures from proteins (Gustafsson et al., 2015). The analysis of lipids, metabolites, endogenous peptides, and intact proteins directly from tissue does not require enzyme treatment. Prior to analysis a small organic acid, referred to as the matrix, is suspended in an acidified solvent buffer and deposited over the sample (Goodwin, 2012; Kaletas et al., 2009). During this process the matrix incorporates the analyte molecules, forming crystals. A focussed laser beam can then be applied to the surface in order to induce analyte ionization. The small organic molecules of the matrix mixture are designed to absorb the energy of the focussed laser beam resulting in an explosive desorption of the matrix and analyte crystals. The protons present in the acidified matrix solution also facilitate the analyte ionization. MALDI is a "soft" ionization technique, meaning many analytes may be analyzed without significant fragmentation.

Following ionization the ions are rapidly accelerated though a strong electric field in the ion source, the ions then enter the TOF tube or "drift" region, allowing separation based on velocity and thus m/z. The charged analyte molecules strike the detector and the number of events within a time period are recorded, providing intensity information. By comparing the measured information to calibration standards, the mass of the detected analyte ion can be calculated. The spatial resolution of MALDI-TOF imaging sits at around 20 μm and can accommodate a molecular mass range of over 100 kDa.

1.2.2 MALDI-FT-ICR

MALDI-FT-ICR is routinely applied to the imaging analysis of lipids, drugs, and metabolites. MALDI is carried out as described earlier with the Fourier transform ion cyclotron resonance (FT-ICR) acting as the mass analyser, where the m/z of analyte ions is determined by the cyclotron frequency of the ions within the fixed magnetic field of the instrument (Scigelova, Hornshaw, Giannakopulos, & Makarov, 2011).

Following ionization the charged molecules are trapped in a Penning trap, which consists of a series of electrodes that use a homogeneous axial magnetic field to capture the molecules radially, and an inhomogeneous quadrupole electric field to confine the molecules axially. Within the Penning trap, a RF electric field is applied in the plane of ion cyclotron motion which excites the ions to move in a larger, synchronous, cyclotron radius. Once the excitation field is removed, the ions are left rotating at their cyclotron frequencies in phase. The ions induce a charge which is detected as an image current by electrode detectors in close proximity. The recorded signal consists of a superposition of sine waves; this information can be used to calculate the masses of the detected ions using a Fourier transform equation. In FT-ICR a superconducting magnet is used to generate a highly stable magnetic field resulting in very high levels of resolution.

1.2.3 SIMS-TOF

SIMS utilizes beams of primary monoatomic (e.g., Bi+, Au+) or polyatomic ($C+_{60}$, $Bi+_3$, $Cs+_n$) ions in a vacuum for ionization (Fletcher, Vickerman, & Winograd, 2011). During SIMS the sample surface is bombarded with a beam of primary ions, causing molecules on the surface of the sample to be released, a process referred to as sputtering. During sputtering energy from the primary ions is transferred to the analyte molecules. Around 1% of the sputtered analyte molecules possess an electric charge; these ions can be detected by a mass analyzer, such as a TOF (as described earlier). The energy of the primary ion beam is typically high compared to the covalent bond energies of the analyte molecules, resulting in their fragmentation. This makes SIMS a "hard" ionization technique, although polyatomic ion beams offer softer ionization than monoatomic ion beams (Bodzon-Kulakowska & Suder, 2016). SIMS requires no matrix and offers higher depth and spatial resolution (<10 μm) as compared to MALDI. However, MALDI is more sensitive as it has access to higher yields of analyte and is able to detect larger molecules (Fletcher, Lockyer, & Vickerman, 2011), with the highest detectable mass range of SIMS sitting at around 1 kDa.

1.2.4 DESI

DESI uses an electrospray source of highly charged aqueous spray droplets under ambient conditions to gently desorb and ionize analyte molecules from the sample surface (Takáts, Wiseman, Gologan, & Cooks, 2004). An advantage of DESI is that it allows for the analysis of a sample in its native state. In the ion source an electrospray emitter is used to generate charged microdroplets from a solvent, which create a thin film on the sample surface that dissolves the analyte molecules. Secondary microdroplets are formed from the kinetic impact of the primary droplets, which contain the highly charged solvent and dissolved analyte. These secondary microdroplets are ionized from the sample surface to the mass spectrometer inlet. A DESI source can be coupled to most standard electrospray mass spectrometers (Hsu & Dorrestein, 2015) where following ionization, the analytes are sent via an ion transfer tube to the mass analyser for detection. During ionization little analyte fragmentation occurs, hence DESI is considered a "soft" ionization technique with a spatial resolution of around 50–100 μm and an upper mass range detection limit of around 2 kDa.

2. PROTEIN MSI IN CANCER RESEARCH

Changes in protein function play a crucial role in the development of cancer. Such alterations can be caused by number of events including altered localization, posttranslational modifications (PTMs), and/or abundance levels. Furthermore, the biology of a tumor depends not only on the cancerous cells but also on their interaction with surrounding stroma, blood vessels, and the immune system (Kriegsmann, Kriegsmann, & Casadonte, 2015). The ability to spatially resolve different proteins while simultaneously acquiring information concerning their relative abundance is particularly important in heterogeneous diseases such as cancer. Therefore, an approach combining high-resolution MS with in situ spatial analysis of the diseased area has the potential to directly identify and quantify proteins and peptides.

When MSI first entered the field of tissue-based research, it was used to analyze the spatial distribution of proteins (Stoeckli, Chaurand, Hallahan, & Caprioli, 2001). Following this ground-breaking research, different imaging methods have been established and applied to a wide range of biological problems. Among the several MSI techniques, MALDI-MSI is one of the most commonly used. MALDI-MSI has the capability of detecting a broad range of compounds present in a tissue section, provided the compound can be ionized and desorbed into gaseous phase during the MALDI process. High-resolution MS, such as MALDI-FT-ICR, has greater strength in the field of low-molecular weight compounds, while MALDI-TOF is used for the analysis of peptides and small proteins up to 25 kDa (Aichler & Walch, 2015). MALDI-MSI of thin tissue sections can result in the detection of over 500 individual protein signals within the mass range of 2–70 kDa that can be directly correlated with specific morphological regions of the tissue (Chaurand, Schwartz, & Caprioli, 2004).

Protocols have been developed for the preparation of tissue sections and matrix application that provide a high level of reproducibility (Gustafsson et al., 2013). The methodology behind protein MSI, via the analysis of their proteolytic peptides, is straightforward; a digestion enzyme is uniformly applied over the entire tissue surface either by a spotting or spray deposition method. Thereafter, using the same application method, internal calibrants and matrix are applied. Data acquisition is then carried out by a MS (Gustafsson et al., 2013), where analyte ionization, mass analysis, and detection occur. This results in the subsequent generation of mass spectrums at regular series of points across the tissue section. The mass spectra produced are then visualized as ion intensity maps; generating two-dimensional distribution maps of the detected mass spectra containing information about the location and the relative intensity of the detected peptides. This visualization is similar in nature to images generated by histological or IHC analysis, but with the advantage that multiple molecules can be examined within a single measurement (Aichler & Walch, 2015) (Fig. 1).

Three-dimensional MALDI-MS images can also be generated by analyzing a number of consecutively cut tissue sections, and following data acquisition, combining the results to generate three-dimensional maps (Andersson, Groseclose, Deutch, & Caprioli, 2008; Sinha et al., 2008).

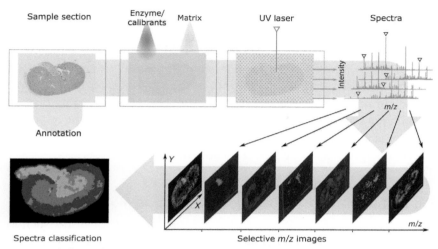

Fig. 1 Work flow of MALDI mass spectrometry imaging.

Recently, methods were developed for the integration of three-dimensional volume reconstructions of spatially resolved MALDI-MSI ion intensity images taken of whole mouse heads with corresponding in vivo data provided by magnetic resonance imaging (MRI) (Sinha et al., 2008). The added spatial dimension provides contextual information to the MALDI-MSI data that allow the interrogation of proteomic relationships within a tissue volume. The work displayed a correlation between the in vivo MRI and proteomic profiles obtained by MALDI-MSI, with both techniques able to distinguish tumor from healthy tissues (Sinha et al., 2008).

Until recently, MALDI-MSI had been carried out almost exclusively on fresh frozen samples as it was believed proteins were inaccessible from FFPE tissue due to the cross-linking caused by the formalin fixation. However, this limitation has been overcome and equal number of proteins can be identified from both fresh frozen and FFPE tissues (Diehl et al., 2015). The use of FFPE tissues in cancer research is a huge advantage, given that the samples are well preserved by the formalin fixation process and can be stored for long periods of time at room temperature. Another attractive feature is that the samples are often well documented as they have been processed by clinical pathology departments, meaning information on patient treatment response, disease progression, and other relevant clinical data have been cataloged (Gorzolka & Walch, 2014). Additionally, FFPE tissues are well suited for the construction of tissue microarrays (TMAs), where needle core biopsies from multiple tissue samples are assembled into a paraffin block. However, fresh frozen TMAs have also been studied in a limited context (Fejzo & Slamon, 2010).

One of the main advantages of MALDI-MSI is the technique's ability to detect changes in disease-associated protein expression prior to histological transformation, making it a useful tool in the identification of early stage disease biomarkers (McDonnell et al., 2010). There have been number of cancer studies that have successfully applied peptide imaging to the discovery of potential biomarkers with the capacity to differentiate tumors of various subtypes, stages, and/or degree of metastasis.

2.1 Distinguishing Tissue Types by Peptide MSI

The ability of peptide MSI to distinguish specific types of tissue, such as regions of healthy, cancer, stroma, and vasculature, has been shown multiple times (Mittal, Klingler-Hoffmann, Arentz, Zhang, et al., 2016). Schwamborn et al. performed MSI on prostate cancer specimens and was able to classify the cancerous tissue from normal tissue with 85% sensitivity and 90% specificity, with an overall cross validation accuracy of 88% (Schwamborn et al., 2007). Moreover, MSI of prostate cancer TMAs on a large scale allowed for the immediate prioritization of MSI signals based on associations with clinic-pathological and molecular data (Steurer et al., 2013). A comparison of these signals with clinic-pathological features revealed statistical association with a favorable phenotype such as grade, stage, and relationship with prolonged time to prostate-specific antigen recurrence (Steurer et al., 2013). Peptide MSI experiments have been able to discriminate normal, preinvasive, and invasive lung tumor tissues with an accuracy of 90% (Rahman et al., 2011), to differentiate normal tissue from oral squamous cell carcinoma (Patel et al., 2009), to discriminate normal tissue from gastric cancer tissues (Kim et al., 2010), and have identified mitogen-activated protein kinase/extracellular signal-regulated kinase kinase 2 (MEKK2) as a marker with the capacity to accurately discern cancerous prostate from normal tissue (Cazares et al., 2009).

One of the most challenging aspects of neurochemistry is the detection of endogenous neuropeptides due to their low in vivo concentrations ranging from pico- to femto-molar levels, which makes it difficult to obtain enough material for quantitative analysis (Andersson, Andren, & Caprioli, 2010; Strand, 2003). MALDI-MSI helps overcome this issue as it allows the direct detection of neuropeptides in discrete regions of a brain sections, enabling the study of the physiological and disease-related metabolic processing of neuropeptides (Andersson et al., 2010). Recently, Andersson et al. has been able to sequence and identify neuropeptides by in situ tandem mass spectrometry (MS/MS) directly off rat brain sections using a MALDI-QTOF mass spectrometer (Andersson et al., 2008).

Furthermore, peptide MALDI-MSI has been employed to differentiate benign regions from malignant in ovarian cancer (Lemaire, Ait Menguellet, et al., 2007) and in thyroid cancer (Pagni et al., 2015). Likewise, biomarkers have been identified that could distinguish Hodgkin's lymphoma from lymphadenitis (Schwamborn et al., 2010).

2.2 Determining Tumor Margins by Peptide MSI

One of the major aims of clinical oncology is to ensure a tumor has been completely removed during surgery to minimize the possibility of disease recurrence (Han et al., 2011). Defining tumor margins also plays a pivotal role in cancer staging, which subsequently influences the administration of treatment. Moreover, the molecular characterization of tumor margins will assist in understanding the process of invasion into surrounding tissues and may help in the clinical management of disease (Agar et al., 2010). MALDI-MSI has been proposed as a valuable tool for gaining information with regards to tumor margins and heterogeneity (Chaurand, Sanders, Jensen, & Caprioli, 2004). Peptide MSI has interestingly revealed histologically normal tissue adjacent to renal cell carcinoma regions, share many molecular characteristics with that of the tumor tissue, providing unprecedented insight into cancer development (Oppenheimer, Mi, Sanders, & Caprioli, 2010). The technique has also been used to identify protein biomarkers for the differentiation of delineated hepatocellular carcinoma from adjacent cirrhotic tissue (Le Faouder et al., 2011). Analysis of ovarian cancer interface zones, the regions between tumor and normal tissues, by MSI revealed a unique peptide profile from both the tumor and normal tissue, and detected plastin 2 and peroxiredoxin 1 as interface markers specific to ovarian cancer (Kang et al., 2010).

2.3 Prediction of Metastasis by Peptide MSI

Metastatic status is often the defining feature in determining if a patient will receive chemotherapy and is currently resolved by the removal and analysis of local lymph nodes, which frequently results in noncancer-related health complications. Identification of primary tumor markers of metastasis would prevent the unnecessary removal of lymph nodes, thereby reducing patient suffering and associated treatment costs, and allow the introduction of more personalized cancer therapies (Casadonte et al., 2014). However, identification of metastasis based on analysis of the primary tumor tissues is challenging, especially when the primary carcinoma site is unknown (Pavlidis & Pentheroudakis, 2016). A statistical classification model based on peptide MSI data has been developed for determining the metastatic status of primary cancers of pancreatic origin and their metastasis to secondary breast cancer with an overall accuracy of 83% (Casadonte et al., 2014). In another study the HER2 status of gastric cancers could be predicted with 90% accuracy using peptide expression patterns originating from breast cancers (Balluff et al., 2010). These studies display the potential of peptide MSI in the profiling of cancers, independent of their site of origin.

The likelihood of lymph node metastasis (LNM) is one of the most important factors to consider while determining the appropriate treatment for a cancer patient. Several studies have applied MALDI-MSI to the prediction of LNM status. Yanagisawa et al. was able to classify primary lung tumors with and without LNM with an accuracy of 85% (Yanagisawa et al., 2003). Different peptide signatures have been identified with the ability to distinguish melanoma (LNM) from control lymph nodes (Hardesty, Kelley, Mi, Low, & Caprioli, 2011). Similarly, the overexpression of S100A10, thioredoxin, and S100A6, have been shown to be highly associated with LNM in papillary thyroid carcinoma (PTC) (Nipp et al., 2011), and it was suggested the three proteins may be used for risk stratification regarding metastatic potential in PTC. Other protein biomarkers have been suggested for the prediction of LNM in several cancers from peptide MSI experiments, such as S100A8 and S100A9 for gastric adenocarcinoma (Choi et al., 2012), COX7A2, TAGLN2, and S100-A10 for monitoring the Barrett's adenocarcinoma development as well as for predicting regional (LNM) and disease outcome (Elsner et al., 2012).

Recently, it has been shown that the LNM status of primary endometrial tumors could be predicted with an accuracy of 88% using a classification system based on tissue peptide markers detected by MALDI-MSI (Mittal, Klingler-Hoffmann, Arentz, Winderbaum, et al., 2016; Winderbaum, Koch, Mittal, & Hoffmann, 2016). In the study FFPE-TMAs comprised of duplicate tumor tissue cores from endometrial cancer patients diagnosed with and without LNM were analyzed by MALDI-TOF-MSI. An overview of the sample preparation and approach used for the proteomic marker identification by MALDI-MSI-(top panel) and LC-MS/MS (bottom panel) is outlined in Fig. 2.

2.4 Analysing Chemoresponse by Peptide MSI

Innate and acquired chemoresistance significantly impact the survival rates of late stage cancer patients. Administration of chemotherapy to patients with specific cancer types has traditionally been standardized, despite the knowledge that response can differ significantly among a cohort of patients due to genetic diversity and tumor heterogeneity. Identification of markers or tests that predict resistance to standard chemotherapy regimens would help save patients from damaging, ineffective treatments, and may help determine if an alternative treatment course may be more beneficial.

IHC is one of the most widely applied proteomic techniques in clinical practice. However, IHC is often unable to distinguish the different isoforms of the same protein that may play different role in the disease development. Recent studies using MSI have been able to distinguish two isoforms of the protein defensin. This is of particular interest given the 2 defensin protein isoforms respond differently in breast cancer tumors when exposed to chemotherapy (Bauer et al., 2010). The finding was discovered during the analysis of a breast cancer patient receiving neoadjuvant paclitaxel and radiation therapy, and was one of the first biomarker discovery studies to utilize MALDI-MSI (Bauer et al., 2010). Mitochondrial defects have also been found and linked to chemotherapy response in oesophageal adenocarcinoma patients using peptide MSI (Aichler et al., 2013), displaying the techniques potential to change classifications of tumors and chemoresponse.

2.5 Identification of Diagnostic and Prognostic Markers by Peptide MSI

A high proportion of cancer patients are diagnosed in the later stages of disease, when the cancer has already metastasized to distant lymph nodes or organs. Detecting cancer in the earlier stages of disease, prior to metastasis, would help increase survival rates and reduce the treatment burden for patients. In a genetically engineered mouse model, peptide MSI was used to detected several discriminative peptide masses that could distinguish between intraepithelial neoplasia, intraductal papillary mucinous neoplasm, and normal pancreatic tissue (Grüner et al., 2012). Using MALDI-MSI to analyze TMAs, significant differences between papillary and clear cell renal cell cancer were found (Steurer, Seddiqi, et al., 2014). Analysis of bladder

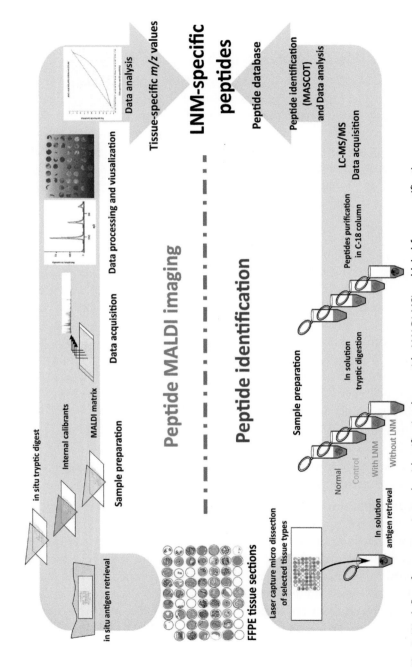

Fig. 2 Work flow for proteomic marker identification by peptide MALDI-MSI and label-free quantification.

cancer TMAs by MSI lead to the identification of multiple signals in the mass spectra associated with unfavorable tumor phenotype and clinical characteristics, such as tumor recurrence, progression, and patient survival time (Steurer, Singer, et al., 2014).

MALDI-MSI has also been used to identify a protein signature that can accurately determine the HER2 status in breast cancer with sensitivity of 83%, for specificity of 92%, and overall accuracy of 89% (Rauser et al., 2010). A seven-protein signature of novel tissue markers was identified as an independent indicator of unfavorable overall survival in intestinal type gastric cancer (Balluff, Rauser, et al., 2011). Other investigators identified histone 4 expression as specific to poorly differentiated gastric cancer tissues by peptide MSI (Morita et al., 2010). A comparison of genomically stable (diploid) and unstable (aneuploid) colon cancer tissues against normal mucosa by MSI identified thymosin beta 4 as an independent prognostic marker for colorectal cancer (Gemoll, Strohkamp, Schillo, Thorns, & Habermann, 2015). Analysis of lung cancer TMAs containing 112 needle core biopsies were used to generate a support vector machine model utilizing MSI information on 73 peptides that could classify adenocarcinomas with an accuracy of 97% and squamous cell carcinomas with an accuracy of 98% (Groseclose, Massion, Chaurand, & Caprioli, 2008).

2.6 Characterisation of Intra- and Intertumor Variability by Peptide MSI

Understanding the spatiotemporal patterns and dynamics of intratumor heterogeneity is crucial for clinical management and designing personalized targeted regimens for cancer patients (Renovanz & Kim, 2014). Balluff et al. showed for the first time that MALDI-MSI of tumor tissues can reveal microscopically indistinct tumor populations that might have adverse impacts in clinical outcome (Balluff et al., 2015). In the proof-of-principal study tumor-specific mass spectra were grouped using an advanced statistically clustering algorithm in order to obtain segmentation maps of molecularly distinct regions. The distinct regions were then statistically compared with patient clinical data to identify *phenotypic* tumor subpopulations (Balluff et al., 2015). This approach revealed that several of the detected tumor subpopulations are associated with different overall survival characteristics of gastric cancer patients ($p=0.025$) and the presence of regional metastasis in patients with breast cancer ($p=0.036$). This approach enables researchers to gain deeper insights into the underlying biological process and changes of tumor subpopulations on a genetic, metabolic, and proteomic level, which may result in novel targeted therapies (Balluff et al., 2015).

The ability of MALDI-MSI to reveal inter- and intratumoral biomolecular heterogeneity of histologically contiguous tumors has been shown, specifically in the capacity to distinguish different types and grades of myxoid sarcomas (Willems et al., 2010). By applying hierarchical clustering to a gastric cancer MALDI-MSI dataset, it has been demonstrated the different histological regions could be distinguished solely on the basis of their protein and peptide expression profiles (Deininger, Ebert, Fütterer, Gerhard, & Röcken, 2008). Ion mobility separation combined with MALDI-MSI has been used to classify prostate cancer TMAs and has identified a number of proteins, including type I collagen and tumor necrosis factor receptor, directly from the tissue that could discriminate different tumor classes (Djidja et al., 2010). Low grade urothelial neoplasia could be separated from high grade urothelial neoplasia with an overall cross validation rate of 97.18% upon analysis of peptide MSI data (Oezdemir et al., 2012). Likewise, based on protein profiling, a grading system for meningioma was developed (Agar et al., 2010). Proteomic signatures associated with breast cancer-activated stromal tissues have been identified by MSI, and hierarchical clustering of the proteomic signals could be used to distinguish the activated intratumoral stroma from the quiescent extratumoral stroma (Dekker et al., 2014). High grade and low grade glioma can also be distinguished by MALDI-MSI (Chaurand et al., 2004).

2.7 Practical Considerations for Proteolytic Peptide MSI: Sample Preparation

MSI of proteolytic peptides in cancer is frequently performed on fresh frozen or FFPE tissues; each sample type has unique advantages and limitations with regards to MSI. The protein component of fresh tissues is easily accessible, especially when compared to FFPE tissues, due to a lack of chemical fixation. Hence, no retrieval steps are required prior to the proteolyic digestion of fresh frozen samples. A disadvantage of fresh frozen tissues, however, is that they require rapid freezing upon collection and storage at extremely low temperatures to inhibit endogenous enzymatic degradation (Fentz, Zornig, Juhl, & David, 2004). Fresh tissues also require solvent washing to remove lipids and other biomolecules that may compete for ionization and interrupt the detection of peptides. Washing steps must be performed carefully, as extensive washing may cause the loss of soluble proteins

(McDonnell et al., 2010; Seeley, Oppenheimer, Mi, Chaurand, & Caprioli, 2008). Two significant advantages of FFPE samples are that they are widely used in modern clinical practice, meaning very large sample banks have often been developed, and that the fixation process allows for storage of the samples at room temperature for indefinite periods of time without loss of morphological information (Wisztorski et al., 2007). The downfall is that formalin fixation results in the formation of methylene bridges between amino residues that causes protein cross-linking (Metz et al., 2004). This cross-linking makes the proteins inaccessible to proteolysis and must be reversed prior to any further preparation steps. Heat-induced antigen retrieval prior to in situ proteolytic digestion is a widely used method for reversing protein cross-links, which allows for the peptide profiling of FFPE tissues that is comparable to fresh frozen tissue analysis (Casadonte & Caprioli, 2011).

2.8 Practical Considerations for Peptide MSI: Spatial Resolution

Currently, MALDI is the most popular ionization method applied in the field of MSI. The success of analyte ionization in a MALDI experiment relies significantly on the matrix used; the type of matrix applied will impact on the mass range of the detectable ions, the intensity of detection, and the minimum spatial resolution achievable (Lemaire et al., 2006). For peptide MALDI-MSI commonly used matrices include α-cycano-4-hydroxycinnamic acid (CHCA) and 2,5-dihydroxybenzoic acid (DHB) (van Hove, Erika, Smith, & Heeren, 2010). The homogeneity of matrix crystallization is a major limiting factor with regard to the spatial resolution that can be achieved by MALDI-MSI, as larger and nonuniform matrix crystals result in lower resolutions. DHB matrix is prone to producing large, needle-like crystals, making CHCA a more suitable matrix as it commonly forms small, homogeneous crystals resulting in higher image resolution (Schwartz, Reyzer, & Caprioli, 2003). The way in which the matrix is applied also has a significant impact on the spatial resolution, with the application of larger size droplets increasing the likely hood of analyte displacement. Instrumentation has been developed that allows matrix solutions to be sprayed or spotted onto the sample surface in a more homogeneous manner compared to manual application (Aerni, Cornett, & Caprioli, 2006). For example, using a matrix spray device CHCA can be applied to form uniform crystals that allow for spatial resolution peptide imaging of <25 μm (McDonnell et al., 2010; Schuerenberg & Deininger, 2010).

Laser specifications also have a significant impact on the spatial resolution achievable. As opposed to the limitations of matrix crystallization, laser beam resolutions can easily reach less than 10 μm, theoretically allowing for peptide imaging at the single cell level (Balluff, Schöne, Höfler, & Walch, 2011). However, reducing the size of the laser beam concurrently reduces the yield of ionized analytes. Higher frequency lasers and overlapping laser shots can counteract the issue of lower ion yields during detection (Jurchen, Rubakhin, & Sweedler, 2005). Thus, currently matrix is the limiting factor for the spatial resolution achievable in MALDI-MSI.

2.9 Practical Considerations for Peptide MSI: Mass Analysers

The type of mass analyser used for MSI will largely determine the mass resolution and mass range of the detectable ions. TOF mass analysers are commonly coupled to MALDI sources and have high acquisition speeds but typically lower mass resolution (Caprioli et al., 1997). Higher mass accuracy can be achieved through the use of internal or external calibrants. Other commonly used mass analysers include FT-ICR and Orbitraps; these instruments have significantly higher mass accuracy and resolution (Taban et al., 2007).

Internal or external calibrants can be used to improve the mass accuracy of acquired spectra during or after data acquisition (Gobom et al., 2002). Calibrants are often comprised of a small number of specific analytes with known masses that span a specific mass range. Within this mass range, the use of calibrants can significantly increase the mass accuracy of the detected analytes. However, the mass error of any analyte molecules that fall outside the mass range of the calibrants cannot be corrected (Coombes, Koomen, Baggerly, Morris, & Kobayashi, 2005).

2.10 Practical Considerations for Peptide MSI: Identification

Currently, MALDI-MSI enables the discrimination of heterogeneous tissue types, but does not allow for the direct identification of the peptide sequences. During data acquisition the intact mass, or the m/z, of the detected peptides is recorded. This allows peptide mass fingerprinting (PMF) to be performed, a process where the intact peptide masses are matched back to a database containing the in silico digested proteome or proteomes of specific species (Pappin, Hojrup, & Bleasby, 1993). Due to a lack of peptide sequence information, often PMF can provide a list of potential peptide identifications but is not able to provide definitive identifications for those peptides with similar masses. However, the higher the mass accuracy of the instrument used the more likely it is a definitive identification will be gained (Horn, Peters, Klock, Meyers, & Brock, 2004). The use of complementary MS techniques, such as matching data to peptides sequenced by traditional liquid chromatography electrospray ionization tandem mass spectrometry (LC-ESI-MS/MS) or by in situ tandem mass spectrometry (MALDI-MS/MS), can help clarify peptide sequence identifications (Lemaire, Desmons, et al., 2007). However, there can be significant difficulties associated with these options also. Multiple sequence matches are often encountered when matching MSI m/z values back to LC-ESI-MS/MS data obtained from the same sample. Generating high quality in situ MS/MS spectrums is often challenging due to the issue that acquisition of very complex peptide mass spectra directly off tissue results in poor fragmentation and high MS/MS complexity, both of which limit direct peptide sequencing from tissues (Gustafsson et al., 2012).

3. LIPID MSI IN CANCER RESEARCH
3.1 Profiling Lipids in Cancer by DESI-MSI

The soft ionization technique of DESI has been developed for the MSI of small molecules and lipids. DESI uses highly charged solvent droplets traveling at high velocities for analyte ionization in an ambient atmosphere. An advantage of DESI for the analysis of lipids and small molecules is that no sample preparation steps are required, unlike MALDI where matrix deposition is a necessity. A further advantage is that a DESI source can be coupled to most standard mass analysers, such as a FT-ICR, Orbitrap, Ion Trap, or a TOF.

In practice, DESI-MSI has been applied to the study of human cancers for the discrimination of tumor subtypes and grades, as well as for the identification of tumor margins (Calligaris et al., 2014; Eberlin et al., 2014). One particular application used DESI-MSI to generate lipid profiles for the classification of human brain tumors (Eberlin et al., 2012). This study was able to classify the subtype, grade, and tumor concentration of 36 human glioma samples via lipidomics imaging in agreement with the histopathology diagnoses (Eberlin et al., 2012). Lipidomic analysis by DESI-MSI has identified the phospholipids phosphatidylserine (PS), phosphatidylinositol (PI), and phosphatidylethanolamine as biomarkers for distinguishing healthy tissue from benign and malignant tumor tissues (Abbassi-Ghadi et al., 2014; Dill et al., 2011, 2009; Eberlin et al., 2012).

Discriminatory lipid signatures between cancerous and normal breast tissue have been detected by DESI-MSI (Calligaris et al., 2014), with the delineation of tumor margins possible through the analysis of PI (18:0/20:4). PI (18:0/20:4) has been shown to be abundant at the tumor center and tumor edge, whereas it was absent or weak in normal tissues. This lipid sharply stratifies the tumor margin in agreement the histopathological staining. Using DESI-MSI, another study found PS (20:4/18:0) and PI (18:0/20:4) are significantly lower in concentration in primary tumor tissues and LNMs when compared to normal lymph node tissues (Abbassi-Ghadi et al., 2014). The same lipid, PI (18:0/20:4), was found to be a biomarker for distinguishing benign and malignant breast tumors (Yang et al., 2015). Collectively, the discussed research shows MSI lipid profiling of cancer tissues can be used for classifying tumor types, with high specificity in differentiating histopathologic grade, and in the mapping of tumor margins.

3.2 Profiling Lipids in Cancer by SIMS-MSI

A less popular ionization method for MSI in the lipidomics field is static SIMS. Generally, a static SIMS source is coupled to a TOF (or QTOF) instrument, and is particularly suitable for the analysis of intact molecules such as lipids and metabolites. It uses a beam of energetic primary ions with either monatomic source, such as Cs^+ and Ga^+ for limited mass range of low hundreds Daltons, or cluster ions such as SF_5^+, Bi_3^+, and Au_3^+ that have extended the mass range to a couple of thousands Daltons (Lanni, Rubakhin, & Sweedler, 2012). The potential of SIMS-TOF imaging for lipidomics has been shown in rat brain sections, a well-established model system for MSI (Benabdellah et al., 2010; Passarelli & Winograd, 2011; Sjovall, Lausmaa, & Johansson, 2004; Sjovall, Lausmaa, Nygren, Carlsson, & Malmberg, 2003). In the field of cancer lipidomics, it is frequently used in cellular imaging, where single cells or a population of cells are isolated after drug treatment for the characterisation of molecules on the cell surface. For example, lipid composition analysis of individual human breast cancer stem cells (CSCs) reported significantly lower expression levels of palmitoleic acids FA (16:1) as compared to nonstem cancer cells (NSCCs), and is thought to have been successfully characterized from complex clinical specimens using TOF-SIMS (Waki et al., 2014).

4. GLYCAN MSI IN CANCER RESEARCH

Glycosylation is the enzymatic process where oligosaccharides, polysaccharides, and carbohydrates (i.e., glycans) are attached to proteins, lipids, or other organic molecules (Ohtsubo & Marth, 2006). In recent years, the word "glycomics" has emerged to describe the study of glycan structures from an organism's "glycome." The mammalian glycome repertoire is comprised of 10 known monosaccharides. These monosaccharides include xylose (Xyl), glucose (Glc), galactose (Gal), mannose (Man), N-acetylglucosamine (GlcNAc), N-acetylgalactosamine (GalNAc), fucose (Fuc), glucuronic acid (GlcA), iduronic acid (IdoA), and N-acetylneuraminic acid (NeuAc or Neu5Ac), a sialic acid (Wuhrer, Deelder, & van der Burgt, 2011). Glycan complexity arises from variation in expression of specific glycosyltransferases, resulting in a diversity of monosaccharide arrangements. Furthermore, modifications such as sulfation, acetylation, or phosphorylation increase the complexity of glycosylation.

The major type of glycosylation is protein glycosylation, where glycans are attached to asparagine residues by their amine group (N-linked) or at serine/threonine residues by their hydroxyl group (O-linked) (North, Hitchen, Haslam, & Dell, 2009). Protein glycosylation is an important PTM which has relevance in many biological processes such as cell signaling, immune responses, extracellular interaction, and cell adhesion (Varki, 1993). In the tumor microenvironment, aberrant protein glycosylation such as the expression of truncated glycans have been well described in various cancers (Dube & Bertozzi, 2005). These structures may resemble glycomic "fingerprints" which discriminate between healthy and cancerous tissues or potentially discriminate between various cancer subtypes (Abbott et al., 2010).

One of the most common glycosylation changes in various cancers is an increase in branching of N-glycans, where large tetra-antennary structures are formed by the expression of N-acetylglucosaminyltransferase V (GnT-V) (Lau & Dennis, 2008). The presence of increased branching antennas creates additional sites for the addition of NeuAc by sialyltransferases in various cancers (Bull et al., 2013). Likewise, terminal modifications on glycan structures such as fucosylation on the cancer cell surface can also give rise to the presence of Lewis and sialyl Lewis antigens (sialyl Lea; [Neu5Acα2-3Galβ1-3(Fucα1-4)GlcNAcβ] and sialyl Lex; [Neu5Acα2-3Galβ1-4(Fucα1-3)

GlcNAcβ]) which have been shown to correlate with tumor progression and metastasis (Nakagoe et al., 2000). It is evident that aberrant glycosylation is indeed a key event in metastasis and therefore the discovery of cancer-specific protein glycosylation markers would be beneficial in the staging and treatment of patients (Drake et al., 2010).

As glycomics gains recognition in this postgenomics era, MS-based methodologies are becoming indispensable and routinely used for the reliable profiling of glycans from clinical samples. Glycan MSI is a recent MS-based development which allows the visualization of the tissue-specific spatial distribution of glycans. N-glycans have been the most comprehensively studied type of protein glycosylation, as they can be easily released from proteins by enzymatic cleavage using PNGase F (Jensen, Karlsson, Kolarich, & Packer, 2012). The majority of these studies have applied optimized N-glycan MALDI-MSI methods to FFPE tissues, with the most recent focus on various cancer tissues.

One of the first N-glycan MALDI-MSI methods to emerge was published in 2013 (Powers et al., 2013). This proof-of-principle paper described the measurement of N-glycans by MALDI-MSI across fresh frozen mouse brain tissue. The m/z values from the sum spectrum were assigned to N-glycans based on further MS structural analysis, for example, high-performance liquid chromatography (HPLC)-MS. The authors concluded that this new approach could lead to novel disease-related targets for biomarker and therapeutic applications. Subsequently, N-glycan MALDI-MSI has been successfully applied to the analysis of glioblastoma tumor xenografts compared to normal brain tissues (Toghi Eshghi et al., 2014), where it was found specific N-glycans associated with distinct tissue regions. N-glycan imaging of FFPE pancreatic and prostate cancers, and a human hepatocellular carcinomas in TMAs found tumor and nontumor regions could be distinguished in the TMA format, enabling high-throughput analysis (Powers et al., 2014).

The use of complimentary MS methods, such as liquid chromatography tandem MS (LC-MS/MS) with MALDI-TOF-MSI, have been developed to enable the mapping of tissue-specific N-glycans at high resolution (Gustafsson et al., 2015). The application of these techniques has resulted in several interesting findings. A panel of over 30 N-glycans has been detected from FFPE human hepatocellular carcinoma tissues (Powers, Holst, Wuhrer, Mehta, & Drake, 2015), from which the N-glycans were extracted and the sialylated species were stabilized by an off-tissue ethylation reaction. N-glycan sialyation linkage-specificity was further found to correlate with the different tissue types analyzed. A novel method for the analysis of N-glycan sialylated species (i.e., α 2–3 and α 2–6) by MALDI-MSI in colon carcinoma found sialylated N-glycans with α 2–3 linkage were observed in stroma, tumor, and necrotic regions, while sialylated N-glycans with α 2–6 linkage were observed in necrotic, collagen-rich, and red blood cell regions (Holst et al., 2016).

The analysis of FFPE stage III ovarian cancer tissues found high mannose structure (Man5 + Man3GlcNAc2) in the tumor regions, while triantennary complex structures (Hex3HexNAc3Deoxyhexose1 + Man3GlcNAc2) were observed in the stromal regions (Fig. 3). This observation was consistent between not only the patients investigated, but also between each N-glycan family. The tissue analyzed in the study had been annotated by a pathologist, and it was found specific glycan structures were differentially detected in specific tissue regions, such as tumor, stroma, and adipose tissue.

In summary, N-glycan MALDI-MSI has developed immensely in recent times, with improvements in areas such as sensitivity and mass accuracy,

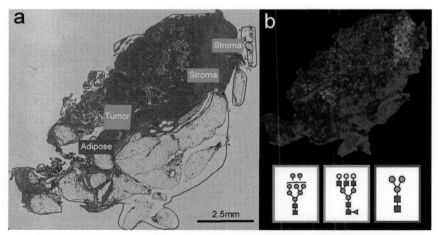

Fig. 3 Glycan analysis of ovarian cancer. The formalin-fixed paraffin-embedded stage III ovarian cancer section was treated with antigen retrieval prior to printing of 15 nL/spot dialyzed PNGase F with 250 μm spacing. 2,5-DHB (20 mg/mL) was sprayed onto the sections and MS spectra were acquired by oversampling at 100 μm intervals using a MALDI-TOF/TOF MS instrument. (A) Haematoxylin and eosin stain of the ovarian cancer section with tumor (*red*), stroma (*green*), and adipose (*blue*) regions annotated by a pathologist. (B) An ion intensity map of *m/z* 1743.7 (*red*), *m/z* 2174.9 (*green*), and *m/z* 933.3 (*blue*) on the ovarian cancer section. *Yellow circle*, galactose (Gal); *green circle*, mannose (Man); *blue square*, N-acetylglucosamine (GlcNAc); *red triangle*, fucose (Fuc).

compositional analysis by complementary techniques (i.e., LC–MS/MS and lectin histochemistry staining), and identification of linkage-specific sialylated species by in situ derivatization methods. However, further work in the field is required to properly answer clinical questions and discover cancer-specific markers. TMAs have not yet been fully utilized in N-glycan MSI experiments, but offer promise in providing researchers and clinicians with high-throughput data for analysis.

5. DRUG IMAGING IN CANCER RESEARCH

Quantification of drugs and their metabolites is crucial in oncology where drug distribution within the tumor tissue is thought to play a pivotal role in response to therapy and could partly explain the variable response rates observed among patients with similar tumor types (Minchinton & Tannock, 2006). Traditionally, autoradiography has been used to visualize the localization of a drug. However, this technique suffers from some significant limitations. One of them is that the synthesis of the radiolabeled drug is an expensive and time-consuming process. Second, the technique does not provide information on the molecular structure. Hence, metabolites either cannot be distinguished from the parent drug or are not detected if the radiolabel is lost in a metabolic process (Kertesz et al., 2008). The other clinically used noninvasive drug imaging techniques are positron emission tomography and MRI. These techniques can provide information on the drug distribution but suffers from the limitation of low spatial resolution. Due to the relatively simple sample preparation steps required for MSI the technique has rapidly emerged as an alternative to the conventional methods of analysis. MSI allows detailed distribution analysis of the parent drug and its metabolites in a single experiment and without any labeling of the targeted compound in contrast to conventional immunohistochemical methods. Such drug characterization in situ, by both spatial and temporal behaviors within tissue compartments, provides a new understanding of the dynamic processes impacting drug uptake and metabolism at the local sites targeted by a drug therapy (Kwon et al., 2015). Multiple MSI techniques including MALDI-MSI, SIMS, and nanostructure initiator mass spectrometry (NIMS) have been used to analyze the distribution of the drug and metabolites. However, MALDI-MSI is the most frequently used method for looking at anticancer drug distribution in tumor samples.

5.1 MALDI-MSI on Tissue Sections

MALDI-MSI was used for the first time in 2003 to directly analyze and image pharmaceutical compounds in intact tissues (Reyzer, Hsieh, Ng, Korfmacher, & Caprioli, 2003). From then, the technique has been applied to the imaging of a number of drugs. Prideaux et al. applied MALDI-MRM-MSI to visualize the distribution of the second line tuberculosis drug moxifloxacin at a range of time points after dosing in tuberculosis-infected rabbits, which was further validated by quantitative LC-MS/MS of lung and granuloma extracts from adjacent biopsies taken from the same animals (Prideaux et al., 2011). Drug distribution within the granulomas was observed to be inhomogeneous, and very low levels were observed in the caseum in comparison to the cellular granuloma regions (Prideaux et al., 2011). Fehniger et al. measured the occurrence of inhaled bronchodilator, ipratropium, within human bronchial biopsies obtained by fiber optic bronchoscopy shortly after dosing exposure and showed the drug is rapidly absorbed into the airway wall (Fehniger et al., 2011). Nilson et al. provided the first evidence that compounds administered by inhaled delivery at standard pharmacological dosage can be quantitatively detected by MALDI-MSI with accuracy and precision (Nilsson et al., 2010). The distribution of the inhaled drug tiotropium was tracked and quantified in the lungs of dosed rats and a concentration gradient (80 fmol–5 pmol) away from the central airways into the lung parenchyma and pleura was observed (Nilsson et al., 2010). Hsieh et al. visualized the spatial distribution of astemizole (withdrawn in most countries due to its side effects) and its primary metabolite in rat brain tissues and showed that the drug alone is likely to be responsible for the associated central nervous system side effects upon elevated exposure (Hsieh, Li, & Korfmacher, 2010). Using MALDI-MSI the specific distribution of unlabeled chloroquine was examined in the retinas of rats and was found to be similar to autoradiograms results reported previously (Yamada, Hidefumi, Shion, Oshikata, & Haramaki, 2011).

Atkinson et al. showed the distribution of the bioreductive anticancer drug AQ4N (banoxatrone), its active metabolite AQ4, and ATP in treated human tumor xenografts (Atkinson, Loadman, Sutton, Patterson, & Clench, 2007). The distribution of ATP was found similar to that of AQ4N, i.e., in regions of abundant ATP there was no evidence of conversion of AQ4N into AQ4, indicating that the cytotoxic metabolite AQ4 is confined to hypoxic regions of the tumor as intended (Atkinson et al., 2007). Using

MALDI-FT-ICR, Cornett et al. imaged the antitumor drug imatinib and two of its metabolites from a mouse brain glioma and showed the metabolites to be more abundant in tumor region compared to normal, which is consistent with other imaging studies of the drug (Cornett, Frappier, & Caprioli, 2008). In order to better understand the penetration of the anticancer drug oxaliplatin, MALDI-MSI of tissue sections from treated rat kidneys was performed and it was observed that the drug and its metabolites were localized exclusively in the kidney cortex, suggesting the drug did not penetrate deeply into the organ (Bouslimani, Bec, Glueckmann, Hirtz, & Larroque, 2010). Sugihara et al. demonstrated the first data on the localization of personalized medicine (vemurafenib) within tumor compartments of malignant melanoma (MM). In a proof-of-concept in vitro study the overexpression and localization of the drug in the MM was shown using MS fragment ion signatures (Sugihara et al., 2014). Using MALDI-MSI, it has been shown the drug irinotecan reaches tumors mainly through microvessels whereas the conversion of the drug into its active form SN-38 in tumors of colorectal cancer was poor compared to normal tissue (Buck et al., 2015). The distribution of tamoxifen in both ER-positive and ER-negative breast cancer tumor tissues as analyzed by MALDI-MSI detected the drug at significantly lower intensities in tumor cells compared with stroma in ER-negative samples (Végvári et al., 2016). Aikawa et al. employed a quantitative MSI method (a combination of MALDI-MSI and LC-MS/MS) in a preclinical mouse model to evaluate the intrabrain distribution of an anticancer drug (Aikawa et al., 2016). While no differences were observed between the mice for the drug concentration, diffuse drug distribution was found in the brain of the knockout mice vs wild mice (Aikawa et al., 2016).

To remove the background signals from matrix degradation observed in MALDI-MSI, the method nanoparticle-assisted laser desorption ionization (n-PALDI) was developed which utilises nanoparticles as matrixes. It was found the protocol could be used to investigate the distribution of anticancer agents in primary tumors and in metastasis, to ascertain whether resistance is related to inadequate drug penetration in poorly vascularized parts of the tumor (Morosi et al., 2013). Using the n-PALDI protocol, the different distribution of paclitaxel in tumor and normal tissues was visualized and related to the dosage-schedules and pathological features of the tumors (Morosi et al., 2013).

5.2 MALDI-MSI on Whole Body Sections

Rohner et al. first demonstrated the MALDI-MSI of drugs in whole body animal sections in 2005 (Rohner, Staab, & Stoeckli, 2005). Since then, a wide range of drugs have been imaged in whole body animal sections, allowing the label-free tracking of both endogenous and exogenous compounds with spatial resolution and molecular specificity (Khatib-Shahidi, Andersson, Herman, Gillespie, & Caprioli, 2006). MALDI-MSI has been used to simultaneously detect drugs and their individual metabolite distributions at various time points across whold body tissue sections (Khatib-Shahidi et al., 2006). Stoeckli et al. measured the distribution of a ^{14}C labeled compound (dosed intratracheally) in whole rat tissue sections using MALDI-MSI and whole body autoradiography (WBA) which displayed good quantitative agreement between both techniques (Stoeckli, Staab, & Schweitzer, 2007). Trim et al. compared the distribution of vinblastine within whole body sections using MALDI-MSI and WBA which displayed MALDI-MSI to be advantageous by separating the drug from an endogenous isobaric lipid (Trim et al., 2008).

5.3 MALDI-MSI on 3D Tissue Cultures

Liu et al. applied MALDI-MSI to the analysis of 3D spheroids in order to assess the distribution of pharmaceuticals and their metabolites. As a proof-of-concept study, MALDI-MSI was applied to in HCT 116 colon carcinoma multicellular spheroids to analyze the distribution of irinotecan, showing the time-dependent penetration of the drug and three metabolites (Liu, Weaver, & Hummon, 2013).

6. DATA ANALYSIS

6.1 Spatial Information

The main advantage of MSI over other methods is the access it provides to spatial information. Possibly the most common approach to leveraging the spatial information in MSI data is simply in displaying results as spatial plots. Fig. 4 shows the results of several analyses of such data. These analyses are blind to the spatial information in the data but their results have spatial patterns that can be easily seen when plotted. Plotting such results next to a stained image of the tissue allows for the molecular information in the MSI data

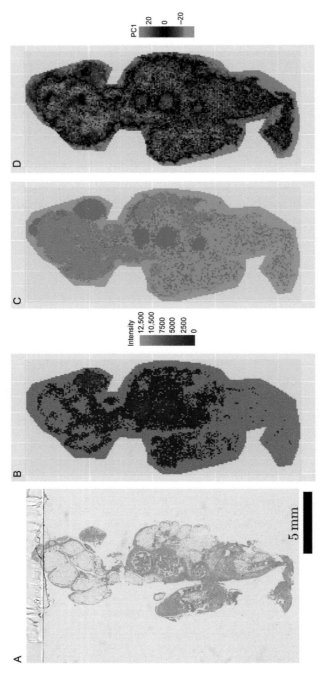

Fig. 4 Results of various analyses of tryptic peptide MALDI-MSI data collected from the same section of several primary serous ovarian tumors embedded in surrounding stromal tissue. (A) H&E stained section of tissue. (B) Intensity of a single peptide peak at $m/z = 1628.8$ confirmed as belong to heterogeneous nuclear ribonucleoprotein A1 which has been shown to be of interest in the past (Chen, Zhang, & Manley, 2010; Lee et al., 2010). (C) Results of 4-means clustering on the log-intensities including all measured m/z values. (D) Results of the first principle component including all measured m/z values on the log-intensities.

to be correlated to the histology of the tissue, thereby allowing biologically relevant use of the spatial information. Such results can even be overlayed directly with stained images (Walch, Rauser, Deininger, & Höfler, 2008), and used to partially automate and aid in histological annotations in an unbiased manner (Cornett et al., 2006). A popular alternative to the k-means clustering shown in Fig. 4 is the so-called semisupervised hierarchical clustering approach applied by Deininger et al. (2008). These semisupervised methods allow the user to manipulate the clustering results until they agree with the observed histology. Jones et al. (2011) demonstrated how multiple approaches to multivariate analysis, such as k-means and principle component analysis (PCA) (Fig. 4), could be combined in an automated manner to produce more reliable partitioning of spatial regions based on the multivariate MSI data.

Spatial information can also be applied to the partitioning of data in a histology driven fashion. To use Fig. 4 as an example, one could use the histology shown in the H&E stain to distinguish different tissue types such as tumor and surrounding stroma, and then by matching the image to the acquired region, partition the spectra into groups representing specific tissue. These groups of spectra could then be compared in an attempt to detect molecular differences between tissue types—this is also a very common approach used in the analysis of MSI data. Using this technique Oppenheimer et al. (2010) demonstrated that histologically healthy tissues within a margin of tumor tissues share molecular characteristics, and hypothesized that this phenomena could be involved in tumor recurrence postresection. Such information could be used by surgeons to identify an appropriate margin to excise the tissue surrounding a tumor.

6.2 Preprocessing: Peak Detection

MSI data have a uniquely complicated structure due to the spatial information it contains, and additionally can be computationally challenging due to the shear amount of information recorded during acquisition. The data therefore require novel approaches to analysis when compared to other, more traditional forms of mass spectrometry (Jones, Deininger, Hogendoorn, Deelder, & McDonnell, 2012). Furthermore, the appropriate approach will vary depending on the application. One approach to addressing the quantity of data is to perform some, preferably computationally fast, data reduction early in the data analysis workflow and then work with the reduced data. This allows for analyses to be performed quickly, but also means that all further analyses will be affected by the choices made in the data reduction step. The most simplistic approach to data reduction is centroiding—essentially reducing spectra to their local maxima. Centroiding is perhaps the most simplistic approach to peak detection. Essentially centroiding involves reducing the acquired spectra to their local maxima, allowing peaks in the spectra to be identified, information relevant to these peaks to be stored and the remainder of the "noise" to be removed. Yang, He, and Yu (2009) provide a good overview of peak detection in the context of MALDI-MS. Wavelet approaches to peak detection, such as those suggested by Du, Kibbe, and Lin (2006), Lange, Gropl, Reinert, Kohlbacher, and Hildebrandt (2006), show some promise but have not yet made it into mainstream software packages, and so do not currently see much use. Regardless, peak detection prior to further analysis is a useful tool as it can reduce the volume of data by as much as three orders of magnitude. However, care must be taken when interpreting final results; it is important to be aware that all analyses will be affected by any peak detection steps and that a quality control analysis of the data following peak detection should be implemented to ensure its performance.

6.3 Classification of FFPE-TMAs and the Importance of Dimension Reduction

The application of MSI to FFPE-TMAs for classification of various diagnostic and prognostic clinical factors is a particularly promising field of research. The ability to use FFPE tissue allows access to large archives including tissue samples and associated patient meta data (Hood et al., 2005). The use of TMAs with MSI facilitates the collection of label-free and untargeted mass data from large numbers of patients (Groseclose et al., 2008), which in turn allows for classification problems to be considered. One of the key obstacles to the classification of such data is that, due to the untargeted nature of MSI, the resulting data are of high-dimensionality. The natural approach to address the dimensionality of such data is to perform some dimension reduction prior to classification. Mascini et al. (2015) proposed the use PCA for dimension reduction prior to classification. PCA is a classical multivariate approach that preserves the maximum amount of variability in the dimension-reduced data. Winderbaum et al. (2016) proposed using a cross correlation analysis (CCA) to rank MSI variables based on their contribution to the highest multivariate correlation to the clinical variable and to reduce the dimension of the data by restricting to only the most highly ranked MSI variables. Either way, dimension reduction prior to classification is crucial. In order to apply these ideas in clinical practice, large cohort studies and/or clinical trials would need to be carried out to validate the performance of these MS-based classifiers. To date most studies have been exploratory, using small samples of patients as a proof of principle, with a study by Steurer et al. (2013) being a notable exception. Dependent on large scale validation studies, classification of MSI data could potentially be used in the individualisation of treatments, predicting factors such as LNM or potentially even chemotherapy response.

7. CONCLUDING REMARKS
7.1 The Future of MSI: Molecular Pathology

For over 100 years clinical pathology has relied upon a description of morphology by trained specialists for tumor classification and grading. The information that can be gained from such an analyses is limited, however, and is often purely descriptive. During the last decade mass spectrometers have found their way from physics and biochemistry research laboratories into clinical pathology laboratories, bringing together researchers from a wide spectrum of disciplines, such as physics, biochemistry, medicine, pathology, and mathematics. A key feature of MSI is the ability to discern molecular signatures of disease with the additional bonus that the features can be directly related back to the tissue morphology. In the context of clinical oncology, the application of MSI has the potential to shift traditional pathology from descriptive morphological analyses to detailed molecular analyses. As such, the introduction of MSI equipment into pathology laboratories will extend their capabilities and revolutionize management and treatment of cancer.

REFERENCES

Abbassi-Ghadi, N., Veselkov, K., Kumar, S., Huang, J., Jones, E., Strittmatter, N., et al. (2014). Discrimination of lymph node metastases using desorption electrospray ionisation-mass spectrometry imaging. *Chemical Communications (Cambridge, England)*, *50*(28), 3661–3664.

Abbott, K. L., Lim, J. M., Wells, L., Benigno, B. B., McDonald, J. F., & Pierce, M. (2010). Identification of candidate biomarkers with cancer-specific glycosylation in the tissue and serum of endometrioid ovarian cancer patients by glycoproteomic analysis. *Proteomics*, *10*(3), 470–481.

Aerni, H.-R., Cornett, D. S., & Caprioli, R. M. (2006). Automated acoustic matrix deposition for MALDI sample preparation. *Analytical Chemistry*, *78*(3), 827–834.

Agar, N. Y., Malcolm, J. G., Mohan, V., Yang, H. W., Johnson, M. D., Tannenbaum, A., et al. (2010). Imaging of meningioma progression by matrix-assisted laser desorption ionization time-of-flight mass spectrometry. *Analytical Chemistry, 82*(7), 2621–2625.

Aichler, M., Elsner, M., Ludyga, N., Feuchtinger, A., Zangen, V., Maier, S. K., et al. (2013). Clinical response to chemotherapy in oesophageal adenocarcinoma patients is linked to defects in mitochondria. *The Journal of Pathology, 230*(4), 410–419.

Aichler, M., & Walch, A. (2015). MALDI Imaging mass spectrometry: Current frontiers and perspectives in pathology research and practice. *Laboratory Investigation, 95*(4), 422–431.

Aikawa, H., Hayashi, M., Ryu, S., Yamashita, M., Ohtsuka, N., Nishidate, M., et al. (2016). Visualizing spatial distribution of alectinib in murine brain using quantitative mass spectrometry imaging. *Scientific Reports, 6*, 23749.

Andersson, M., Andren, P., & Caprioli, R. M. (2010). MALDI imaging and profiling mass spectrometry in neuroproteomics. In O. Alzate (Ed.), *Neuroproteomics*. Boca Raton, FL: CRC Press/Taylor & Francis.

Andersson, M., Groseclose, M. R., Deutch, A. Y., & Caprioli, R. M. (2008). Imaging mass spectrometry of proteins and peptides: 3D volume reconstruction. *Nature Methods, 5*(1), 101–108.

Atkinson, S. J., Loadman, P. M., Sutton, C., Patterson, L. H., & Clench, M. R. (2007). Examination of the distribution of the bioreductive drug AQ4N and its active metabolite AQ4 in solid tumours by imaging matrix-assisted laser desorption/ionisation mass spectrometry. *Rapid Communications in Mass Spectrometry, 21*(7), 1271–1276.

Balluff, B., Elsner, M., Kowarsch, A., Rauser, S., Meding, S., Schuhmacher, C., et al. (2010). Classification of HER2/neu status in gastric cancer using a breast-cancer derived proteome classifier. *Journal of Proteome Research, 9*(12), 6317–6322.

Balluff, B., Frese, C. K., Maier, S. K., Schone, C., Kuster, B., Schmitt, M., et al. (2015). De novo discovery of phenotypic intratumour heterogeneity using imaging mass spectrometry. *The Journal of Pathology, 235*(1), 3–13.

Balluff, B., Rauser, S., Meding, S., Elsner, M., Schone, C., Feuchtinger, A., et al. (2011). MALDI imaging identifies prognostic seven-protein signature of novel tissue markers in intestinal-type gastric cancer. *The American Journal of Pathology, 179*(6), 2720–2729.

Balluff, B., Schöne, C., Höfler, H., & Walch, A. (2011). MALDI imaging mass spectrometry for direct tissue analysis: Technological advancements and recent applications. *Histochemistry and Cell Biology, 136*(3), 227–244.

Bauer, J. A., Chakravarthy, A. B., Rosenbluth, J. M., Mi, D., Seeley, E. H., De Matos Granja-Ingram, N., et al. (2010). Identification of markers of taxane sensitivity using proteomic and genomic analyses of breast tumors from patients receiving neoadjuvant paclitaxel and radiation. *Clinical Cancer Research, 16*(2), 681–690.

Benabdellah, F., Seyer, A., Quinton, L., Touboul, D., Brunelle, A., & Laprevote, O. (2010). Mass spectrometry imaging of rat brain sections: Nanomolar sensitivity with MALDI versus nanometer resolution by TOF-SIMS. *Analytical and Bioanalytical Chemistry, 396*(1), 151–162.

Bodzon-Kulakowska, A., & Suder, P. (2016). Imaging mass spectrometry: Instrumentation, applications, and combination with other visualization techniques. *Mass Spectrometry Reviews, 35*(1), 147–169.

Bouslimani, A., Bec, N., Glueckmann, M., Hirtz, C., & Larroque, C. (2010). Matrix-assisted laser desorption/ionization imaging mass spectrometry of oxaliplatin derivatives in heated intraoperative chemotherapy (HIPEC)-like treated rat kidney. *Rapid Communications in Mass Spectrometry, 24*(4), 415–421.

Buck, A., Halbritter, S., Spath, C., Feuchtinger, A., Aichler, M., Zitzelsberger, H., et al. (2015). Distribution and quantification of irinotecan and its active metabolite SN-38 in colon cancer murine model systems using MALDI MSI. *Analytical and Bioanalytical Chemistry*, *407*(8), 2107–2116.

Büll, C., Boltje, T. J., Wassink, M., de Graaf, A. M., van Delft, F. L., den Brok, M. H., et al. (2013). Targeting aberrant sialylation in cancer cells using a fluorinated sialic acid analog impairs adhesion, migration, and in vivo tumor growth. *Molecular Cancer Therapeutics*, *12*(10), 1935–1946.

Calligaris, D., Caragacianu, D., Liu, X., Norton, I., Thompson, C. J., Richardson, A. L., et al. (2014). Application of desorption electrospray ionization mass spectrometry imaging in breast cancer margin analysis. *Proceedings of the National Academy of Sciences of the United States of America*, *111*(42), 15184–15189.

Caprioli, R. M., Farmer, T. B., & Gile, J. (1997). Molecular imaging of biological samples: Localization of peptides and proteins using MALDI-TOF MS. *Analytical Chemistry*, *69*(23), 4751–4760.

Casadonte, R., & Caprioli, R. M. (2011). Proteomic analysis of formalin-fixed paraffin-embedded tissue by MALDI imaging mass spectrometry. *Nature Protocols*, *6*(11), 1695–1709.

Casadonte, R., Kriegsmann, M., Zweynert, F., Friedrich, K., Bretton, G., Otto, M., et al. (2014). Imaging mass spectrometry to discriminate breast from pancreatic cancer metastasis in formalin-fixed paraffin-embedded tissues. *Proteomics*, *14*(7–8), 956–964.

Cazares, L. H., Troyer, D., Mendrinos, S., Lance, R. A., Nyalwidhe, J. O., Beydoun, H. A., et al. (2009). Imaging mass spectrometry of a specific fragment of mitogen-activated protein kinase/extracellular signal-regulated kinase kinase kinase 2 discriminates cancer from uninvolved prostate tissue. *Clinical Cancer Research*, *15*(17), 5541–5551.

Chaurand, P., Sanders, M. E., Jensen, R. A., & Caprioli, R. M. (2004). Proteomics in diagnostic pathology: Profiling and imaging proteins directly in tissue sections. *The American Journal of Pathology*, *165*(4), 1057–1068.

Chaurand, P., Schwartz, S. A., & Caprioli, R. M. (2004). Assessing protein patterns in disease using imaging mass spectrometry. *Journal of Proteome Research*, *3*(2), 245–252.

Chen, M., Zhang, J., & Manley, J. L. (2010). Turning on a fuel switch of cancer: hnRNP proteins regulate alternative splicing of pyruvate kinase mRNA. *Cancer Research*, *70*(22), 8977–8980.

Cho, Y. T., Su, H., Wu, W. J., Wu, D. C., Hou, M. F., Kuo, C. H., et al. (2015). Biomarker characterization by MALDI-TOF/MS. *Advances in Clinical Chemistry*, *69*, 209–254.

Choi, J. H., Shin, N. R., Moon, H. J., Kwon, C. H., Kim, G. H., Song, G. A., et al. (2012). Identification of S100A8 and S100A9 as negative regulators for lymph node metastasis of gastric adenocarcinoma. *Histology and Histopathology*, *27*(11), 1439–1448.

Coombes, K. R., Koomen, J. M., Baggerly, K. A., Morris, J. S., & Kobayashi, R. (2005). Understanding the characteristics of mass spectrometry data through the use of simulation. *Cancer Informatics*, *1*, 41–52.

Cornett, D. S., Frappier, S. L., & Caprioli, R. M. (2008). MALDI-FTICR imaging mass spectrometry of drugs and metabolites in tissue. *Analytical Chemistry*, *80*(14), 5648–5653.

Cornett, D. S., Mobley, J. A., Dias, E. C., Andersson, M., Arteaga, C. L., Sanders, M. E., et al. (2006). A novel histology-directed strategy for MALDI-MS tissue profiling that improves throughput and cellular specificity in human breast cancer. *Molecular & Cellular Proteomics*, *5*(10), 1975–1983.

Crecelius, A. C., Cornett, D. S., Caprioli, R. M., Williams, B., Dawant, B. M., & Bodenheimer, B. (2005). Three-dimensional visualization of protein expression in mouse brain structures using imaging mass spectrometry. *Journal of the American Society for Mass Spectrometry*, *16*(7), 1093–1099.

de Hoffmann, E., & Stroobant, V. (2007). *Mass spectrometry: Principles and applications* (3rd ed.). England: Wiley.

Deininger, S.-O., Ebert, M. P., Fütterer, A., Gerhard, M., & Röcken, C. (2008). MALDI imaging combined with hierarchical clustering as a new tool for the interpretation of complex human cancers. *Journal of Proteome Research*, 7(12), 5230–5236.

Dekker, T. J. A., Balluff, B. D., Jones, E. A., Schöne, C. D., Schmitt, M., Aubele, M., et al. (2014). Multicenter matrix-assisted laser desorption/ionization mass spectrometry imaging (MALDI MSI) identifies proteomic differences in breast-cancer-associated stroma. *Journal of Proteome Research*, 13(11), 4730–4738.

Diehl, H. C., Beine, B., Elm, J., Trede, D., Ahrens, M., Eisenacher, M., et al. (2015). The challenge of on-tissue digestion for MALDI MSI—A comparison of different protocols to improve imaging experiments. *Analytical and Bioanalytical Chemistry*, 407(8), 2223–2243.

Dill, A. L., Eberlin, L. S., Costa, A. B., Zheng, C., Ifa, D. R., Cheng, L., et al. (2011). Multivariate statistical identification of human bladder carcinomas using ambient ionization imaging mass spectrometry. *Chemistry*, 17(10), 2897–2902.

Dill, A. L., Ifa, D. R., Manicke, N. E., Costa, A. B., Ramos-Vara, J. A., Knapp, D. W., et al. (2009). Lipid profiles of canine invasive transitional cell carcinoma of the urinary bladder and adjacent normal tissue by desorption electrospray ionization imaging mass spectrometry. *Analytical Chemistry*, 81(21), 8758–8764.

Djidja, M.-C., Claude, E., Snel, M. F., Francese, S., Scriven, P., Carolan, V., et al. (2010). Novel molecular tumour classification using MALDI–mass spectrometry imaging of tissue micro-array. *Analytical and Bioanalytical Chemistry*, 397(2), 587–601.

Drake, P. M., Cho, W., Li, B., Prakobphol, A., Johansen, E., Anderson, N. L., et al. (2010). Sweetening the pot: adding glycosylation to the biomarker discovery equation. *Clinical Chemistry*, 56(2), 223–236.

Du, P., Kibbe, W. A., & Lin, S. M. (2006). Improved peak detection in mass spectrum by incorporating continuous wavelet transform-based pattern matching. *Bioinformatics*, 22(17), 2059–2065.

Dube, D. H., & Bertozzi, C. R. (2005). Glycans in cancer and inflammation—Potential for therapeutics and diagnostics. *Nature Reviews. Drug Discovery*, 4(6), 477–488.

Eberlin, L. S., Norton, I., Dill, A. L., Golby, A. J., Ligon, K. L., Santagata, S., et al. (2012). Classifying human brain tumors by lipid imaging with mass spectrometry. *Cancer Research*, 72(3), 645–654.

Eberlin, L. S., Tibshirani, R. J., Zhang, J., Longacre, T. A., Berry, G. J., Bingham, D. B., et al. (2014). Molecular assessment of surgical-resection margins of gastric cancer by mass-spectrometric imaging. *Proceedings of the National Academy of Sciences of the United States of America*, 111(7), 2436–2441.

Elsner, M., Rauser, S., Maier, S., Schöne, C., Balluff, B., Meding, S., et al. (2012). MALDI imaging mass spectrometry reveals COX7A2, TAGLN2 and S100-A10 as novel prognostic markers in Barrett's adenocarcinoma. *Journal of Proteomics*, 75(15), 4693–4704.

Fehniger, T. E., Vegvari, A., Rezeli, M., Prikk, K., Ross, P., Dahlback, M., et al. (2011). Direct demonstration of tissue uptake of an inhaled drug: Proof-of-principle study using matrix-assisted laser desorption ionization mass spectrometry imaging. *Analytical Chemistry*, 83(21), 8329–8336.

Fejzo, M. S., & Slamon, D. J. (2010). Tissue microarrays from frozen tissues-OCT technique. *Methods in Molecular Biology*, 664, 73–80.

Fentz, J. S., Zornig, C., Juhl, H. H., & David, K. A. (2004). Tissue ischemia time affects gene and protein expression patterns within minutes following surgical tumor excision. *Biotechniques*, 36(6), 1030–1037.

Fletcher, J. S., Lockyer, N. P., & Vickerman, J. C. (2011). Developments in molecular SIMS depth profiling and 3D imaging of biological systems using polyatomic primary ions. *Mass Spectrometry Reviews*, 30(1), 142–174.

Fletcher, J. S., Vickerman, J. C., & Winograd, N. (2011). Label free biochemical 2D and 3D imaging using secondary ion mass spectrometry. *Current Opinion in Chemical Biology*, *15*(5), 733–740.

Gemoll, T., Strohkamp, S., Schillo, K., Thorns, C., & Habermann, J. K. (2015). MALDI-imaging reveals thymosin beta-4 as an independent prognostic marker for colorectal cancer. *Oncotarget*, *6*(41), 43869–43880.

Gobom, J., Mueller, M., Egelhofer, V., Theiss, D., Lehrach, H., & Nordhoff, E. (2002). A calibration method that simplifies and improves accurate determination of peptide molecular masses by MALDI-TOF MS. *Analytical Chemistry*, *74*(15), 3915–3923.

Goodwin, R. J. (2012). Sample preparation for mass spectrometry imaging: Small mistakes can lead to big consequences. *Journal of Proteomics*, *75*(16), 4893–4911.

Gorzolka, K., & Walch, A. (2014). MALDI mass spectrometry imaging of formalin-fixed paraffin-embedded tissues in clinical research. *Histology and Histopathology*, *29*(11), 1365–1376.

Groseclose, M. R., Massion, P. P., Chaurand, P., & Caprioli, R. M. (2008). High-throughput proteomic analysis of formalin-fixed paraffin-embedded tissue microarrays using MALDI imaging mass spectrometry. *Proteomics*, *8*(18), 3715–3724.

Grüner, B. M., Hahne, H., Mazur, P. K., Trajkovic-Arsic, M., Maier, S., Esposito, I., et al. (2012). MALDI imaging mass spectrometry for in situ proteomic analysis of preneoplastic lesions in pancreatic cancer. *PLoS One*, *7*(6), e39424.

Gustafsson, O. J., Briggs, M. T., Condina, M. R., Winderbaum, L. J., Pelzing, M., McColl, S. R., et al. (2015). MALDI imaging mass spectrometry of N-linked glycans on formalin-fixed paraffin-embedded murine kidney. *Analytical and Bioanalytical Chemistry*, *407*(8), 2127–2139.

Gustafsson, J. O., Eddes, J. S., Meding, S., Koudelka, T., Oehler, M. K., McColl, S. R., et al. (2012). Internal calibrants allow high accuracy peptide matching between MALDI imaging MS and LC-MS/MS. *Journal of Proteomics*, *75*(16), 5093–5105.

Gustafsson, O. J., Eddes, J. S., Meding, S., McColl, S. R., Oehler, M. K., & Hoffmann, P. (2013). Matrix-assisted laser desorption/ionization imaging protocol for in situ characterization of tryptic peptide identity and distribution in formalin-fixed tissue. *Rapid Communications in Mass Spectrometry*, *27*(6), 655–670.

Han, E. C., Lee, Y.-S., Liao, W.-S., Liu, Y.-C., Liao, H.-Y., & Jeng, L.-B. (2011). Direct tissue analysis by MALDI-TOF mass spectrometry in human hepatocellular carcinoma. *Clinica Chimica Acta*, *412*(3–4), 230–239.

Hardesty, W. M., Kelley, M. C., Mi, D., Low, R. L., & Caprioli, R. M. (2011). Protein signatures for survival and recurrence in metastatic melanoma. *Journal of Proteomics*, *74*(7), 1002–1014.

Holst, S., Heijs, B., de Haan, N., van Zeijl, R. J., Briaire-de Bruijn, I. H., van Pelt, G. W., et al. (2016). Linkage-specific in situ sialic acid derivatization for N-glycan mass spectrometry imaging of formalin-fixed paraffin-embedded tissues. *Analytical Chemistry*, *88*(11), 5904–5913.

Hood, B. L., Darfler, M. M., Guiel, T. G., Furusato, B., Lucas, D. A., Ringeisen, B. R., et al. (2005). Proteomic analysis of formalin-fixed prostate cancer tissue. *Molecular & Cellular Proteomics*, *4*(11), 1741–1753.

Horn, D. M., Peters, E. C., Klock, H., Meyers, A., & Brock, A. (2004). Improved protein identification using automated high mass measurement accuracy MALDI FT-ICR MS peptide mass fingerprinting. *International Journal of Mass Spectrometry*, *238*(2), 189–196.

Hsieh, Y., Li, F., & Korfmacher, W. A. (2010). Mapping pharmaceuticals in rat brain sections using MALDI imaging mass spectrometry. *Methods in Molecular Biology*, *656*, 147–158.

Hsu, C. C., & Dorrestein, P. C. (2015). Visualizing life with ambient mass spectrometry. *Current Opinion in Biotechnology*, *31*, 24–34.

Jensen, P. H., Karlsson, N. G., Kolarich, D., & Packer, N. H. (2012). Structural analysis of N- and O-glycans released from glycoproteins. *Nature Protocols*, 7(7), 1299–1310.

Jones, E. A., Deininger, S.-O., Hogendoorn, P. C., Deelder, A. M., & McDonnell, L. A. (2012). Imaging mass spectrometry statistical analysis. *Journal of Proteomics*, 75(16), 4962–4989.

Jones, E. A., van Remoortere, A., van Zeijl, R. J. M., Hogendoorn, P. C. W., Bovée, J. V. M. G., Deelder, A. M., et al. (2011). Multiple statistical analysis techniques corroborate intratumor heterogeneity in imaging mass spectrometry datasets of myxofibrosarcoma. *PLoS One*, 6(9), e24913.

Jurchen, J. C., Rubakhin, S. S., & Sweedler, J. V. (2005). MALDI-MS imaging of features smaller than the size of the laser beam. *Journal of the American Society for Mass Spectrometry*, 16(10), 1654–1659.

Kaletas, B. K., van der Wiel, I. M., Stauber, J., Guzel, C., Kros, J. M., Luider, T. M., et al. (2009). Sample preparation issues for tissue imaging by imaging MS. *Proteomics*, 9(10), 2622–2633.

Kang, S., Shim, H. S., Lee, J. S., Kim, D. S., Kim, H. Y., Hong, S. H., et al. (2010). Molecular proteomics imaging of tumor interfaces by mass spectrometry. *Journal of Proteome Research*, 9(2), 1157–1164.

Kertesz, V., Berkel, V., Gary, J., Vavrek, M., Koeplinger, K. A., Schneider, B. B., et al. (2008). Comparison of drug distribution images from whole-body thin tissue sections obtained using desorption electrospray ionization tandem mass spectrometry and autoradiography. *Analytical Chemistry*, 80(13), 5168–5177.

Khatib-Shahidi, S., Andersson, M., Herman, J. L., Gillespie, T. A., & Caprioli, R. M. (2006). Direct molecular analysis of whole-body animal tissue sections by imaging MALDI mass spectrometry. *Analytical Chemistry*, 78(18), 6448–6456.

Kim, H. K., Reyzer, M. L., Choi, I. J., Kim, C. G., Kim, H. S., Oshima, A., et al. (2010). Gastric cancer-specific protein profile identified using endoscopic biopsy samples via MALDI mass spectrometry. *Journal of Proteome Research*, 9(8), 4123–4130.

Klerk, L. A., Maarten Altelaar, A. F., Froesch, M., McDonnell, L. A., & Heeren, R. M. A. (2009). Fast and automated large-area imaging MALDI mass spectrometry in microprobe and microscope mode. *International Journal of Mass Spectrometry*, 285(1–2), 19–25.

Kriegsmann, J., Kriegsmann, M., & Casadonte, R. (2015). MALDI TOF imaging mass spectrometry in clinical pathology: A valuable tool for cancer diagnostics (review). *International Journal of Oncology*, 46(3), 893–906.

Kwon, H. J., Kim, Y., Sugihara, Y., Baldetorp, B., Welinder, C., Watanabe, K., et al. (2015). Drug compound characterization by mass spectrometry imaging in cancer tissue. *Archives of Pharmacal Research*, 38(9), 1718–1727.

Lange, E., Gropl, C., Reinert, K., Kohlbacher, O., & Hildebrandt, A. (2006). High-accuracy peak picking of proteomics data using wavelet techniques. *Pacific Symposium on Biocomputing*, 11, 243–254. Paper presented at the Pacific Symposium on Biocomputing.

Lanni, E. J., Rubakhin, S. S., & Sweedler, J. V. (2012). Mass spectrometry imaging and profiling of single cells. *Journal of Proteomics*, 75(16), 5036–5051.

Lau, K. S., & Dennis, J. W. (2008). N-Glycans in cancer progression. *Glycobiology*, 18(10), 750–760.

Le Faouder, J., Laouirem, S., Chapelle, M., Albuquerque, M., Belghiti, J., Degos, F., et al. (2011). Imaging mass spectrometry provides fingerprints for distinguishing hepatocellular carcinoma from cirrhosis. *Journal of Proteome Research*, 10(8), 3755–3765.

Lee, D. H., Chung, K., Song, J.-A., Kim, T.-h., Kang, H., Huh, J. H., et al. (2010). Proteomic identification of paclitaxel-resistance associated hnRNP A2 and GDI 2 proteins in human ovarian cancer cells. *Journal of Proteome Research*, 9(11), 5668–5676.

Lemaire, R., Ait Menguellet, S., Stauber, J., Marchaudon, V., Lucot, J.-P., Collinet, P., et al. (2007). Specific MALDI imaging and profiling for biomarker hunting and validation: Fragment of the 11S proteasome activator complex, Reg alpha fragment, is a new potential ovary cancer biomarker. *Journal of Proteome Research*, 6(11), 4127–4134.

Lemaire, R., Desmons, A., Tabet, J. C., Day, R., Salzet, M., & Fournier, I. (2007). Direct analysis and MALDI imaging of formalin-fixed, paraffin-embedded tissue sections. *Journal of Proteome Research*, 6(4), 1295–1305.

Lemaire, R., Tabet, J. C., Ducoroy, P., Hendra, J. B., Salzet, M., & Fournier, I. (2006). Solid ionic matrixes for direct tissue analysis and MALDI imaging. *Analytical Chemistry*, 78(3), 809–819.

Liu, X., Weaver, E. M., & Hummon, A. B. (2013). Evaluation of therapeutics in three-dimensional cell culture systems by MALDI imaging mass spectrometry. *Analytical Chemistry*, 85(13), 6295–6302.

Luxembourg, S. L., Mize, T. H., McDonnell, L. A., & Heeren, R. M. (2004). High-spatial resolution mass spectrometric imaging of peptide and protein distributions on a surface. *Analytical Chemistry*, 76(18), 5339–5344.

Mascini, N. E., Eijkel, G. B., ter Brugge, P., Jonkers, J., Wesseling, J., & Heeren, R. M. A. (2015). The use of mass spectrometry imaging to predict treatment response of patient-derived xenograft models of triple-negative breast cancer. *Journal of Proteome Research*, 14(2), 1069–1075.

McDonnell, L. A., Corthals, G. L., Willems, S. M., van Remoortere, A., van Zeijl, R. J. M., & Deelder, A. M. (2010). Peptide and protein imaging mass spectrometry in cancer research. *Journal of Proteomics*, 73(10), 1921–1944.

Metz, B., Kersten, G. F. A., Hoogerhout, P., Brugghe, H. F., Timmermans, H. A. M., De Jong, A. D., et al. (2004). Identification of formaldehyde-induced modifications in proteins reactions with model peptides. *Journal of Biological Chemistry*, 279(8), 6235–6243.

Minchinton, A. I., & Tannock, I. F. (2006). Drug penetration in solid tumours. *Nature Reviews. Cancer*, 6(8), 583–592. http://dx.doi.org/10.1038/nrc1893.

Mittal, P., Klingler-Hoffmann, M., Arentz, G., Winderbaum, L., Lokman, N. A., Zhang, C., et al. (2016). Maldi imaging of primary endometrial cancers reveals proteins associated with lymph node metastasis. *Proteomics*, 16, 1793–1801.

Mittal, P., Klingler-Hoffmann, M., Arentz, G., Zhang, C., Kaur, G., Oehler, M. K., et al. (2016). Proteomics of endometrial cancer diagnosis, treatment, and prognosis. *Proteomics. Clinical Applications*, 10(3), 217–229.

Morita, Y., Ikegami, K., Goto-Inoue, N., Hayasaka, T., Zaima, N., Tanaka, H., et al. (2010). Imaging mass spectrometry of gastric carcinoma in formalin-fixed paraffin-embedded tissue microarray. *Cancer Science*, 101(1), 267–273.

Morosi, L., Spinelli, P., Zucchetti, M., Pretto, F., Carrà, A., D'Incalci, M., et al. (2013). Determination of paclitaxel distribution in solid tumors by nano-particle assisted laser desorption ionization mass spectrometry imaging. *PLoS One*, 8(8), e72532.

Nakagoe, T., Fukushima, K., Nanashima, A., Sawai, T., Tsuji, T., Jibiki, M., et al. (2000). Expression of Lewis(a), sialyl Lewis(a), Lewis(x) and sialyl Lewis(x) antigens as prognostic factors in patients with colorectal cancer. *Canadian Journal of Gastroenterology*, 14(9), 753–760.

Nilsson, A., Fehniger, T. E., Gustavsson, L., Andersson, M., Kenne, K., Marko-Varga, G., et al. (2010). Fine mapping the spatial distribution and concentration of unlabeled drugs within tissue micro-compartments using imaging mass spectrometry. *PLoS One*, 5(7), e11411.

Nipp, M., Elsner, M., Balluff, B., Meding, S., Sarioglu, H., Ueffing, M., et al. (2011). S100-A10, thioredoxin, and S100-A6 as biomarkers of papillary thyroid carcinoma with lymph node metastasis identified by MALDI imaging. *Journal of Molecular Medicine*, 90(2), 163–174.

North, S. J., Hitchen, P. G., Haslam, S. M., & Dell, A. (2009). Mass spectrometry in the analysis of N-linked and O-linked glycans. *Current Opinion in Structural Biology*, 19(5), 498–506.

Oezdemir, R. F., Gaisa, N. T., Lindemann-Docter, K., Gostek, S., Weiskirchen, R., Ahrens, M., et al. (2012). Proteomic tissue profiling for the improvement of grading of noninvasive papillary urothelial neoplasia. *Clinical Biochemistry, 45*(1–2), 7–11.

Ohtsubo, K., & Marth, J. D. (2006). Glycosylation in cellular mechanisms of health and disease. *Cell, 126*(5), 855–867.

Oppenheimer, S. R., Mi, D., Sanders, M. E., & Caprioli, R. M. (2010). Molecular analysis of tumor margins by MALDI mass spectrometry in renal carcinoma. *Journal of Proteome Research, 9*(5), 2182–2190.

Pagni, F., Mainini, V., Garancini, M., Bono, F., Vanzati, A., Giardini, V., et al. (2015). Proteomics for the diagnosis of thyroid lesions: Preliminary report. *Cytopathology, 26*(5), 318–324.

Pappin, D. J. C., Hojrup, P., & Bleasby, A. J. (1993). Rapid identification of proteins by peptide-mass fingerprinting. *Current Biology, 3*(6), 327–332.

Passarelli, M. K., & Winograd, N. (2011). Lipid imaging with time-of-flight secondary ion mass spectrometry (ToF-SIMS). *Biochimica et Biophysica Acta, 1811*(11), 976–990.

Patel, S. A., Barnes, A., Loftus, N., Martin, R., Sloan, P., Thakker, N., et al. (2009). Imaging mass spectrometry using chemical inkjet printing reveals differential protein expression in human oral squamous cell carcinoma. *Analyst, 134*(2), 301–307. http://dx.doi.org/10.1039/B812533C.

Pavlidis, N., & Pentheroudakis, G. (2016). Cancer of unknown primary site. *The Lancet, 379*(9824), 1428–1435.

Powers, T. W., Holst, S., Wuhrer, M., Mehta, A. S., & Drake, R. R. (2015). Two-dimensional N-glycan distribution mapping of hepatocellular carcinoma tissues by MALDI-imaging mass spectrometry. *Biomolecules, 4*, 2554–2572.

Powers, T. W., Jones, E. E., Betesh, L. R., Romano, P. R., Gao, P., Copland, J. A., et al. (2013). Matrix assisted laser desorption ionization imaging mass spectrometry workflow for spatial profiling analysis of N-linked glycan expression in tissues. *Analytical Chemistry, 85*(20), 9799–9806.

Powers, T. W., Neely, B. A., Shao, Y., Tang, H., Troyer, D. A., Mehta, A. S., et al. (2014). MALDI imaging mass spectrometry profiling of N-glycans in formalin-fixed paraffin embedded clinical tissue blocks and tissue microarrays. *PLoS One, 9*(9), e106255.

Prideaux, B., Dartois, V., Staab, D., Weiner, D. M., Goh, A., Via, L. E., et al. (2011). High-sensitivity MALDI-MRM-MS imaging of moxifloxacin distribution in tuberculosis-infected rabbit lungs and granulomatous lesions. *Analytical Chemistry, 83*(6), 2112–2118.

Rahman, S. M., Gonzalez, A. L., Li, M., Seeley, E. H., Zimmerman, L. J., Zhang, X. J., et al. (2011). Lung cancer diagnosis from proteomic analysis of preinvasive lesions. *Cancer Research, 71*(8), 3009–3017.

Rauser, S., Marquardt, C., Balluff, B., Deininger, S.-O., Albers, C., Belau, E., et al. (2010). Classification of HER2 receptor status in breast cancer tissues by MALDI imaging mass spectrometry. *Journal of Proteome Research, 9*(4), 1854–1863.

Renovanz, M., & Kim, E. L. (2014). Intratumoral heterogeneity, its contribution to therapy resistance and methodological caveats to assessment. *Frontiers in Oncology, 4*, 142.

Reyzer, M. L., Hsieh, Y., Ng, K., Korfmacher, W. A., & Caprioli, R. M. (2003). Direct analysis of drug candidates in tissue by matrix-assisted laser desorption/ionization mass spectrometry. *Journal of Mass Spectrometry, 38*(10), 1081–1092.

Rodrigo, M. A., Zitka, O., Krizkova, S., Moulick, A., Adam, V., & Kizek, R. (2014). MALDI-TOF MS as evolving cancer diagnostic tool: A review. *Journal of Pharmaceutical and Biomedical Analysis, 95*, 245–255.

Rohner, T. C., Staab, D., & Stoeckli, M. (2005). MALDI mass spectrometric imaging of biological tissue sections. *Mechanisms of Ageing and Development, 126*(1), 177–185.

Ronci, M., Bonanno, E., Colantoni, A., Pieroni, L., Di Ilio, C., Spagnoli, L. G., et al. (2008). Protein unlocking procedures of formalin-fixed paraffin-embedded tissues: Application to MALDI-TOF imaging MS investigations. *Proteomics, 8*(18), 3702–3714.

Schuerenberg, M., & Deininger, S.-O. (2010). Matrix application with ImagePrep. *Imaging mass spectrometry* (pp. 87–91). New York, USA: Springer.

Schwamborn, K., & Caprioli, R. M. (2010). Molecular imaging by mass spectrometry—Looking beyond classical histology. *Nature Reviews. Cancer, 10*(9), 639–646. http://dx.doi.org/10.1038/nrc2917.

Schwamborn, K., Krieg, R. C., Jirak, P., Ott, G., Knuchel, R., Rosenwald, A., et al. (2010). Application of MALDI imaging for the diagnosis of classical Hodgkin lymphoma. *Journal of Cancer Research and Clinical Oncology, 136*(11), 1651–1655.

Schwamborn, K., Krieg, R. C., Reska, M., Jakse, G., Knuechel, R., & Wellmann, A. (2007). Identifying prostate carcinoma by MALDI-Imaging. *International Journal of Molecular Medicine, 20*(2), 155–159.

Schwartz, S. A., Reyzer, M. L., & Caprioli, R. M. (2003). Direct tissue analysis using matrix-assisted laser desorption/ionization mass spectrometry: Practical aspects of sample preparation. *Journal of Mass Spectrometry, 38*(7), 699–708.

Scigelova, M., Hornshaw, M., Giannakopulos, A., & Makarov, A. (2011). Fourier transform mass spectrometry. *Molecular & Cellular Proteomics, 10*(7). M111.009431.

Seeley, E. H., Oppenheimer, S. R., Mi, D., Chaurand, P., & Caprioli, R. M. (2008). Enhancement of protein sensitivity for MALDI imaging mass spectrometry after chemical treatment of tissue sections. *Journal of the American Society for Mass Spectrometry, 19*(8), 1069–1077.

Signor, L., & Boeri Erba, E. (2013). Matrix-assisted laser desorption/ionization time of flight (MALDI-TOF) mass spectrometric analysis of intact proteins larger than 100 kDa. *Journal of Visualized Experiments, 79*, e506351.

Sinha, T. K., Khatib-Shahidi, S., Yankeelov, T. E., Mapara, K., Ehtesham, M., Cornett, D. S., et al. (2008). Integrating spatially resolved three-dimensional MALDI IMS with in vivo magnetic resonance imaging. *Nature Methods, 5*(1), 57–59. http://dx.doi.org/10.1038/nmeth1147.

Sjovall, P., Lausmaa, J., & Johansson, B. (2004). Mass spectrometric imaging of lipids in brain tissue. *Analytical Chemistry, 76*(15), 4271–4278.

Sjovall, P., Lausmaa, J., Nygren, H., Carlsson, L., & Malmberg, P. (2003). Imaging of membrane lipids in single cells by imprint-imaging time-of-flight secondary ion mass spectrometry. *Analytical Chemistry, 75*(14), 3429–3434.

Steurer, S., Borkowski, C., Odinga, S., Buchholz, M., Koop, C., Huland, H., et al. (2013). MALDI mass spectrometric imaging based identification of clinically relevant signals in prostate cancer using large-scale tissue microarrays. *International Journal of Cancer, 133*(4), 920–928.

Steurer, S., Seddiqi, A. S., Singer, J. M., Bahar, A. S., Eichelberg, C., Rink, M., et al. (2014). MALDI imaging on tissue microarrays identifies molecular features associated with renal cell cancer phenotype. *Anticancer Research, 34*(5), 2255–2261.

Steurer, S., Singer, J. M., Rink, M., Chun, F., Dahlem, R., Simon, R., et al. (2014). MALDI imaging-based identification of prognostically relevant signals in bladder cancer using large-scale tissue microarrays. *Urologic Oncology, 32*(8), 1225–1233.

Stoeckli, M., Chaurand, P., Hallahan, D. E., & Caprioli, R. M. (2001). Imaging mass spectrometry: A new technology for the analysis of protein expression in mammalian tissues. *Nature Medicine, 7*(4), 493–496.

Stoeckli, M., Staab, D., & Schweitzer, A. (2007). Compound and metabolite distribution measured by MALDI mass spectrometric imaging in whole-body tissue sections. *International Journal of Mass Spectrometry, 260*(2–3), 195–202.

Strand, F. L. (2003). Neuropeptides: General characteristics and neuropharmaceutical potential in treating CNS disorders. *Progress in Drug Research, 61*, 1–37.

Sugihara, Y., Végvári, Á., Welinder, C., Jönsson, G., Ingvar, C., Lundgren, L., et al. (2014). A new look at drugs targeting malignant melanoma—An application for mass spectrometry imaging. *Proteomics*, *14*(17–18), 1963–1970.

Taban, I. M., Altelaar, A. F., van der Burgt, Y. E., McDonnell, L. A., Heeren, R. M., Fuchser, J., et al. (2007). Imaging of peptides in the rat brain using MALDI-FTICR mass spectrometry. *Journal of the American Society for Mass Spectrometry*, *18*(1), 145–151.

Takáts, Z., Wiseman, J. M., Gologan, B., & Cooks, R. G. (2004). Mass spectrometry sampling under ambient conditions with desorption electrospray ionization. *Science*, *306*(5695), 471–473.

Toghi Eshghi, S., Yang, S., Wang, X., Shah, P., Li, X., & Zhang, H. (2014). Imaging of N-linked glycans from formalin-fixed paraffin-embedded tissue sections using MALDI mass spectrometry. *ACS Chemical Biology*, *9*(9), 2149–2156.

Trim, P. J., Henson, C. M., Avery, J. L., McEwen, A., Snel, M. F., Claude, E., et al. (2008). Matrix-assisted laser desorption/ionization-ion mobility separation-mass spectrometry imaging of vinblastine in whole body tissue sections. *Analytical Chemistry*, *80*(22), 8628–8634.

van Hove, A., Erika, R., Smith, D. F., & Heeren, R. M. A. (2010). A concise review of mass spectrometry imaging. *Journal of Chromatography. A*, *1217*(25), 3946–3954.

Varki, A. (1993). Biological roles of oligosaccharides: All of the theories are correct. *Glycobiology*, *3*(2), 97–130.

Végvári, Á., Shavkunov, A. S., Fehniger, T. E., Grabau, D., Niméus, E., & Marko-Varga, G. (2016). Localization of tamoxifen in human breast cancer tumors by MALDI mass spectrometry imaging. *Clinical and Translational Medicine*, *5*(1), 1–8.

Waki, M., Ide, Y., Ishizaki, I., Nagata, Y., Masaki, N., Sugiyama, E., et al. (2014). Single-cell time-of-flight secondary ion mass spectrometry reveals that human breast cancer stem cells have significantly lower content of palmitoleic acid compared to their counterpart non-stem cancer cells. *Biochimie*, *107*(Pt. A), 73–77.

Walch, A., Rauser, S., Deininger, S.-O., & Höfler, H. (2008). MALDI imaging mass spectrometry for direct tissue analysis: A new frontier for molecular histology. *Histochemistry and Cell Biology*, *130*(3), 421–434.

Willems, S. M., van Remoortere, A., van Zeijl, R., Deelder, A. M., McDonnell, L. A., & Hogendoorn, P. C. W. (2010). Imaging mass spectrometry of myxoid sarcomas identifies proteins and lipids specific to tumour type and grade, and reveals biochemical intra-tumour heterogeneity. *The Journal of Pathology*, *222*(4), 400–409.

Winderbaum, L., Koch, I., Mittal, P., & Hoffmann, P. (2016). Classification of MALDI-MS imaging data of tissue microarrays using canonical correlation analysis based variable selection. *Proteomics*, *16*, 1731–1735.

Wisztorski, M., Lemaire, R., Stauber, J., Menguelet, S. A., Croix, D., Mathé, O. J., et al. (2007). New developments in MALDI imaging for pathology proteomic studies. *Current Pharmaceutical Design*, *13*(32), 3317–3324.

Wuhrer, M., Deelder, A. M., & van der Burgt, Y. E. (2011). Mass spectrometric glycan rearrangements. *Mass Spectrometry Reviews*, *30*(4), 664–680.

Yamada, Y., Hidefumi, K., Shion, H., Oshikata, M., & Haramaki, Y. (2011). Distribution of chloroquine in ocular tissue of pigmented rat using matrix-assisted laser desorption/ionization imaging quadrupole time-of-flight tandem mass spectrometry. *Rapid Communications in Mass Spectrometry*, *25*(11), 1600–1608.

Yanagisawa, K., Yu, S., Xu, B. J., Massion, P. P., Larsen, P. H., White, B. C., et al. (2003). Proteomic patterns of tumour subsets in non-small-cell lung cancer. *The Lancet*, *362*(9382), 433–439.

Yang, L., Cui, X., Zhang, N., Li, M., Bai, Y., Han, X., et al. (2015). Comprehensive lipid profiling of plasma in patients with benign breast tumor and breast cancer reveals novel biomarkers. *Analytical and Bioanalytical Chemistry*, *407*(17), 5065–5077.

Yang, C., He, Z., & Yu, W. (2009). Comparison of public peak detection algorithms for MALDI mass spectrometry data analysis. *BMC Bioinformatics*, *10*(1), 4.

CHAPTER THREE

Assessing the Potential of Metal-Assisted Imaging Mass Spectrometry in Cancer Research

M. Dufresne*, N.H. Patterson[†], N. Lauzon*, P. Chaurand*,[1]

*Université de Montréal, Montreal, QC, Canada
[†]Mass Spectrometry Research Center, Vanderbilt University School of Medicine, Nashville, TN, United States
[1]Corresponding author: e-mail address: pierre.chaurand@umontreal.ca

Contents

1. Introduction	68
2. Material	69
3. Silver-Assisted IMS	70
3.1 Preparation of Silver-Coated Glass Slides	70
3.2 Tissue Deposition on Silver- or ITO-Coated Slides	70
3.3 Silver Deposition on Tissue Section	70
3.4 LDI Data Acquisition of a Silver-Coated Tissue Section	73
3.5 Fast Optimization of New Tissue Sections	74
4. Gold-Assisted IMS	75
4.1 Preparation of Gold-Coated Glass Slides	75
4.2 Tissue Deposition on ITO- or Gold-Coated Slides	75
4.3 Sodium Salt Deposition	75
4.4 Gold Deposition on Tissue Section	77
4.5 LDI Data Acquisition of a CBS-Au-Coated Tissue Section	77
5. Applications to Cancer Research	78
6. Concluding Remarks	82
References	83

Abstract

In the last decade, imaging mass spectrometry (IMS) has been the primary tool for biomolecular imaging. While it is possible to map a wide range of biomolecules using matrix-assisted laser desorption/ionization IMS ranging from high-molecular-weight proteins to small metabolites, more often than not only the most abundant easily ionisable species are detected. To better understand complex diseases such as cancer more specific and sensitive methods need to be developed to enable the detection of lower abundance molecules but also molecules that have yet to be imaged by IMS. In recent years, a big shift has occurred in the imaging community from developing wide reaching methods to developing targeted ones which increases sensitivity through the use of more specific sample preparations. This has been primarily marked by the advent of solvent-free matrix deposition methods for polar lipids, chemical derivatization for hormones and metabolites, and the use of alternative ionization agents for neutral lipids. In this chapter, we discuss two of the latest sample preparations which exploit the use of alternative ionization agents to enable the detection of certain classes of neutral lipids along with free fatty acids by high-sensitivity IMS as demonstrated within our lab.

1. INTRODUCTION

Since its discovery, imaging mass spectrometry (IMS) of tissue sections has made possible visualization of many classes biomolecules while directly linking these to the underlying histological features (Stoeckli, Chaurand, Hallahan, & Caprioli, 2001). This new histological information is completely complementary to current histochemical methods and has the ability to simultaneously detect hundreds or even thousands of compounds from a single sample. The wide variety of biomolecules which can be imaged ranges from high mass proteins to their digested peptides down to low-molecular-weight lipids and metabolites making IMS one of the most prevalent molecular imaging techniques (Reyzer & Caprioli, 2007; Schwamborn & Caprioli, 2010). The technology has been primarily driven by the need to understand disease progression at the molecular level. This is especially true for cancer studies, where the understanding of the molecular composition in the early stages of the disease can be instrumental in diagnosing, prognosticating, and/or assessing treatment of the cancer. To this end, development in sample handling and processing remains a key battleground in unraveling the molecular distribution in these samples.

Sample preparation remains the most important part of an IMS experiment. It is during this step that the user decides which type of biomolecule will be detected by the instrument. However, the sample preparation approach chosen may also limit the achievable imaging lateral resolution (Thomas & Chaurand, 2014). With matrix-assisted laser desorption/ionization (MALDI) this is often controlled by the choice of matrix along with the method by which the matrix is deposited (Berry et al., 2011; Shanta et al., 2011; Thomas, Charbonneau, Fournaise, & Chaurand, 2012). While most current sample preparations are capable of imaging multiple types of biomolecules simultaneously, they often lack the specificity and sensitivity to visualize low abundance or/and hard to ionize species (Chaurand, Schwartz, & Caprioli, 2004; MacAleese, Stauber, & Heeren, 2009). This has recently driven the development of highly specific and sensitive sample preparation methods that mainly focus on alternative ionization agents or chemical derivatization (Cobice et al., 2013; Jun et al., 2010; Kallback, Shariatgorji, Nilsson, & Andren, 2012).

The road forward in IMS is likely to be toward more specific and sensitive methods rather than universal approaches of the past decades. The aim of this chapter is to present some of the latest sample preparation methods developed in our laboratory which increase both specificity and sensitivity for specific classes of biocompounds such as neutral lipids that are either poorly detected or undetected by current MALDI IMS procedures. The first methodology presented relies on the affinity of cationized silver toward olefins forming argentinated $[M+Ag]^+$ molecular ions, and it is used here to enable the detection and imaging of intact cholesterol and free fatty acids (FAs) (Dufresne, Thomas, Breault-Turcot, Masson, & Chaurand, 2013). The second procedure uses sodium cationization to specifically reveal triacylglycerols (TAGs) and cholesterol esters with the help of a UV absorbing gold nanolayer (Dufresne, Masson, & Chaurand, 2016). These techniques will be referred below as silver- or gold-assisted LDI (Ag- and Au-LDI, respectively). Highly detailed protocols are given to facilitate the implementation of these two sample preparation methods in other laboratories. These methods have previously been successfully applied to a transgenic mouse model of Alzheimer's disease (Hamilton et al., 2015), forensic fingermarks (Lauzon, Dufresne, Chauhan, & Chaurand, 2015), and a human atherosclerotic plaque (Patterson et al., 2016). In this chapter, the two methods are being applied to human neuroendocrine tumors metastasized to the liver and liver cholangiocarcinoma highlighting their imaging capabilities in providing meaningful and complementary biological information.

2. MATERIAL

The system required for the deposition of metallic silver or gold layers is either a metal evaporation system or, as presented in this chapter, a metal sputtering system (308R sputter coater, Cressington Scientific Instruments Ltd., Watford, England). Silver, gold, and chrome targets used (target size depends on the system used) for sputtering must be at least of 99.5% purity (ESPI Metals, Ashland OR, USA). Standard microscope glass slides (25 × 75 mm) can be obtained from most generic scientific vendors. Salt layer deposition by liquid interface can be performed with any MALDI matrix spray system such as the SunCollect (SunChrom Wissenschaftliche Geräte GmbH, Friedrichsdorf, Germany) or, as used in this chapter, a TM-Sprayer (HTX Technologies, Chapel Hill, NC). Sodium bicarbonate, sodium carbonate, sodium acetate (NaAc), HPLC grade acetonitrile (ACN), and HPLC grade water can be purchased from most generic chemical vendors. Indium tin oxide-coated microscope glass slides (25 × 75 mm) can be purchased from Delta Technologies (Loveland CO, USA). Finally, an imaging capable LDI (time-of-flight) mass spectrometer (Ultraflextreme from Bruker Daltonics, Billerica, MA, USA) was used for IMS data acquisition.

3. SILVER-ASSISTED IMS
3.1 Preparation of Silver-Coated Glass Slides

Silver-coated microscope slide preparation is as follows (if ITO- or gold-coated slides are used, go to Section 3.2). Using the metal sputtering system, an initial 2- to 3-nm thick film of chrome (30 s of deposition time) is deposited over the glass slide to enhance silver adhesion to the surface. If the silver layer detaches once the tissue section is deposited on the slide than increasing the thickness of the chrome layer will increase the adhesion of the silver to the glass slide. Adding more chrome will only increase the adhesion without any adverse effect on the analysis. A silver layer is then sputter deposited over the chrome film until a thickness of roughly 100 nm (2 min 30 s of deposition time) is reached. Further increasing the silver layer thickness will not provide better IMS analyte detection but it will not negatively affect the analysis.

3.2 Tissue Deposition on Silver- or ITO-Coated Slides

Fresh frozen or formalin fixed tissues can be cut at thicknesses between 5 and 40 μm using a cryostat (fresh frozen) or a vibratome (formalin fixed) and mounted on the metal-coated slides. Due to the electrical conductivity of the silver layer, some samples with thickness higher than 40 μm may be used if thinner section cannot be obtained. More detailed procedures for fresh tissue handling, sectioning, and thaw-mounting can be found elsewhere (Kaletas et al., 2009; Thomas & Chaurand, 2014).

3.3 Silver Deposition on Tissue Section

(A) Controlling silver thickness during sputtering. Most advanced sputtering systems like the one used in this chapter have up to five individual instrumental parameters which needs to be considered. If the system does not allow the user to control all five parameters, a procedure for building a calibration curve on your own system is given in part B. The position of the head containing the sputtered metal block over the tissue section is paramount in controlling the rate of deposition over the tissue section. In the most advanced systems, the sputtering head can be moved along three axes: x (horizontal head support), y (vertical head support), and z (head rotation) (Fig. 1). To insure reproducibility, all axes should first be optimized and kept constant thereafter. The x axis should be extended as long as possible to prevent heating of the tissue section due to the argon plasma generated in the sputtering system. The head being further away will also help in generating a more homogenous metallic surface over the tissue section. In our case, a distance of ~9 cm was used, and no signal degradation has been observed from the sample over 4 years of regular use. Setting of the y and z value is the trickier part. When sputtering metals over a surface, one must be aware that the deposition rate directly under the sputtering head is greater than on its edges. This means that putting a plate directly under the sputtering head will lead to a thicker layer of metal in the middle of the plate and thinner layers on the edges, which later on may lead to significant inhomogeneity and imaging artifacts.

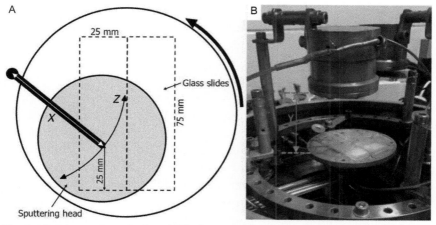

Fig. 1 (A) Scheme (*top view*) and (B) photograph (*side view*) of silver sputter system. See text for details. *Reproduced with permission from Thomas, A., Patterson, N. H., Dufresne, M., & Chaurand, P. (2016) In R. Cramer (Ed.),* Advances in MALDI and laser-induced soft ionization mass spectrometry *(pp. 129–147). Copyright 2016 Springer International Publishing.*

To prevent this effect, either the sputtering head (if y and z axis can be controlled) or the sample slide can be offset along with using a rotating sample stage. In our case, the sputtering head is offset by centering the head at one-half of the radius of the sample stage (Fig. 1). With this offset and a sample rotation of 25 rpm, a homogenous metal layer over the tissue sections (≤ 2 nm variation) can be reached. The argon partial pressure used in the sputtering system controls the rugosity (surface roughness) and to some extent the rate of metal deposition. In the system used, a partial pressure of 0.02 mbar was optimized to achieve high homogeneity and as smooth as possible silver layer. Finally, the electrical current applied on the sputtering head to produce the argon plasma is the last optimizable parameter. We set this value to the available maximum of our system (80 mA) since no change in MS signal could be observed when optimizing this parameter. Using the maximum value only increases the speed at which the silver is deposited over the sample. It is also important to note that this last parameter is linear meaning that if the system is set to 40 mA then doubling the deposition time would yield the same final thickness.

(B) Developing a calibration curve for sputtering system optimization. To generate a thickness calibration curve you will need access to a precise thickness measuring instrument such as an AFM microscope. To measure the deposited thickness over a 25×75 mm-glass slide, we first mount a 22×22 mm-cover slip glass slide in the middle of the slide before silver sputtering. After a given time of silver deposition the cover slip is removed generating three to four edges that can be used to measure the silver layer thickness. This procedure is repeated for several deposition times to generate an instrument-specific calibration curve. It is very important that the depositions are not serial since the sputtering head tends to heat up after 1–2 min of use. This heating affects the deposition rate of the metal and can lead to a nonlinear calibration curve. To prevent this we usually let the head cool down for 20–30 min between depositions. Another point to note is that the edge created by the cover slip over the slide will have higher aggregation rate. To work around this artifact, we image all three to four edges in triplicates using the imaging function of the AFM microscope while ensuring that the AFM image goes as far in as possible over the silver layer while starting over the glass slide. With this you should be able to see the edge and simply measure the thickness using the points pass the edge resulting in a more reliable measurement. With this procedure, a calibration curve can be produced with any type of sputtering or evaporating instruments with its own set of parameters. This procedure can be time-consuming, but it only needs to be done once for a given set of parameters.

(C) Optimization of the silver layer thickness over the tissue section. Our optimization of silver sputtering revealed that 30 s of silver deposition is the optimal thickness for nonfatty tissue sections deposited on a silver slide. We consider tissue sections to be nonfatty if only small amount of adipocytes ($\leq 10\%$) are present or if the tissue is devoid of visceral fat. Visceral fat has a tendency of migrating inside the tissue sections during the thaw mounting process and if possible should be removed or trimmed from the frozen tissue block prior to sectioning to minimize imaging artifacts. It is important not to deposit too thick of a silver layer since from experience it will result in lower ion yields most probably because thicker layers hinder the desorption/ionization of the underlying analytes.

3.4 LDI Data Acquisition of a Silver-Coated Tissue Section

When all the sample preparation steps are done we can begin the analyses of the tissue section by LDI IMS. If the sample cannot be analyzed on the same day, we suggest inserting the sample in a sealed bag under an inert atmosphere (N_2) and storing it in either a $-20°C$ or a $-80°C$ freezer to prevent analyte degradation (Patterson, Thomas, & Chaurand, 2014). Once the sample is in the instrument the laser fluency must be optimized. Typically, the laser fluency needed for proper ionization using silver is below the one used for α-cyano-4-hydroxycinnamic acid (CHCA) or 2,5-dihydroxybenzoic acid DHB MALDI matrices but above the fluency needed for 1,5-diaminonaphthalene (DAN) matrix. Precise adjustment of the fluency is critical since most analytes are highly prone to in-source fragmentation under LDI conditions. This is especially true when analyzing FAs since high laser fluencies irradiating the sample may induce fragmentation of either phospholipids or TAGs from the tissue section into neutral FAs. These can then be ionized by silver and contaminate the free FA signals present in the sample.

Using silver as the ionization agent also has an advantage when it comes to mass calibration. Silver will produce a series of clusters from Ag_2^+ to Ag_9^+ under LDI conditions in a mass window from m/z 200 to 1100. These clusters can be used for postprocessing internal calibration of entire IMS data sets. Ag-LDI IMS detects FAs, cholesterols, and some high abundance

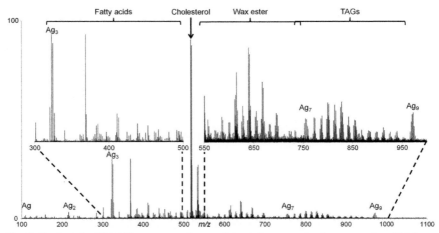

Fig. 2 Typical Ag-LDI MS spectrum from a latent fingermark after summing 500 laser shots. To date, latent fingermarks present the widest variety of lipids species which can be detected as silver adduct by Ag-LDI such as FAs, cholesterol, wax esters, and TAGs. Silver clusters are typically detected throughout the entire studied mass range and when used as internal standards can provide high mass accuracy (~5 ppm).

TAGs from thin tissue sections down to 5 μm lateral resolution. Using other types of samples, such as fingermarks, the detection of other classes of molecules like wax esters and a greater variety of TAGs is possible as shown in Fig. 2 (Lauzon et al., 2015). All in all, most olefin containing molecules have the capability to ionize with Ag^+ by forming adducts through a Lewis acid–base interaction (Christie, 1987; Dobson, Christie, & Nikolovadamyanova, 1995).

3.5 Fast Optimization of New Tissue Sections

When considering the vast differences between various types of cancers and the organs where these originate or have metastasized, it becomes very important to have a way of optimizing the necessary silver layer in a timely manner for each cancer sample which may be very fatty (such as breast cancer or lymph nodes) or of unknown fat content. Since for most tissue sections a typical thickness of 14 ± 2 nm of silver is used, this can be used as a starting point to slowly optimize the silver thickness using only one tissue sample. At this point, mass spectra can be acquired over the different histologies present in the tissue specimen (guided by a serial section stained with H&E) and observe how the relative intensities of the silver clusters (mentioned in Section 3.4) behave compared to the analyte signals. To ensure that enough silver is present on the section for proper

Assessing the Potential of Metal-Assisted IMS 75

desorption/ionization, both the second and third clusters (Ag_2^+ and Ag_3^+) must be roughly equal or higher in intensity than all other observed signals (see Fig. 2). This ensures that the most abundant signal(s) does not ionically suppress other signals by preferentially forming adducts with the available gas phase Ag^+ due to their overwhelming concentration. If there is an ion signal which has a higher intensity than the two clusters, then one can simply insert the sample back into the sputtering system and add additional silver layers at an increment rate of 10 s (~5 nm) of sputtering until the appropriate ion intensities are observed. This may need to be done multiple times for very fatty samples, but it will greatly limit ionic suppression. As for the maximum possible imaging resolution, we have demonstrated that this method can be used down to 5 μm with the instrumentation available in our lab. It is however possible that with better laser focus on target higher spatial resolutions can be achieved.

4. GOLD-ASSISTED IMS

4.1 Preparation of Gold-Coated Glass Slides

Gold-coated glass slides preparation is paramount to this procedure since gold surfaces offer a much better tissue adhesion than ITO slides. We found that the procedure described later often leads to the tissue detaching from the ITO surface due to the very high salt concentration used, while this does not happen with gold-coated slides. As with silver, an initial film of chrome of 2–3 nm (30 s) is deposited over the glass slide to enhance gold adhesion to the surface. A ~150-nm gold layer is then deposited over the chrome surface. Gold precoated slides can also be bought commercially, but it is cheaper to make these if a sputtering system is available (~1$ per slides when homemade).

4.2 Tissue Deposition on ITO- or Gold-Coated Slides

The procedure for tissue sectioning toward Au-LDI IMS is identical to the one described in Section 3.2.

4.3 Sodium Salt Deposition

The initial step for this method is to dope the tissue section with a sodium salt to privilege the formation of Na^+ adducts ($[M+Na^+]$ molecular ion) necessary for TAG and cholesterol ester ionization. One may assume that there is enough endogenous sodium present in fresh frozen tissue sections, but it has been found to be insufficient since the addition of an external sodium source dramatically increases the observed signals for TAGs (Dufresne et al., 2016). If you are using formalin fixed tissue sections and are using PBS washes during the fixation and/or rinsing steps, you can go directly to the next section since processing with PBS incorporates more than enough sodium in the section for successful Au-LDI IMS analyses. For fresh frozen tissue sections the amount of sodium salt is proportional to the tissue thickness. All initial optimizations were carried out on 14-μm thick tissue sections, but we found that the amount of salt is linear with the thickness such that a 20-μm thick tissue section requires ~40% more sodium salt for optimal analyte detection. For this step a sprayer unit for MALDI imaging is highly recommended.

We use a first-generation TM-Sprayer unit but other sprayers may be used by simply developing a method that deposits the same optimal density of sodium salt solution. The optimal salt solution was found to be a 1:1 aqueous mixture of 85 mM carbonate buffer with 250 mM of sodium acetate combined with ACN with a 7:3 ratio (CBS). Once properly sprayed, this should yield a layer with a density of 675 μg/cm^2 of salt over the tissue section. To measure the salt density, simply weigh three times a 22 × 22 mm-cover slip before and after the salt deposition. This step can be repeated three times with different cover slips for accuracy. We weigh each cover slip three times since the added weight is quite small. We tested several salts and most of these are either too hygroscopic leading very long drying time (>30 s) which leads to analyte delocalization and hinder high spatial resolution imaging. It is important to note that both the concentration sprayed and the solvent mixture compositions are instrument specific. In our case, higher salt concentrations can partially clog the TM-Sprayer unit causing the spray to deviate from the sample. As for the solvent composition, the mix of water and ACN gave the fastest drying time (<6 s) with the highest amount of ACN added without the formation of a two phase solution due to the high salt concentration present in solution. These two parameters could and should be optimized for other sprayers to insure that the proper salt density on tissue is reached. We highly advise using water without ACN if your system can spray such solution without compromising the drying time since water does not solubilize TAGs or cholesterol esters and should not induce delocalization.

4.4 Gold Deposition on Tissue Section

For optimization and generating a calibration curve with a sputtering system, refer to Section 3.3A and B. In Au-LDI IMS, because of its strong absorbance at the wavelength used for desorption/ionization (355 nm), gold serves here as the desorption agent assisting the sodium salt in detecting TAGs and cholesterol esters. The optimization of the gold layer thickness after sodium salt deposition revealed that a layer of 28 ± 3 nm (40 s) of gold using 40 mA (instead of 80 mA, see Section 3.3A) yields the highest intensity signals for both target compound class. As previously mentioned, reducing the electrical current results in a proportional reduction in the sputtering rate of the metal. Since we reduced the current by half we double the deposition time (40 s at 40 mA rather than 20 s at 80 mA). For gold deposition, we decided to reduce the current because short deposition times (with our instrument <30 s) are less reproducible and yield less homogenous surfaces. This can be problematic when dealing with multiple samples or when a single 25×75 mm-plate contains multiple sections.

4.5 LDI Data Acquisition of a CBS-Au-Coated Tissue Section

Once the gold layer is deposited over the CBS-coated tissue section, LDI IMS analyses are possible. As with Ag-LDI, if the samples cannot be analyzed on the same day we suggest storage under nitrogen atmosphere in a freezer (see Section 3.4). Laser fluency is once again the first parameter which needs to be optimized. In the case of the CBS-Au method, laser fluency is quite similar to that used for CHCA and DHB matrices. We have however observed that the usable range of fluency is larger than with other LDI methods. When analyzing high fat tissues, the laser fluency can be significantly diminished often to fluencies typically used with DAN matrix and the opposite can often happen with lean tissues. This method is also capable of internal postacquisition recalibration using the gold clusters present in all mass spectra. To this end we usually acquire mass spectra between m/z 100 and 1100, which gives us access to Au clusters 1 through 3 covering the bottom half of the mass range (see Fig. 3). Adding one of the detected TAGs to the calibration list extends the calibration range into the upper half of the mass range. Au-LDI IMS is possible at high spatial resolution down to 10 μm. Higher spatial resolutions have not been tested, but the salt crystal size measured by transmission electron microscopy were found to be ≤ 3 μm and sputtered gold layers only becomes

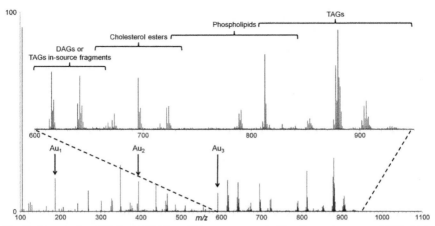

Fig. 3 Composite CBS-Au-LDI MS profile spectrum resulting from the sum of 250 shots acquired from a rabbit adrenal gland and 250 shots acquired from a mouse liver. This composite mass spectrum displays many of the neutral lipids species that can be detected by the CBS-Au LDI such as the ambiguous diacylglycerols or TAG in-source fragments signals along with cholesterol esters and intact TAGs expressed as sodium adduct. Gold cluster signals (covering the first half of the MS spectrum) and a TAG signal at higher MW (after its identity has been confirmed by MS/MS) can be used for internal calibration.

heterogeneous on the nanometer scale (Dufresne et al., 2016). Therefore, these very fine salt crystals along with the highly homogenous gold layer should enable at least a 5 μm of lateral resolution.

5. APPLICATIONS TO CANCER RESEARCH

Understanding the chemical composition of the various histologies present in a progressing cancer is primordial in the development of novel therapeutic treatment of such disease. For one, our lab has a special interest in the study of liver metastases since metastatic cancers often are the most hazardous to our health. Liver metastases are a good starting point as most cases receive simple resection surgery after response to chemotherapy or in some cases simple resection if the tumor is estimated small enough at the time of diagnosis. Due to this, large amounts of liver metastasis resections are available for research purposes in various biobanks often accompanied by the patient's medical history. These medical data can significantly help analyze the large amounts of data produced by IMS technology.

The first example presented here is a human neuroendocrine metastasis to the liver. Fig. 4 shows four selected ion images acquired by Ag-LDI or

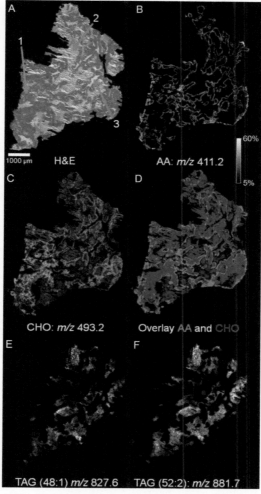

Fig. 4 Ag-LDI and CBS-Au-LDI IMS from a human neuroendocrine liver metastasis at 50 μm lateral resolution. (A) H&E staining of serial section. (1) *Deep purple* staining corresponding to cancer cell proliferation regions. (2) *Pink* region mostly devoid of nuclei corresponding to necrotic areas. (3) Tumor stroma region which is surrounded by cancer cells. (B) and (C) Ag-LDI ion images for arachidonic acid and cholesterol, respectively. (D) Overlay of both (B) and (C) ion images showing that the arachidonic acid (*red* signal) comes from tumor areas surrounding the cholesterol signals (*blue* signal). Arachidonic acid is expressed in high abundance at the edge between the cancer cell proliferation regions and the necrotic areas. This area is characterized by apoptotic tumor cells. (E) and (F) Two TAG ion images acquired by CBS-Au-LDI MS corresponding to the necrotic area found in this liver metastasis.

CBS-Au-LDI IMS from serial tissue sections. As it is often the case with ion images generated by metal-assisted LDI IMS, most families of molecules tend to have highly similar distributions within a tissue section. In this case, Fig. 4C shows that cholesterol was found to primarily correlate with the cancer cell proliferation regions highlighted in deep purple on the H&E presented Fig. 4A. Upon Ag-LDI IMS analysis, cholesterol distribution within a given tissue sample is often observed to correlate with high cellular density tissue regions such as the granular region in the brain cerebellum (Dufresne et al., 2013) or in this case, the cancer cell proliferation region, both displaying high quantities of cells per surface unit. This can be rationalized when taking into account that cholesterol plays a critical role in cellular wall structure. Lower signal intensity for cholesterol was also found in the stroma regions located in the middle of the proliferation regions but very little to no cholesterol signal was observed in the necrotic areas. Fig. 4B shows the ion image for arachidonic acid which is localized in higher abundance in the cancer cell apoptotic regions located between the cancer proliferation and the necrotic regions. This is better visualized in an overlay compiled from the cholesterol and arachidonic ion images (Fig. 4D). All of the six other detected FA signals (16:0, 16:1, 18:1, 18:2, 20:3, and 22:6) gave similar abundance distributions among the four observable histologies with the highest concentration within the apoptotic regions and with some lower expression within the necrotic areas. This last region visualized in bright pink in the H&E staining Fig. 4A did not correlate well with any of the Ag-LDI data but correlated almost perfectly with CBS-Au-LDI data. The signals with the best correlation to the necrotic areas were the 16 different TAGs that could be observed in this region. One short chain TAG (Fig. 4E) with a total of 48 carbons and 1 unsaturation (48:1) as well as the most abundant TAG signal (Fig. 4F) with 52 carbons and 2 unsaturations (52:2) are presented to highlight these necrotic areas. Usually, TAGs are found to mostly correlate with adipocytes, but in this case these very large necrotic areas do not present any adipocyte (which would be highlighted in the H&E) but only display vast amounts of cellular debris.

The second example presented is a human cholangiocarcinoma or bile duct liver cancer (Fig. 5). In this case, Ag-LDI analysis of cholesterol once again showed a good correlation to the cancer proliferation regions, but in this case the proliferation regions are much smaller and much more dispersed, which makes the ion image seems much less structured (Fig. 5A–C). Along with cholesterol, six FAs were detected with the distribution of arachidonic acid being presented in Fig. 5B. This time FAs

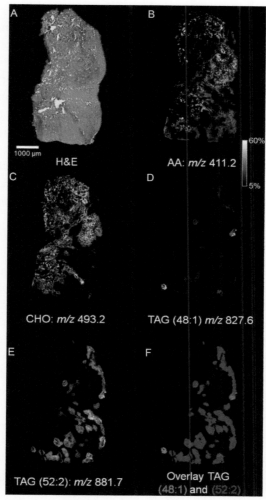

Fig. 5 Ag-LDI and CBS-Au-LDI IMS from a human cholangiocarcinoma tumor at 50 μm lateral resolution. (A) H&E staining of the next serial section. (B) and (C) Ag-LDI ion images for arachidonic acid and cholesterol, respectively. (D) and (E) CBS-Au-LDI ion images for two TAG signals corresponding to specific lipid deposits found in the necrotic regions of this liver metastasis. (F) Overlay of (D) and (E) showing a very heterogeneous distribution between TAG (52:2) in *blue* and the smaller TAG (48:1) in *red*.

appeared to be more evenly distributed around the proliferation regions without being specific to the apoptotic regions like in the previous case. We can also see more signals in the large necrotic areas at the right side of the specimen compared to the previous sample. We also see that the distribution in the necrotic areas has distinct circular features, where the signals for FAs are quite lower in intensity. These circular regions are adipocyte bundles, which are better highlighted by the CBS-Au-LDI results (Fig. 5D–E). The ion images for both triglycerides not only correlate very well with these adipocytes but also show the differential distribution between the short FA chain TAGs and longer FA chain TAGs. We have yet to understand why the smaller chain TAGs have such a distinct distribution but this is a good example the types of features which has remained unobservable with previous MALDI-based IMS or other histological methods. Fig. 5F shows the overlap images of both TAG (48:1) and TAG (52:2), which highlights in pink the regions, where all eight detected short chain TAGs can be found and in blue the regions where the six longer chain TAGs are observed.

6. CONCLUDING REMARKS

Current MALDI IMS approaches are designed for broad molecular analysis, where the end goal is often seen as many species as possible in a single experiment. These methods offer the possibility of selecting wide classes of biomolecules by controlling the matrix deposition step and the type of solvent involved. This is especially true for low-molecular-weight compounds like lipids, where the advent of solvent-free approaches enabled high spatial resolution IMS of common phospholipids. Unfortunately, these wide reaching methods often only visualize the most abundant species. This becomes an issue when considering complex samples such as tumors. To better understand all of these different forms of cancer at the molecular level, one must move away from these wide reaching methods toward more specific and more sensitive approaches which can enable the imaging of low abundance or hard to ionize molecules with significant biological output. One approach to introduce specificity in IMS is to exploit the affinity of metal cations for certain functional groups found in the targeted biomolecules.

One such interaction is found between olefin containing molecules and cationized silver atoms. As detailed earlier, we have developed a simple sample preparation protocol to analyze cholesterol, FAs, along with many other types of olefin containing molecules directly from thin tissue sections by Ag-LDI IMS. This solvent-free procedure yields a highly homogenous silver metallic layer over the tissue section enabling the detection of biomolecules as silver adducts with lateral resolution reaching cellular dimensions (∼5 µm) (Dufresne et al., 2013). A second metal cation that can be used for ionization is sodium. Sodium is known to ionize a wide range of biomolecules in previously reported MALDI and LDI IMS methods; however, such an event is not always desirable. One solution to give more specificity to sodium cation ionization is through the use of a gold metallic layer instead of using conventional MALDI matrices. This enabled us to specifically visualize TAGs with an increase in signal of more than 30 folds compared to DHB matrix along with an increase in the number of detected TAG species from 7 to 25 (Dufresne et al., 2016). This method was also designed to enable high spatial resolution IMS down s0 µm. The advent of these targeted methods offering better specificity and sensitivity in IMS, which are complementary to the more traditional MALDI-based IMS approaches, will help us better understand the biological footprints left by various diseases and conditions such as the multiple forms of a cancer.

REFERENCES

Berry, K. A. Z., et al. (2011). MALDI imaging of lipid biochemistry in tissues by mass spectrometry. *Chemical Reviews, 111*, 6491–6512. http://dx.doi.org/10.1021/cr200280p.

Chaurand, P., Schwartz, S. A., & Caprioli, R. M. (2004). Profiling and imaging proteins in tissue sections by MS. *Analytical Chemistry, 76*, 86A–93A. http://dx.doi.org/10.1021/ac0415197.

Christie, W. W. (1987). A stable silver-loaded column for the separation of lipids by high-performance liquid-chromatography. *Journal of High Resolution Chromatography & Chromatography Communications, 10*, 148–150. http://dx.doi.org/10.1002/jhrc.1240100309.

Cobice, D. F., et al. (2013). Mass spectrometry imaging for dissecting steroid intracrinology within target tissues. *Analytical Chemistry, 85*, 11576–11584. http://dx.doi.org/10.1021/ac402777k.

Dobson, G., Christie, W. W., & Nikolovadamyanova, B. (1995). Silver ion chromatography of lipids and fatty-acids. *Journal of Chromatography B, Biomedical Applications, 671*, 197–222. http://dx.doi.org/10.1016/0378-4347(95)00157-e.

Dufresne, M., Masson, J. F., & Chaurand, P. (2016). Sodium-doped gold-assisted laser desorption ionization for enhanced imaging mass spectrometry of triacylglycerols from thin tissue sections. *Analytical Chemistry, 88*, 6018–6025. http://dx.doi.org/10.1021/acs.analchem.6b01141.

Dufresne, M., Thomas, A., Breault-Turcot, J., Masson, J.-F., & Chaurand, P. (2013). Silver-assisted laser desorption ionization for high spatial resolution imaging mass spectrometry of olefins from thin tissue sections. *Analytical Chemistry, 85*, 3318–3324. http://dx.doi.org/10.1021/ac3037415.

Hamilton, L. K., et al. (2015). Aberrant lipid metabolism in the forebrain niche suppresses adult neural stem cell proliferation in an animal model of Alzheimer's disease. *Cell Stem Cell, 17*, 397–411. http://dx.doi.org/10.1016/j.stem.2015.08.001.

Jun, J. H., et al. (2010). High-spatial and high-mass resolution imaging of surface metabolites of Arabidopsis thaliana by laser desorption-ionization mass spectrometry using colloidal silver. *Analytical Chemistry, 82*, 3255–3265. http://dx.doi.org/10.1021/ac902990p.

Kaletas, B. K., et al. (2009). Sample preparation issues for tissue imaging by imaging MS. *Proteomics, 9*, 2622–2633. http://dx.doi.org/10.1002/pmic.200800364.

Kallback, P., Shariatgorji, M., Nilsson, A., & Andren, P. E. (2012). Novel mass spectrometry imaging software assisting labeled normalization and quantitation of drugs and neuropeptides directly in tissue sections. *Journal of Proteomics*, *75*, 4941–4951. http://dx.doi.org/10.1016/j.jprot.2012.07.034.

Lauzon, N., Dufresne, M., Chauhan, V., & Chaurand, P. (2015). Development of laser desorption imaging mass spectrometry methods to investigate the molecular composition of latent fingermarks. *Journal of the American Society for Mass Spectrometry*, *26*, 878–886. http://dx.doi.org/10.1007/s13361-015-1123-0.

MacAleese, L., Stauber, J., & Heeren, R. M. A. (2009). Perspectives for imaging mass spectrometry in the proteomics landscape. *Proteomics*, *9*, 819–834. http://dx.doi.org/10.1002/pmic.200800363.

Patterson, N. H., Thomas, A., & Chaurand, P. (2014). Monitoring time-dependent degradation of phospholipids in sectioned tissues by MALDI imaging mass spectrometry. *Journal of Mass Spectrometry: JMS*, *49*, 622–627. http://dx.doi.org/10.1002/jms.3382.

Patterson, N. H., et al. (2016). Three-dimensional imaging MS of lipids in atherosclerotic plaques: Open-source methods for reconstruction and analysis. *Proteomics*, *16*, 1642–1651. http://dx.doi.org/10.1002/pmic.201500490.

Reyzer, M. L., & Caprioli, R. M. (2007). MALDI-MS-based imaging of small molecules and proteins in tissues. *Current Opinion in Chemical Biology*, *11*, 29–35. http://dx.doi.org/10.1016/j.cbpa.2006.11.035.

Schwamborn, K., & Caprioli, R. M. (2010). MALDI imaging mass spectrometry—Painting molecular pictures. *Molecular Oncology*, *4*, 529–538. http://dx.doi.org/10.1016/j.molonc.2010.09.002.

Shanta, S. R., et al. (2011). Binary matrix for MALDI imaging mass spectrometry of phospholipids in both ion modes. *Analytical Chemistry*, *83*, 1252–1259. http://dx.doi.org/10.1021/ac1029659.

Stoeckli, M., Chaurand, P., Hallahan, D. E., & Caprioli, R. M. (2001). Imaging mass spectrometry: A new technology for the analysis of protein expression in mammalian tissues. *Nature Medicine*, *7*, 493–496. http://dx.doi.org/10.1038/86573.

Thomas, A., Charbonneau, J. L., Fournaise, E., & Chaurand, P. (2012). Sublimation of new matrix candidates for high spatial resolution imaging mass spectrometry of lipids: Enhanced information in both positive and negative polarities after 1,5-diaminonapthalene deposition. *Analytical Chemistry*, *84*, 2048–2054. http://dx.doi.org/10.1021/ac2033547.

Thomas, A., & Chaurand, P. (2014). Advances in tissue section preparation for MALDI imaging MS. *Bioanalysis*, *6*, 967–982. http://dx.doi.org/10.4155/bio.14.63.

CHAPTER FOUR

MALDI Mass Spectrometry Imaging of *N*-Linked Glycans in Cancer Tissues

R.R. Drake[1], T.W. Powers, E.E. Jones, E. Bruner, A.S. Mehta, P.M. Angel

Medical University of South Carolina, Charleston, SC, United States
[1]Corresponding author: e-mail address: draker@musc.edu

Contents

1. Introduction   86
2. Glycosylation and Cancer   87
      2.1 Function and Types of Glycosylation   87
      2.2 N-Linked Glycan Biosynthesis   89
      2.3 N-Linked Glycans and Cancer   89
3. Methodology for N-Linked Glycan Detection by MALDI Imaging   92
      3.1 Tissue Sources   92
      3.2 Glycan Visualization in Tissues: Lectins and Anticarbohydrate Antibodies   95
      3.3 Histochemistry Stains   96
      3.4 Peptide *N*-Glycosidase F   97
      3.5 Matrix and Instrumentation Choices for *N*-Glycans   98
      3.6 Structural Confirmation   100
4. *N*-Glycan Distribution Linked With Histopathology   101
      4.1 Major Structural Classes   101
      4.2 High-Mannose *N*-Glycans   102
      4.3 Nontumor Stroma and Normal Tissue Glycans   103
      4.4 *N*-Glycan Branching and Sialylation   104
      4.5 Fucosylation and the Glycan Isomer Problem   105
5. Emerging Applications   106
      5.1 Combined Glycan and Peptide MS Imaging   106
      5.2 Custom Multitumor TMA and Other Enzymes   108
      5.3 Linkage to Genomic Studies   110
      5.4 Potential Clinical Diagnostic Applications of *N*-Glycan MSI Data   110
6. Summary   111

References   112

Abstract

Glycosylated proteins account for a majority of the posttranslation modifications of cell surface, secreted, and circulating proteins. Within the tumor microenvironment, the presence of immune cells, extracellular matrix proteins, cell surface receptors, and interactions between stroma and tumor cells are all processes mediated by glycan binding and recognition reactions. Changes in glycosylation during tumorigenesis are well documented to occur and affect all of these associated adhesion and regulatory functions. A MALDI imaging mass spectrometry (MALDI-IMS) workflow for profiling N-linked glycan distributions in fresh/frozen tissues and formalin-fixed paraffin-embedded tissues has recently been developed. The key to the approach is the application of a molecular coating of peptide-N-glycosidase to tissues, an enzyme that cleaves asparagine-linked glycans from their protein carrier. The released N-linked glycans can then be analyzed by MALDI-IMS directly on tissue. Generally 40 or more individual glycan structures are routinely detected, and when combined with histopathology localizations, tumor-specific glycans are readily grouped relative to nontumor regions and other structural features. This technique is a recent development and new approach in glycobiology and mass spectrometry imaging research methodology; thus, potential uses such as tumor-specific glycan biomarker panels and other applications are discussed.

1. INTRODUCTION

MALDI imaging mass spectrometry (MALDI-IMS) profiling of N-linked glycan distributions in fresh/frozen tissues, formalin-fixed paraffin-embedded (FFPE) tissue blocks and tissue microarrays (TMAs) is a recent development and new approach in glycobiology research methodology. Reported initially in frozen kidney tissues (Powers et al., 2013) and quickly followed by application to different FFPE cancer tissues (Powers et al., 2014), this methodology is particularly relevant for cancer tissues, as most known cancer biomarkers are glycoproteins or carbohydrate antigens. The basic approach is to spray a molecular coating of the enzyme that releases N-linked carbohydrates from carrier proteins, peptide N-glycosidase F (PNGaseF), on-tissues, followed by matrix application, and MALDI-IMS. Beyond the initial publications, eight other reports from multiple laboratories have been published for FFPE cancer tissues (Drake, Jones, Powers, & Nyalwidhe, 2015; Everest-Dass et al., 2016; Heijs et al., 2016; Holst et al., 2016; Powers, Holst, Wuhrer, Mehta, & Drake, 2015) and other noncancer FFPE tissue types (Briggs et al., 2016; Gustafsson et al., 2015; Toghi Eshghi et al., 2014). Compared to other biomolecules targeted by MALDI-IMS-like lipids, metabolites, and proteins/peptides, IMS of N-glycans has some unique analytical aspects. One major consideration for N-glycan imaging is that FFPE tissues are the best starting material, which is a particular advantage due to the vast repositories of clinical FFPE tissues, primarily from cancer patients, archived worldwide. Similarly, tissue microarrays derived from multiple FFPE tissues can also be efficiently used (Powers et al., 2015, 2014). The use of the enzyme, PNGaseF to release the N-glycans means that the derived IMS signal is highly specific to only released N-glycan patterns from tissue. Many N-glycan structures are known and can be confirmed by collision-induced dissociation (CID) directly from tissues or following extraction of tissue (Holst et al., 2016; Powers et al., 2015). For most cancer samples analyzed thus far, the numbers of glycans detected per tissue sample are manageable, generally 40–60 per sample. This in turn facilitates the generation of different glycan panels associated with specific histopathology features and tissue subregions useful for cancer biomarker assessment. These localized regions of interest can be further targeted for identification of the carrier glycoproteins.

The focus of this chapter is to provide an overview of the methods and challenges associated with the emerging area of N-glycan tissue imaging by mass spectrometry as applied to cancer. A background of the significance of glycosylation in cancer development and progression is provided, as well as a summary of other methods used to evaluate glycan expression in cancer tissues. Using a colorectal adenocarcinoma FFPE tissue that has been evaluated for N-glycan and peptide MALDI-IMS, examples of the type of glycan data obtained from this tissue are provided to illustrate the strengths of the approach, as well as highlight the challenges and limitations of the method. Application of the method to tissue microarrays has been covered in previous studies (Powers et al., 2015, 2014), but data from an example custom tissue microarray that is being used for continued method development are provided. Future applications of the approach are numerous and are discussed in context with other large data "omics" analyses and clinical diagnostics.

2. GLYCOSYLATION AND CANCER
2.1 Function and Types of Glycosylation

The functions and regulation of glycoproteins carrying N-linked and O-linked sugar chains have been extensively reviewed (Kudelka, Ju, Heimburg-Molinaro, & Cummings, 2015; Moremen, Tiemeyer, & Nairn, 2012; Varki, 2016). It is estimated that over 50% of human proteins are glycosylated, making it one of the most common and complex posttranslational modifications. There are over 300 metabolic enzymes, glycosyltransferases and glycosidases, involved in glycan biosynthesis and processing (Zoldoš,

Novokmet, Bečeheli, & Lauc, 2013). Glycoproteins account for ∼80% of the proteins located at the cell surface and in the extracellular environment, and serve as binding ligands for cell adhesion, extracellular matrix molecules, signaling receptors, immune cells, lectins, and pathogens (Varki, 2016). Glycans present on newly synthesized glycoproteins also aid in basic protein folding, intracellular transport, and secretion processes.

There are 10 monosaccharide units from which mammalian glycans are constructed, but additional diversity can be achieved by further modification of those monosaccharides, like sulfation (Moremen et al., 2012). Structurally, hexose (Hex) monosaccharides consist of glucose (Glu), galactose (Gal), and mannose (Man) residues. N-acetylhexosamine (HexNAc) monosaccharides consist of both N-acetylglucosamine (GlcNAc) and N-acetylgalactosamine (GalNAc). The remaining monosaccharides include fucose (dHex, Fuc), N-acetylneuraminic acid (NeuAc), glucuronic acid, iduronic acid, and xylose. The glycans will be further discussed in this chapter, and their representative symbols are listed in Fig. 1A.

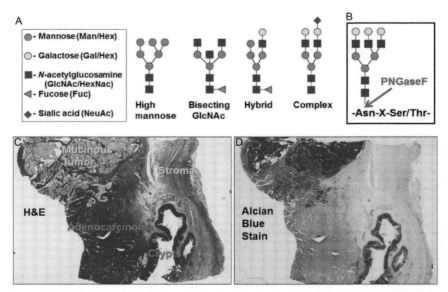

Fig. 1 (A) Five glycans and their symbols, used throughout the chapter, and four N-glycans representative of different structure classes. (B) Site of action of peptide N-glycosidase F (PNGaseF). (C) Hematoxylin and eosin stain (H&E) of a colon cancer FFPE tissue slice. Indicated are areas of adenocarcinoma (*red font*), normal crypt epithelial cells (*blue font*), mucinous tumor, and stroma (*yellow font*). (D) Alcian blue stain of the colon tumor.

2.2 N-Linked Glycan Biosynthesis

The generation of the many possible N-glycan structures that are attached to glycoproteins is the result of a series of sequential glycan addition and subtraction reactions mediated by specific glycosidases and glycosyltransferases (reviewed in Rini, Esko, & Varki, 2009; Stanley, Schachter, & Taniguchi, 2009). Briefly, the addition of N-glycans to proteins occurs cotranslationally in the endoplasmic reticulum (ER), whereby a Glc3Man9GlcNAc2-P-P-dolichol donor substrate is transferred to an asparagine residue on the recipient protein by an oligosaccharyltransferase complex. The N-glycan consensus sequence is N-X-S/T, where X is any amino acid other than proline. After transfer, the glycan is trimmed sequentially in the ER and cis-Golgi by glucosidases and mannosidases to Man5GlcNAc2, which serves as the precursor to complex and hybrid glycan structures (see Fig. 1). Glycans that are not processed or incompletely processed by the mannosidases are termed high-mannose glycans, containing 5–9 mannose residues (Man5–9GlcNAc2) (Fig. 1A). To generate complex glycans, Man5GlcNAc2 is acted upon by N-acetylglucosaminyltransferases I and II to generate an initial biantennary complex glycan. Triantennary and tetraantennary complex glycans are generated from the activity of N-acetylglucosaminyltransferases IV and V to add additional branches. N-acetylglucosaminyltransferase III is an additional enzyme that transfers a bisecting GlcNAc residue onto the complex glycan (Fig. 1A). In the trans-Golgi, further maturation may take place to provide additional glycan diversity. Among the most common additions are the transfer of β-linked galactose residues to the nonreducing end, the addition of α1,6-linked fucose residues to the GlcNAc directly bound to the asparagine (termed core fucosylation), and the addition of sialic acids and fucose residues on the branched chains. Different examples of these N-glycan structures are shown in Figs. 1–7.

2.3 N-Linked Glycans and Cancer

Alterations and changes in cell surface glycosylation during tumorigenesis are well documented and have been extensively reviewed (Christiansen et al., 2014; Kudelka et al., 2015; Pinho & Reis, 2015; Taniguchi & Kizuka, 2015). This is underscored by the fact that the majority of current FDA-approved tumor markers are glycoproteins or glycan antigens, including PSA, and also CA19-9 (Adamczyk, Tharmalingam, & Rudd, 2012; Ruhaak, Miyamoto, & Lebrilla, 2013). Broadly for cancers, increased

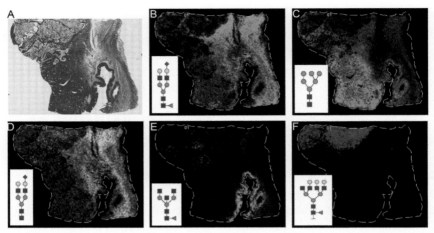

Fig. 2 Examples of regiospecificity of different *N*-glycans detected by MALDI-FT-ICR imaging mass spectrometry. The colon tumor FFPE tissue was antigen retrieved and incubated with PNGaseF, and *N*-glycans detected by MALDI-IMS as previously described (Powers et al., 2014). (A) H&E stain; (B) a stroma glycan (*green*), Hex5HexNAc4Fuc1NeuAc (+2Na), $m/z = 2122.810$; (C) a high-mannose tumor glycan (*red*), Man6, $m/z = 1419.515$; (D) a stroma glycan (*yellow*), Hex5HexNAc4NeuAc, $m/z = 1976.745$ (+2Na); (E) a crypt glycan (*aqua blue*) Hex3HexNAc5Fuc1, $m/z = 1688.647$; (F) a mucinous tumor glycan (*pink*), Hex6HexNAc6Fuc1, $m/z = 2377.798$.

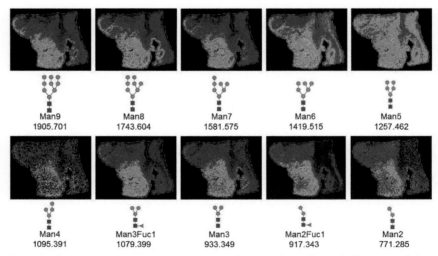

Fig. 3 High-mannose and paucimannose *N*-glycans. The indicated glycans and their masses were detected primarily in the adenocarcinoma regions of the colon tumor.

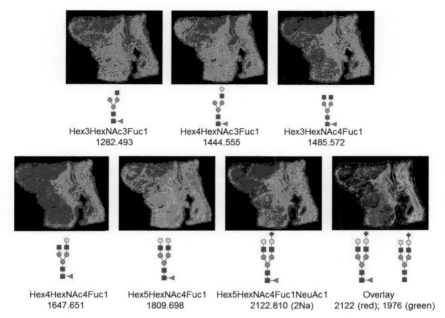

Fig. 4 Core fucosylated, biantennary *N*-glycans. The indicated glycans and their masses were detected primarily in the stroma regions and adenocarcinoma regions of the colon tumor. The *bottom left panel* shows an overlay of two stroma glycans at 2122 (*red*) and 1976 (*green*). All masses include one sodium, except where indicated.

expression of *N*-acetylglucosaminyl transferase V (GnT-V) transcripts in tumors leads to an increase in β1,6-GlcNAc branching in N-linked structures necessary for larger tri- and tetraantennary structures associated with the metastatic phenotype of multiple cancer types (Miwa, Song, Alvarez, Cummings, & Stanley, 2012; Pinho & Reis, 2015; Schultz, Swindall, & Bellis, 2012; Taniguchi & Kizuka, 2015). Structurally, increased branching of glycans in cancers is typified by increased detection of sialyl Lewis X and sialyl Lewis A antigens, as well increases in polylactosamine modifications. These structures in turn are recognized by selectins and other carbohydrate lectins expressed on different tissues involved with immune cell binding, and in the case of extravasation, binding to cells in distant organ/tissue sites. Conversely, the presence of bisecting GlcNAc residues from the activity of *N*-acetylglucosaminyl transferase III (GnT-III) alters the structural conformation of the glycan chains and tend to limit branching (Miwa et al., 2012; Pinho & Reis, 2015; Taniguchi & Kizuka, 2015) and therefore tumor progression. Within the tumor microenvironment, the presence of immune

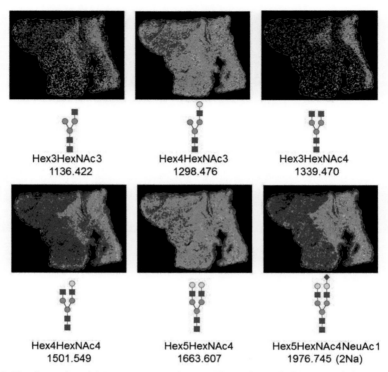

Fig. 5 Nonfucosylated, biantennary N-glycans. The indicated glycans and their masses were detected primarily in the stroma regions and adenocarcinoma regions of the colon tumor. All masses include one sodium, except where indicated.

cells, extracellular matrix proteins, cell surface receptors, and interactions between stroma and tumor cells are all processes mediated by glycan-binding and recognition reactions. For example, increased levels of α2,6 sialylation on integrins have been implicated in metastasis (Schultz et al., 2012; Uemura et al., 2009).

3. METHODOLOGY FOR N-LINKED GLYCAN DETECTION BY MALDI IMAGING

3.1 Tissue Sources

Most imaging mass spectrometry techniques that are attempting to detect protein, lipid, and metabolite targets rely on the availability of fresh frozen tissues. The original N-glycan MALDI-IMS study described the technique for use in frozen tissues (Powers et al., 2013). In subsequent studies, it was

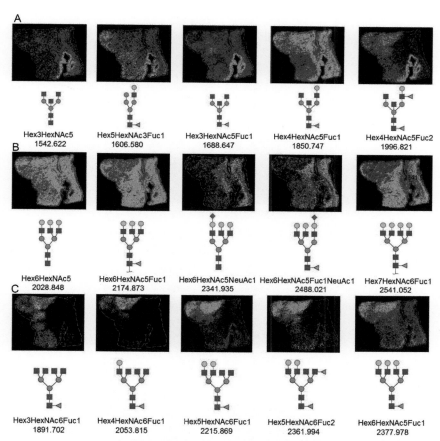

Fig. 6 Other glycan examples. (A) Hybrid and bisecting N-acetylglucosamine glycans present primarily in the epithelial crypt cell region. (B) Tri- and tetraantennary N-glycan examples in the stroma and adenocarcinoma regions. (C) Tetraantennary N-glycans that are present primarily in the mucinous tumor region. All structures shown are representative of glycan composition, but not specific linkages or individual branch modifications.

observed that use of FFPE tissues processed by standard histochemistry workflows, i.e., xylene washes to remove paraffin and antigen retrieval to reverse cross-links, yielded more N-glycan signal intensities as well as more target analytes (Powers et al., 2014). The reasons for this difference are not completely understood, but one major factor is the significant processing of FFPE tissues. Formalin readily reacts with many biological molecules including lipids, proteins, RNA, and DNA. The process of the reaction not only forms cross-links between tissue content but also modifies almost

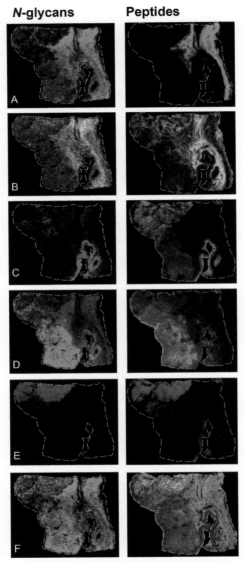

Fig. 7 Illustration of colocalization of N-glycans and peptides detected by MALDI-FT-ICR imaging mass spectrometry. The same glycans from Fig. 2 are included in the indicated panels, and a corresponding peptide image. For peptides, tissue was sprayed with trypsin and digested, followed by MALDI-IMS similar to published protocols (Heijs et al., 2016). (A) Stroma (*green*) glycan $m/z=2122.810$, peptide $m/z=1055.491$. (B) Stroma (*yellow*), glycan $m/z=1976.745$ (+2Na), peptide $m/z=886.439$; (C) Epithelial crypt (*aqua blue*) glycan, $m/z=1688.647$, peptide $m/z=911.444$; (D) adenocarcinoma (*red*), glycan $m/z=1419.515$, peptide $m/z=890.449$; (E) mucinous tumor (*pink*), glycan $m/z=2377.798$, peptide $m/z=931.431$; (F) glycan overlay of the shown mucinous tumor glycan (*pink*), adenocarcinoma glycan (*red*), and stroma glycan (*green*); peptide overlay of each peptide mass and color shown in (A)–(E).

all end groups in biological molecules such as primary and secondary amines, amides, hydroxyls, and sulfhydryls (Dapson, 2007). In FFPE tissue, such molecules, therefore, become undetectable through massive cross-linking. After undergoing formalin fixation, FFPE tissues are processed for paraffin embedding. This procedure involves heat-mediated solvent exchanges of ethanol, xylenes, and molten paraffin through the tissue, further washing out metabolites, lipids, and molecules that may not have been cross-linked (Scalia et al., 2016). In particular, for N-glycans, which generally do not have any free amino groups and are not cross-linked, the extensive processing most likely serves to expose the N-linked glycan moieties on the remaining glycoproteins, allowing PNGaseF digestion to be more efficient. The methods certainly can be applied to frozen tissues, and hypothesize that in these tissues the N-glycans are less exposed to enzyme, and that cleaved glycans may still be bound to binding partners and/or shielded from ionization. The remainder of the chapter will focus on use of FFPE tissues as the starting material.

3.2 Glycan Visualization in Tissues: Lectins and Anticarbohydrate Antibodies

The most common methods used to evaluate N-glycans and other carbohydrates in tissues are the use of specific anticarbohydrate antibodies and lectins. These reagents can be used for standard histochemistry types of analysis with FFPE tissues. Antibodies to anticarbohydrate antigens are frequently derived as tumor antigens associated with known blood group antigen designations (Heimburg-Molinaro et al., 2011; Kudelka et al., 2015; Ravn & Dabelsteen, 2000). A well-known example carbohydrate antigen is that of CA19-9, and its measurement is used clinically in the prognosis for pancreatic and other gastrointestinal cancers. CA19-9 antibody recognizes a four sugar sialyl-Lewis-A carbohydrate motif comprised of GlcNAc-Gal, α1-4 Fuc, and α2,3 sialic acid. Another carbohydrate antibody that has been described associated with prostate cancer detection is the F77 antigen, which recognizes blood group H antigen-related Lewis Y glycan structures with α1,2-Fuc (Nonaka et al., 2014). Most carbohydrate-targeted antibodies recognize a sugar structural epitopes of 1–4 sugars. While they can be effective for tissue immunohistochemistry staining, they may be less useful for defining the type of glycan class or carrier; i.e., the detected glycan could be present on glycolipids, O-linked glycoproteins, or N-linked glycoproteins, or possibly all three, depending on the tissue.

Carbohydrate-binding proteins termed lectins, obtained from plants, bacteria, fungi, and animals, have been used for many decades to evaluate structural features of glycoproteins in many biomedical applications (Hirabayashi, 2004; Lis & Sharon, 1998). Like the carbohydrate-antigen antibodies, lectins recognize specific structural motifs of 2–4 sugar linkages. A frequent limitation to their use, especially for tissue histochemistry applications, are their poor affinity-binding constants, generally in the low micromolar range but can be millimolar (Dam, Roy, Das, Oscarson, & Brewer, 2000; Lis & Sharon, 1998). Because they frequently possess multivalent-binding capabilities, they are frequently used for affinity capture purposes and chromatography separations (Dam et al., 2000; Drake et al., 2006; Lis & Sharon, 1998). For histochemistry purposes in tissues, they can be used to stain for known carbohydrate motifs using workflows similar to standard immunohistochemistry procedures in FFPE tissues. They cannot be used to distinguish whether the glycan motif that is bound by a given lectin is present on a glycolipid, O-linked glycoprotein, or N-linked glycoprotein. The use of imaging MS for N-glycans addresses this issue of identifying specific individual N-glycans and their localization within a tissue, but does not replace the use of lectins or anticarbohydrate antibodies. There are many potential synergistic uses of all three detection modalities in tissue that warrant further exploration.

3.3 Histochemistry Stains

There are many histological stains that target functional groups on carbohydrates through binding or chemical reactions to produce carbohydrate-specific staining (Myers, Fredenburgh, & Grizzle, 2008; Reid, Owen, Kruk, & Maitland, 1990). Alcian blue (AB) and periodic acid schiff (PAS) are two of the most commonly used carbohydrate stains. AB is a cationic, water soluble dye that has a blue color due to copper bound in the molecule (Scott, Quintarelli, & Dellovo, 1964). AB binding to specific functional groups of carbohydrates is controlled by the pH of the solution. In acidic solution (pH 2.5), AB binds to negatively charged sulfate and carboxylate groups of chondroitin sulfate, dermatan sulfate, heparin sulfate, hyaluronic acid, as well as sialo- and sulfomucins (Mowry, 1956; Myers et al., 2008; Scott et al., 1964). At pH of 1.0 or below, sulfated mucins are stained (Spicer, 1960). Specificity of the stain is shown using controls that have been exposed to acidified methanolic solution or digested with hyaluronidase (Myers et al., 2008; Spicer, 1960). AB is frequently used in pathologic evaluation of epithelial and connective tissue tumors (Kindblom & Angervall, 1975; Myers et al., 2008). An example of this stain for a colon cancer FFPE tissue, in comparison to an adjacent slice hematoxylin and eosin stain, is shown in Fig. 1C and D.

The PAS stain is used to target hyaluronans and acid mucopolysaccharides (proteoglycans and glycosaminoglycans). Unlike AB, PAS is a chemical reaction dependent on structural features of the monosaccharide unit. Periodic acid oxidizes 1,2 glycol groups to aldehydes, followed by a colorimetric reaction of the aldehydes with the Schiff reagent to produce a dark pink color (McManus & Cason, 1950). PAS may be used at different concentrations to stain target carbohydrate components. Mild PAS (0.01% PAS for 10 min at 4°C) selectively oxidizes nonsubstituted OH groups of sialic acid (Richter & Makovitzky, 2006), while a 0.5% solution of PAS stains glycogen and mucins deep purple (Gomori, 1946).

AB and PAS are often combined to show pathological changes in distribution of mucin glycoproteins (Myers et al., 2008). Most mucin glycoproteins are heavily O-glycosylated, but do frequently contain one or two N-linked glycan chains. AB is used to stain sialo and sulfo components of carbohydrates blue (used at pH 2.5), while the PAS stain targets the neutral components dark pink. An example for using AB with PAS was reported in colorectal tissues (Jain, Mondal, Sinha, Mukhopadhyay, & Chakraborty, 2014). Although not necessarily specific to N-glycans, this type of histochemical stain is useful in assessing the overall carbohydrate abundance in a given tissue and can be used to correlate with N-glycan image distribution within that tissue.

3.4 Peptide N-Glycosidase F

A key aspect of N-glycan imaging is the selectivity of the peptide N-glycosidase F (PNGaseF) enzyme to cleave the N-linked glycans attached to asparagines in the protein carrier (Fig. 1B). The asparagine is deamidated to an aspartic acid, and this change within the N-X-S/T motif can be used to identify peptides that had been glycosylated. To facilitate use of this enzyme for the scale required for tissue imaging studies, the entire PNGaseF gene (P21163.2) from the genome of *Flavobacterium meningosepticum* was cloned, expressed, and purified in the Mehta laboratory as described (Powers et al., 2013). A histidine epitope tag was included to facilitate purification, and it is stored with minimal buffer components and no other additives. Many commercially available preparations of PNGaseF have low amounts of enzyme and contain additives like glycerol or detergents, which can lead to ion suppression during ionization. The recombinant enzyme has proven to be stable and highly reactive when 20 μg are sprayed robotically onto tissue and incubated over a 2 to 4-h period at 37°C in a humidified chamber (Powers et al., 2013, 2014).

Prior to application of PNGAseF, waxed FFPE tissue sections are heat treated at 60°C for 1 h. This step is essential for denaturing the protein content and substantially increases *N*-glycan signal. Tissue is then dewaxed using typical xylene washing steps, followed by high-temperature antigen retrieval in a citraconic anhydride buffer. It is important to note that placing sections on positively charged slides or poly-L-lysine-coated slides limits sample loss occurring through the antigen retrieval process. Other standard antigen retrieval buffers like Tris can also be used. Parameters often need to be optimized per tissue type (e.g., breast cancer vs prostate cancer). Continued optimization of each of these parameters is still ongoing.

3.5 Matrix and Instrumentation Choices for *N*-Glycans

Classically, the matrix dihydroxybenzoic acid (DHB) has been used as for MALDI analysis of carbohydrates (Harvey, 2015). For MALDI-IMS of *N*-glycans, however, both DHB and alpha-cyano-4-hydroxycinnamic acid (CHCA) have been used with success (Holst et al., 2016; Powers et al., 2015, 2013, 2014). When deciding which matrix, a consideration is the type of instrument that will be used for the imaging experiment. Most MALDI time-of-flight (TOF) instruments have higher pressure sources that tend to produce intense matrix clusters so that matrices are selected to limit interference with signals of interest. DHB has been the choice when using TOF instruments for many years. MALDI-IMS of *N*-glycans has been done on a Fourier transform ion cyclotron resonance (FT-ICR) mass spectrometers (Heijs et al., 2016; Powers et al., 2015, 2013, 2014). This instrument has an intermediate pressure source, with limited high-mass matrix cluster peaks. Here, CHCA appears to result in more intense *N*-glycan signal from tissue.

When applying matrix, thickness of the tissue section must be considered as the ratio of matrix to tissue influences efficient production of signal (Yang & Caprioli, 2011). Thick applications of matrix can also result in increased production of matrix clusters, obscuring target analytes. On the other hand, not enough matrix results in poor ionization and loss of signal. Our group has routinely used FFPE tissue sections 5–7 μm thick, and apply matrix at 0.002154 mg/mm^2. For FFPE tissue sections of 3 μm thickness, 0.001725 mg/mm^2 works well. Currently, matrices for MALDI-IMS are applied by robotic sprayers which increase throughput and provide robust, consistent results (Angel & Caprioli, 2013; Schuerenberg & Deininger, 2010). The example N-glycan images shown in this chapter were all processed using a robotic sprayer to spray the PNGaseF and apply CHCA matrix. In unpublished evaluations of other matrix molecules for N-glycan imaging of tissues, no matrix has yet outperformed CHCA or DHB. A formal analysis of different matrix molecules is still warranted, as well as evaluations of glycan data obtained in negative ion mode.

Primary instruments that have been used for N-glycan imaging are MALDI-TOFs and MALDI-FT-ICRs. Each has certain advantages. TOF instrument provides high-throughput but lowers mass resolution capabilities. A situation that frequently occurs on-tissue is that the second isotope from one peak may be near-isobaric to another. For N-glycans, this is very common in tumor tissues for complex glycans that may have two fucoses (292 m.u.) vs a single sialic acid (291 m.u.). An example is Isotope 2 of Hex5dHex1HexNAc4NeuAc1+1Na at m/z 2101.7426 vs Isotope 1 of Hex5dHex3HexNAc4+1Na at m/z 2101.7551, suggesting a needed resolving power of ~168,000 (assuming equal height at 50% peak height). MALDI-FT-ICRs can easily detect this difference with imaging data typically acquired at estimated resolving powers of 85,000–160,000. FT-ICRs are generally lower throughput, but the advantage is the higher mass resolution and increased sensitivity when compared to MALDI-TOF instruments.

MALDI sources frequently generate structural fragments due to in source decay. In a MALDI spot comparison of carbohydrates, this can be leveraged as pseudo MS3 for structural elucidation (Mechref, Novotny, & Krishnan, 2003). In imaging studies, this becomes a problem as it is impossible to determine an intact structure produced by the biology vs the fragmented structure induced by instrumentation. Sialic acids are a particularly challenging problem as they readily cleave under even low source energy. Vendors for MALDI-FT-ICR instruments have been aware of this problem for analysis of carbohydrates and have built in a cooling nitrogen gas stream into the source, limiting sialic acid cleavage by in source decay (O'Connor & Costello, 2001; O'Connor, Mirgorordskaya, & Costello, 2002).

Although high mass accuracy is useful in distinguishing between near-isobaric species, this cannot determine glycoforms. Ion mobility mass spectrometry incorporates postsource gas-phase separation to separate glycoforms based on collision cross-section. Ion mobility has been used extensively for carbohydrate characterization (Fenn & McLean, 2011; Gray et al., 2016) in MALDI spot or in liquid chromatography analyses. A few reports of on-tissue imaging analyses show that on fresh frozen tissue, ion mobility can be used to separate and determine structural characteristics of carbohydrates apart from other biological molecules (McLean, Fenn, & Enders, 2010). In preliminary studies, an ion mobility imaging approach has been applied to PNGaseF digested FFPE tissues (data not shown). The approach easily distinguishes distinct glycoforms in tumor and non-tumor regions, as well as potential separation of glycan isomers.

3.6 Structural Confirmation

Studies on structural information for N-glycans are performed parallel with imaging experiments, as a combination of on-tissue and off-tissue experiments. The on-tissue MSI experiments provide structural information by tandem mass spectrometry, enzymatic approaches, or chemical derivatization in a way that preserves localization of the N-glycan species. For tumor tissues, this aids in confirming that a specific N-glycoform has nontumor or tumor-specific localization. Tandem mass spectrometry is performed either on the same tissue section as was used for the imaging experiment or on a serial tissue section (Holst et al., 2016; Powers et al., 2015). Serial tissue sections used for tandem mass spectrometry are prepared the same way as for an imaging experiment. Most commonly, the same mass spectrometry source used in the imaging experiment is employed as this eliminates issues of differential ionization. For MALDI-MSI, on-tissue tandem mass spectrometry experiments are generally directed at specific N-glycan species that have been shown to be altered in N-glycan imaging profiles from large cohorts. Tertiary and larger N-glycan structure may be difficult to fragment on-tissue by MALDI MS/MS. The use of differential exoglycosidases in combination with PNGaseF has been used to uncover and define large structures that are not as accessible on-tissue by MALDI-MS/MS methods (Powers et al., 2013). On-tissue chemical derivatization, used to increase detection sensitivity and stabilize moieties such as sialic acid, is applied to tissue sections serial to the tissue section used for the imaging experiment. The use of chemicals in on-tissue derivatization is challenging since chemicals that result in ion suppression must be removed in a way that does not delocalize the N-glycan species. Recently, an ethyl esterification method for high-throughput profiling of sialic acid containing N-glycans (Reiding, Blank,

Kuijper, Deelder, & Wuhrer, 2014) was modified for application to tissue sections (Holst et al., 2016). Application to leiomyosarcome and colon carcinoma tissue sections showed localization of α2,3 and α2,6 linkages shifts in sialic acid content and stabilized the potentially labile sialic acid moieties (Holst et al., 2016).

Off-tissue approaches to defining N-glycan structures first release N-glycans from tissue sections and collect released N-glycans from the tissue as a supernatant. Derivatization methods are then applied to improve fragmentation and detection, followed by chromatography to separate and quantify N-glycoforms (Briggs et al., 2016; Gustafsson et al., 2015; Powers et al., 2013). An advantage to this approach is that virtually any chemical derivatization approach may be used since the sample will be cleaned up by the following chromatography approaches. A disadvantage is that the data no longer have region-specific information. This disadvantage may be partly overcome by microdissection of tissue sections performed prior to releasing the N-glycans (Briggs et al., 2016). Nano LC-MS/MS strategies paired with IMS allow high sensitive detection and fragmentation of molecules up to intact proteins from low microliter extractions of molecules from tissue sections (Schey, Anderson, & Rose, 2013), but this type of strategy has not yet been applied to the imaged N-glycome. LC-MS/MS experiments could also be performed on larger amounts of the same tissue that was used for tissue imaging, as has been done for identification of proteins from melanoma or gastric cancers (Angel & Caprioli, 2013; Balluff et al., 2011; Hardesty, Kelley, Mi, Low, & Caprioli, 2011). The use of this strategy would allow extensive cataloging of N-glycome structural information but without region- and cell-specific information. Both top-down and bottom-up approaches should be considered for structural characterization of N-glycans parallel with imaging experiments.

4. N-GLYCAN DISTRIBUTION LINKED WITH HISTOPATHOLOGY

4.1 Major Structural Classes

In the analysis of the different types of N-glycans that are detected in the tissues, a structural biosynthetic and/or catabolic theme has been observed that is linked to their histopathological localizations. For reference throughout this chapter, we have selected to highlight a FFPE human colon adenocarcinoma tissue that has many distinct features, including regions of adenocarcinoma, mucinous tumor, adjacent normal colon crypts, and different nontumor stroma regions. Imaging of this tissue for N-glycans by FT-ICR-MALDI, and different histological stains of it, will be presented throughout as examples of the type of information that can be obtained with the approach. In this section, examples of five main structural classes will be shown in the same colon tissues. The representative distributions of these five N-glycan groups are highlighted in Fig. 2 and discussed individually in the next sections.

4.2 High-Mannose N-Glycans

N-glycans with high-mannose chains represent an early stage in biosynthetic processing, as described in Fig. 3. If glycans are detected in tissues with primarily only mannose ($n=1-9$) and two N-acetlyglucosamines, these would be considered to be intracellular intermediates residing in the ER or Golgi membranes in normal tissues. As these structures are degraded by mannosidases to only three mannose residues and then further processed to complex N-glycans, these structures would not be expected to be on the cell surface. However, in tumor tissues this is frequently not the case, as high mannose containing glycans and other biosynthetic intermediates (also called paucimannose glycans) are found on secreted glycoproteins and on the cell surface (Loke, Kolarich, Packer, & Thaysen-Andersen, 2016; Nyalwidhe et al., 2013). Specific mannose recognizing C-type lectin receptors are also known to bind to them. These mannose structures are also immunogenic, and specific autoantibodies to high-mannose structures can be detected in sera from cancer and infectious disease patients (Wang, 2012; Wang et al., 2013). Whether high-mannose glycans are secreted or transported to the cell surface to represent a programmed response in tumors, an immune response to the tumor, or are reflective of aberrant transport regulation in cancer cells remains to be determined.

In the published MALDI imaging studies of N-glycans detected in tumor tissues, the high-mannose structures are routinely detected in abundance (Everest-Dass et al., 2016; Heijs et al., 2016; Holst et al., 2016; Powers et al., 2015, 2014). An example of this is shown in Fig. 3, where the number of mannose residues is decreasing from top left to bottom right, reflecting mannosidase processing. The tumor-specific localization of these structures from high-mannose (Man4–9) and paucimannose (Man2–3, ±Fuc) classes is readily apparent. For most tissue types, the paucimannose and Man6–9 structures are most abundant, while the Man4 is usually the lowest in abundance. In this tissue example, distribution of the Man5 glycan is both tumor-localized and located in nontumor regions. Overall, we have found the distribution of Man5 to be primarily in tumor regions, with a presence in nontumor locations to be highly variable. As a class, the high-mannose and paucimannose structures represent easily detectable tumor-localized glycans using MALDI imaging approaches.

4.3 Nontumor Stroma and Normal Tissue Glycans

Complex N-glycans are derived from the Man3GlcNAc2 precursor and built sequentially by the activity of individual glycosyltransferases that add multiple N-acetylglucosamine, galactose, fucose, and sialic acid sugars. Branching of the biantennary chains to include tri- or tetraantennary structures, if present, also occurs during this process (discussed further in Section 4.4). To illustrate this biosynthetic pattern and the role of colocalization within stroma features, images of the colorectal tissue for biantennary structures with (Fig. 4) or without (Fig. 5) a core fucose modification are shown. The structures shown in these two figures represent the most commonly detected N-glycans in tissue MS imaging studies (Holst et al., 2016; Powers et al., 2014), typified by Hex5HexNAc4Fuc1 ($m/z=1809$; Fig. 4) and Hex5HexNAc4 ($m/z=1663$; Fig. 5). The singly sialylated versions of these structures ($m/z=2122$ and 1976, detected as doubly sodiated) are also abundant and share the same overall tissue distributions. In many FFPE tumor tissues, these classes of glycans tend to provide distinct distribution patterns from each other in the nontumor stroma regions, with some overlap, and vary significantly regarding distribution in different tumor types. These are useful target glycans to determine their distribution empirically with each new tissue analyzed, as they are generally highly abundant. This is illustrated in the overlay panel image in Fig. 4 for the sialylated $m/z=2122$ and 1976 glycans. The doubly sialylated N-glycans also have these same distributions but are generally detected at much lower levels (images not shown). The presence of the core fucose appears to be one of the major factors for differential localization relative to the biantennary structures lacking the fucose. The presence of the core fucose of the indicated structures can be confirmed by CID directly from tissues but requires significant amounts of tissue to perform the experiment (Powers et al., 2015).

An example of this phenomenon is shown in colorectal tumor tissue, where the presence of significant epithelial crypt structures is indicative of normal colon tissue regions. As shown in Fig. 6A, there was both a structural and localization uniqueness to the most abundant N-glycans detected in the crypt epithelial cells. The common structural theme was the presence of N-glycans with putatively bisecting N-acetylglucosamine residues, including both fucosylated and nonfucosylated examples. Also shown is an example of a hybrid N-glycan structure at m/z 1794. These structures have been reported to be decreased in colon tumor tissues (Holst et al., 2016), consistent with their nontumor localization shown in Fig. 6A. This is consistent with the activity of GnT-III, the glycosyltransferase responsible for catalyzing the addition of the bisecting GlcNAc, being primarily associated with expression reflecting an antitumorigenesis activity (Taniguchi & Kizuka, 2015).

4.4 N-Glycan Branching and Sialylation

Branching of complex biantennary glycans to tri- and tetraantennary structures is catalyzed by GnT-IV and GnT-V and is frequently associated with more tumorigenic processes. Fig. 6B shows the localization of the basic triantennary structures with or without a fucose, as well as a common tetraantennary glycan with a single fucose of $m/z = 2539$. Additionally, the triantennary structures with a single sialic acid are also shown. The distribution of the two nonfucosylated triantenarry structures at $m/z = 2028$ and 2341 is similar to the abundant nonfucosylated biantennary glycan distributions shown for $m/z = 1663$ and 1976 in Figs. 4 and 5. The fucosylated triantennary glycan at $m/z = 2174$ and tetraantennary glycan at $m/z = 2541$ have a largely adenocarcinoma tumor localization in the adenocarcinoma region of the colon tissue, analogous to the high-mannose glycan distributions. Like that of high-mannose glycans, a tumor-centric distribution of these two glycans is similar in many other tissue types and is usually abundant in signal intensity, such that they are useful targets to examine in any new tumor tissues being analyzed.

A complication in the detection of sialylated glycans by MALDI is their well-known lability during the ionization process. Holst et al. have demonstrated that stabilization of sialylated N-glycans in situ by ethylation reactions increases their detection and overall signal intensities. The Solarix series MALDI-FT-ICR instruments use a different source configuration allowing softer ionization and a cooling gas which facilitates detection of sialylated glycans, as can be seen in the example images used in this chapter. If a more standard MALDI instrument is used for N-glycan imaging, without any type of sialic acid stabilization, more nonsialylated complex glycans will be detected (e.g., $m/z = 1809$ and 1663) than is reflective of the actual amount of sialylated glycan present. At least for the biantennary structures with single or disialylated glycans, this may not be that critical regarding tissue localizations, as these sialylated structures generally colocalize with the nonsialylated precursors. This is not ideal and would have to be empirically determined for each tissue analyzed. Additionally, even with the capabilities of the MALDI-FT-ICR source configuration, the presence of up to two sialic acids is generally the current limit to what N-glycans can be detected. For example, it cannot be concluded whether 3–4 sialic acid structures of the $m/z = 2174$ or 2539 glycans are present in the model colon tissue. An emerging solution is that of the in situ ethylation modification of sialic acid residues prior to PNGaseF release (Holst et al., 2016). The ethylation modification has the added benefit of reacting differently with the two possible isomers of sialic acid, which are added as terminal residues in either α-2,3 or α-2,6 configurations. Ethylation of α-2,6-linked sialic acids results in a mass shift of 28 m.u., while an α-2,3 linkage will result in a lactonization reaction and decrease of 18 m.u. relative to the nonmodified parent glycan. These shifts are readily detectable by any MALDI instrument. This is an emerging technique that is still being refined methodologically and represents a significant improvement in the ability to study tissue sialylation for not only N-glycan structures, but other O-linked and glycosphingolipid species.

4.5 Fucosylation and the Glycan Isomer Problem

In the mucinous tumor region, specific classes of complex glycans were detected, representing branched and fucosylated structures (Fig. 6C). There is a clear structural theme of tetraantennary glycans incorporating different numbers of galactose residues, as well as fucose. The presence of core fucose in the multifucosylated structures is assumed but has not been confirmed. The five structures shown in Fig. 6C were selected to illustrate this biosynthetic scheme, but there were many other glycans detected specifically in this region representing tri- and tetraantennary structures with multiple fucose residues not shown, overall highlighting a role for branching and multifucosylation as indicators of mucinous tumor.

While the high-resolution capability of the MALDI-FT-ICR instrument can be used to determine the structural composition of N-glycans, there are multiple anomeric details that are lacking in regards to the specificity of the linkages between each sugar. For example, fucose residues can be attached via α-1,2 (FUT1,2), α-1,3 (FUT3,4,6,7,9,10,11), or α-1,4 (FUT3,4) linkages for outer arm modifications, or α-1,6 for core fucosylation linkages (FUT8). There are no differences in masses for these fucose linkages, and beyond the core fucose modification for simple biantennary complex glycans that can be confirmed by CID, the presence of two or more fucose residues precludes identification of the specific linkages or which arm of the glycan the attachment occurs. For sialylation, the ethylation modifications does directly address isomer identification, and for biantennary structures, the α-2,3 and α-2,6 linkages can be determined. For tri- or tetraantennary glycans with 1–2 terminal sialic acid residues, it is not possible to determine which arm of the structure carries the modifications. Even with ethylation, only the numbers of total α-2,3 and α-2,6 sialic acid linkages can be inferred, and not their specific structural arm positions. Biologically, the locations of the terminal sialic acid residues are important as they serve as attachment sites for proteins, and their negative charges also affect folding and activity of their carrier protein. An emerging analytical approach that could assist with the challenges of determining fucosylation and sialylation isomers is ion mobility mass spectrometry, an analytical technique that measures the mobility of gas-phase ions through an electric field in the presence of a buffer gas. The mobility of ions is based on their charge, shape, and size, and the method is increasingly being applied to the separation of glycan isomers (Gray et al., 2016). In relation to glycosylated analytes, DESI and MALDI ionization coupled to ion mobility separations were recently reported for tissue imaging of multisialylated gangliosides and other glycosphingolipids (Škrášková et al., 2016). This should be an equally robust approach for N-linked glycan imaging, as well as new MSI comparisons of glycosphingolipid distributions (Angel, Spraggins, Baldwin, & Caprioli, 2012; Jones et al., 2014) with N-glycans.

5. EMERGING APPLICATIONS

5.1 Combined Glycan and Peptide MS Imaging

The N-glycan classes described herein were all released by PNGaseF from a protein carrier. Using the glycan distribution maps obtained by MALDI imaging to identify the glycoprotein carriers of glycans of interest is one potential experimental extension of the approach. The most direct approach would be to target regions of interest for digestion with trypsin, followed by extraction and enrichment of glycopeptides from these regions. While great progress has been made in direct glycopeptide analysis to determine both peptide and glycan sequences using high-resolution tandem MS (Banazadeh, Veillon, Wooding, Zabet, & Mechref, 2016; Thaysen-Andersen & Packer, 2014), there is still a requirement for large amounts of sample earlier what can be obtained by isolation of small, focal regions of tissue. A current alternative approach is to use MS imaging of peptides in situ, combined with extracted peptide sequencing identification, to map protein expression to tissue localization. This distribution map of the peptides can be combined with the N-glycan map to correlate areas of overlap. Additionally, having the protein identification allows bioinformatic analysis to determine which glycoproteins are present in a given area. A recent study from Heijs et al. (2016) describes a sequential version of this approach, one in which a FFPE tissue is first treated with PNGaseF to release N-glycans, which are then imaged, followed by trypsin digestion of the same tissue and peptide imaging. An adjacent tissue slice is treated similarly, but glycans and peptides are extracted for tandem MS analysis and peptide sequencing. In one data example, a colorectal cancer FFPE tissue treated with sequential PNGaseF and trypsin digestions compared to trypsin alone increased the number of unique proteins identified from 236 (corresponding to 948 unique peptides) to 509 proteins (corresponding to 1334 unique peptides) (Heijs et al., 2016). The sequential digestion improved protein identification, most likely due to removal of N-glycans that would otherwise block trypsin access. Cleavage of N-glycans by PNGaseF results in a deamidation of the asparagine residue to aspartic acid, a modification easily determined in LC-MS/MS peptide analysis, but that can also potentially arise as an artifact without careful experimental preparation (Hao, Ren, Alpert, & Sze, 2011). There were 20 deamidated peptides identified in the colorectal cancer tissue using the sequential PNGaseF/trypsin approach, compared to none detected with trypsin alone (Heijs et al., 2016).

To further illustrate this, an adjacent tissue slice of the example colorectal cancer specimen shown in Figs. 2–6 was treated with trypsin and peptide profiling done by MALDI imaging. As shown in Fig. 7, peptide masses indicative of the same tissue regions highlighted in the preceding glycan examples were selected. The high spatial correlation between adenocarcinoma and mucinous glycan and peptide localizations is evident, as well as the two stroma and crypt areas. The glycan examples shown are the same as those in Fig. 2, and peptide identification from this tissue by tandem MS is ongoing. Although still correlative, and with protein ID as described, protein localizations can be validated with immunohistochemistry confirmation on tissues.

As these are generally going to be abundant proteins, it is likely that appropriate antibody reagents are available. This abundance also lends itself to clinical biomarker assay development. From the glycomic side of characterizing the glycoprotein carriers, knowing the structural class of the types of glycans of interest can allow use of affinity enrichment of glycopeptides or glycoproteins using lectins or anticarbohydrate epitope antibodies. It is possible that using targeted enrichment of tissue regions based on the presence of glycans of interest, combined with affinity enrichment based on glycan structure, could lead to improved amounts of glycopeptides amenable to current tandem MS analysis workflows. Overall, the approach described by Heijs et al. represents an effective MALDI-IMS proteomic–glycomic characterization strategy and provides a template for continuing to refine these types of novel analyses.

5.2 Custom Multitumor TMA and Other Enzymes

In order to facilitate method development for N-glycan MSI applications, we have generated two custom TMAs comprised of multiple tissue core pairs of nontumor and tumor regions representing 18 different tumor types (pancreatic, colon, prostate, breast, lung, melanoma, sarcoma, head/neck, kidney, liver, glioma, ovarian, thyroid, uterine, cervical, testicular, gastric, bladder). An example of MSI N-glycan data from one TMA (pancreatic, colon, prostate, breast, kidney, liver, thyroid, bladder) is shown for a high-mannose Man6 glycan ($m/z = 1419$) in Fig. 8. In Fig. 8A, the intensities are shown for the entire TMA, with data normalized across all of the tissue cores. The tumor types present in each row are listed in the figure legend. This allows a comparison of relative expression of the abundance of each individual glycan across the different tumor types. In Fig. 8B, cores from each individual tumor type were analyzed as distinct image files, allowing normalization to be done for each tumor independently. For this high-mannose glycan, there is consistently higher tumor levels for colon, prostate, lung, breast, and bladder cancers, while the other tumor types are variable (Fig. 8B). Data from each of the TMA core pairs are being compared systematically to what N-glycans are detected in the source FFPE blocks. The goal is to generate a reference database of expected MSI N-glycans for each tumor tissue type. This will also provide a framework including ethylated sialic acid structures as well as isomer information as it is obtained. Other images from different platforms like DESI/MALDI ion mobility could also be included.

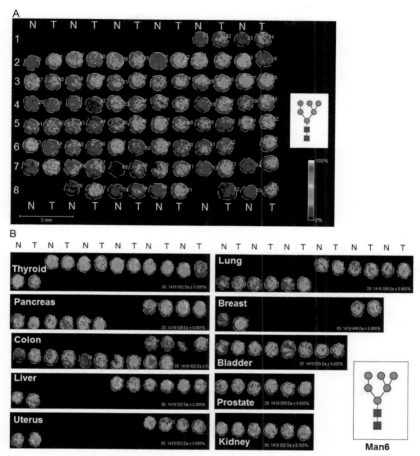

Fig. 8 Example N-glycan MALDI-MSI data for a tissue microarray representing eight different tumor types. (A) MALDI-IMS intensity of the Man6 glycan across each tissue core. Nontumor (N) regions and tumor (T) regions are shown for each tumor type: Rows 1 and 2—uterus (6 cores); Rows 2 and 3—thyroid (12 cores); Row 3—prostate (6 cores); Rows 3 and 4—pancreas (10 cores); Rows 4 and 5—lung (12 cores); Rows 5 and 6—liver (8 cores); Row 6—kidney (6 cores); Rows 6 and 7—colon (14 cores); Rows 7 and 8—breast (6 cores); Row 8—bladder (8 cores). (B) Each tissue type in the microarray analyzed as an individual imaging file. Shown is data for the Man6 N-glycan.

Other glycosidase enzymes specific to galactose (galactosidase), fucose (fucosidase), and sialic acid (sialidase or neuraminidase) resides could be applied to tissue in different combinations with PNGaseF to aid in structural confirmations. Many of these glycosidases are linkage specific, thus in particular for fucose modifications, their use could provide added structural/isomeric information. Use of other enzymes with modified PNGaseF substrate specificity, for example, an enzyme that only cleaves N-glycans with core fucose residues, could also find utility in the MSI N-glycan workflows.

5.3 Linkage to Genomic Studies

Because different glycosyltransferases and glycosidases determine the sequential biosynthesis of N-glycans, the gene expression levels of these enzyme's function could potentially regulate the composition of the N-glycome in each region. An extensive study of glycogene transcripts in different mouse organs, compared with the most abundant N-glycans detected in each tissue, illustrated that the levels of many transcripts (but not all) can be correlated with the levels of specific glycan structures (Nairn et al., 2008). The glycan structural series presented in Figs. 3–6 strongly suggests specific regulatory control of sequential glycosyltransferase networks linked with individual histopathology regions, consistent with transcriptional level control. Several studies have been published that have evaluated the transcriptome profiles of glycan biosynthesis genes to predict what glycan structures would be present (Kawano, Hashimoto, Miyama, Goto, & Kanehisa, 2005; Suga, Yamanishi, Hashimoto, Goto, & Kanehisa, 2007), including glycosylation reaction network analysis strategies (Liu & Neelamegham, 2014). Combining these approaches with the specific glycan structures identified in the N-glycan tissue imaging profiles has not been reported. The abundance of genomic data available via the Cancer Genome Atlas (TCGA) network for many types of cancer represents a wealth of N-glycan gene expression profiling data. Linking the N-glycan imaging maps to tissues with corresponding transcriptomic data represents a direct opportunity to more fully link glycome and genomic datasets to better understand the regulation and function of N-glycans in different cancer types.

5.4 Potential Clinical Diagnostic Applications of N-Glycan MSI Data

There are several clinical diagnostic directions that N-glycan imaging of FFPE samples has made possible, as the panels of N-glycans obtained can be used to develop classification algorithms. One of the more direct possibilities is to use a new MALDI-MSI platform, the rapifleX MALDI Tissuetyper™ TOF instrument (from Bruker Daltonics, Billerica, MA) (Ogrinc Potočnik, Porta, Becker, Heeren, & Ellis, 2015). This system was designed to significantly decrease the time required to acquire and process tissue imaging data, essentially by fivefold or more. Additionally, the laser spot size and raster distances can be done routinely at 10–20 µm resolutions, while maintaining the rapid analysis times. Pilot MALDI-MSI of N-glycans from FFPE tissues has been successful on this new instrument and will be reported separately. These improved acquisition times bring the analysis of surgery-derived biopsy tissues and other standard pathology samples by MALDI-IMS into a feasible time frame for clinical laboratory assays.

Knowing the glycan structures and their tissue locations also facilitates development of specific lectin-based assays for direct lectin histochemistry tissue assays. Although less available, use of anticarbohydrate epitope antibodies could also be used, as fluorescent or light microscopy visualization of cancer biomarkers in tissue is standard in the clinical pathology laboratory. The potential down side to their use is that the lectins or antibodies recognize glycan structural motifs and classes which could also be present on glycosphingolipids or O-glycosylated proteins. Specificity for detection of N-glycan glycoproteins could be lost or confounding.

6. SUMMARY

The capability of profiling and identifying multiple N-glycan species in their native locations within the tumor microenvironment of FFPE tissue samples provides a long missing spatial component in our understanding of the role of N-glycans in tumor development and function. In the field of imaging mass spectrometry, targeting N-glycans is a recent development, and as highlighted herein, there is a wealth of new information that can be obtained for any cancer type. We expect that the lessons learned from these efforts will facilitate development of MSI strategies for O-linked glycans, which lacks an enzyme mediator to release them, as well as heparin and chondroitin sulfate glycosaminoglycans (Shao, Shi, Phillips, & Zaia, 2013). The use of FFPE tissues as the primary source and the defined number of detectable N-glycan species greatly facilitates the potential use of the method, or information derived from it, in subsequent clinical assay measurements related to development of new tumor diagnostics. The limitations of the current MSI approaches for N-glycans are largely addressable, and new innovations in methodology and instrumentation, as well as a focus on imaging of other glycan classes, will continue to aid in the evolution of this approach.

REFERENCES

Adamczyk, B., Tharmalingam, T., & Rudd, P. M. (2012). Glycans as cancer biomarkers. *Biochimica et Biophysica Acta, 1820*(9), 1347–1353.

Angel, P. M., & Caprioli, R. M. (2013). Matrix-assisted laser desorption ionization imaging mass spectrometry: In situ molecular mapping. *Biochemistry, 52*(22), 3818–3828.

Angel, P. M., Spraggins, J. M., Baldwin, H. S., & Caprioli, R. (2012). Enhanced sensitivity for high spatial resolution lipid analysis by negative ion mode matrix assisted laser desorption ionization imaging mass spectrometry. *Analytical Chemistry, 84*(3), 1557–1564.

Balluff, B., Rauser, S., Meding, S., Elsner, M., Schöne, C., Feuchtinger, A., et al. (2011). MALDI imaging identifies prognostic seven-protein signature of novel tissue markers in intestinal-type gastric cancer. *The American Journal of Pathology, 179*(6), 2720–2729.

Banazadeh, A., Veillon, L., Wooding, K. M., Zabet, M., & Mechref, Y. (2016). Recent advances in mass spectrometric analysis of glycoproteins. *Electrophoresis.* http://dx.doi.org/10.1002/elps.201600357. Epub ahead of print, PMID 27757981.

Briggs, M. T., Kuliwaba, J. S., Muratovic, D., Everest-Dass, A. V., Packer, N. H., Findlay, D. M., et al. (2016). MALDI mass spectrometry imaging of N-glycans on tibial cartilage and subchondral bone proteins in knee osteoarthritis. *Proteomics, 16*(11–12), 1736–1741.

Christiansen, M. N., Chik, J., Lee, L., Anugraham, M., Abrahams, J. L., & Packer, N. H. (2014). Cell surface protein glycosylation in cancer. *Proteomics, 14*(4–5), 525–546.

Dam, T. K., Roy, R., Das, S. K., Oscarson, S., & Brewer, C. F. (2000). Binding of multivalent carbohydrates to concanavalin A and Dioclea grandiflora lectin: Thermodynamic analysis of the "multivalency effect". *The Journal of Biological Chemistry, 275*, 14223–14230.

Dapson, R. W. (2007). Macromolecular changes caused by formalin fixation and antigen retrieval. *Biotechnic & Histochemistry, 82*(3), 133–140.

Drake, R. R., Jones, E. E., Powers, T. W., & Nyalwidhe, J. O. (2015). Altered glycosylation in prostate cancer. *Advances in Cancer Research, 126*, 345–382.

Drake, R. R., Schwegler, E. E., Malik, G., Diaz, J., Block, T., Mehta, A., et al. (2006). Lectin capture strategies combined with mass spectrometry for the discovery of serum glycoprotein biomarkers. *Molecular and Cellular Proteomics, 5*(10), 1957–1967.

Everest-Dass, A. V., Briggs, M. T., Kaur, G., Oehler, M. K., Hoffmann, P., & Packer, N. H. (2016). N-glycan MALDI imaging mass spectrometry on formalin-fixed paraffin-embedded tissue enables the delineation of ovarian cancer tissues. *Molecular and Cellular Proteomics, 15*(9), 3003–3016.

Fenn, L. S., & McLean, J. A. (2011). Structural resolution of carbohydrate positional and structural isomers based on gas-phase ion mobility-mass spectrometry. *Physical Chemistry Chemical Physics, 13*(6), 2196–2205.

Gomori, G. (1946). A new histochemical test for glycogen and mucin. *American Journal of Clinical Pathology, 10*, 177–179.

Gray, C. J., Thomas, B., Upton, R., Migas, L. G., Eyers, C. E., Barran, P. E., et al. (2016). Applications of ion mobility mass spectrometry for high throughput, high resolution glycan analysis. *Biochimica et Biophysica Acta, 1860*(8), 1688–1709.

Gustafsson, O. J., Briggs, M. T., Condina, M. R., Winderbaum, L. J., Pelzing, M., McColl, S. R., et al. (2015). MALDI imaging mass spectrometry of N-linked glycans on formalin-fixed paraffin-embedded murine kidney. *Analytical Bioanalytical Chemistry, 407*, 2127–2139.

Hao, P., Ren, Y., Alpert, A. J., & Sze, S. K. (2011). Detection, evaluation and minimization of nonenzymatic deamidation in proteomic sample preparation. *Molecular and Cellular Proteomics, 10*(10). O111.009381.

Hardesty, W. M., Kelley, M. C., Mi, D., Low, R. L., & Caprioli, R. M. (2011). Protein signatures for survival and recurrence in metastatic melanoma. *Journal of Proteomics*, *74*(7), 1002–1014.

Harvey, D. J. (2015). Analysis of carbohydrates and glycoconjugates by matrix-assisted laser desorption/ionization mass spectrometry: An update for 2009–2010. *Mass Spectrometry Reviews*, *34*(3), 268–422.

Heijs, B., Holst, S., Briaire-de Bruijn, I. H., van Pelt, G. W., de Ru, A. H., van Veelen, P. A., et al. (2016). Multimodal mass spectrometry imaging of N-glycans and proteins from the same tissue section. *Analytical Chemistry*, *88*, 7745–7753.

Heimburg-Molinaro, J., Lum, M., Vijay, G., Jain, M., Almogren, A., & Rittenhouse-Olson, K. (2011). Cancer vaccines and carbohydrate epitopes. *Vaccine*, *29*(48), 8802–8826.

Hirabayashi, J. (2004). Lectin-based structural glycomics: Glycoproteomics and glycan profiling. *Glycoconjugate Journal*, *21*, 35–40.

Holst, S., Heijs, B., de Haan, N., van Zeijl, R. J., Briaire-de Bruijn, I. H., van Pelt, G. W., et al. (2016). Linkage-specific in situ sialic acid derivatization for N-glycan mass spectrometry imaging of formalin-fixed paraffin-embedded tissues. *Analytical Chemistry*, *88*(11), 5904–5913.

Jain, P., Mondal, S. K., Sinha, S. K., Mukhopadhyay, M., & Chakraborty, I. (2014). Diagnostic and prognostic significance of different mucin expression, preoperative CEA, and CA-125 in colorectal carcinoma: A clinicopathological study. *Journal of Natural Science, Biology, and Medicine*, *5*, 404–408.

Jones, E. E., Dworski, S., Canals, D., Casas, J., Fabrias, G., Schoenling, D., et al. (2014). On-tissue localization of ceramides and other sphingolipids by MALDI mass spectrometry imaging. *Analytical Chemistry*, *86*(16), 8303–8311.

Kawano, S., Hashimoto, K., Miyama, T., Goto, S., & Kanehisa, M. (2005). Prediction of glycan structures from gene expression data based on glycosyltransferase reactions. *Bioinformatics*, *21*, 3976–3982.

Kindblom, L. G., & Angervall, L. (1975). Histochemical characterization of mucosubstances in bone and soft tissue tumors. *Cancer*, *36*, 985–994.

Kudelka, M. R., Ju, T., Heimburg-Molinaro, J., & Cummings, R. D. (2015). Simple sugars to complex disease-mucin-type o-glycans in cancer. *Advances in Cancer Research*, *126*, 53–135.

Lis, H., & Sharon, N. (1998). Lectins: Carbohydrate-specific proteins that mediate cellular recognition. *Chemistry Reviews*, *98*, 637–674.

Liu, G., & Neelamegham, S. (2014). A computational framework for the automated construction of glycosylation reaction networks. *PloS One*, *9*(6), e100939.

Loke, I., Kolarich, D., Packer, N. H., & Thaysen-Andersen, M. (2016). Emerging roles of protein mannosylation in inflammation and infection. *Molecular Aspects of Medicine*, *51*, 31–55.

McLean, J. A., Fenn, L. S., & Enders, J. R. (2010). Structurally selective imaging mass spectrometry by imaging ion mobility-mass spectrometry. *Methods in Molecular Biology*, *656*, 363–383.

McManus, J. F., & Cason, J. E. (1950). Carbohydrate histochemistry studied by acetylation techniques: I. Periodic acid methods. *The Journal of Experimental Medicine*, *91*(6), 651–654.

Mechref, Y., Novotny, M. V., & Krishnan, C. (2003). Structural characterization of oligosaccharides using MALDI-TOF/TOF tandem mass spectrometry. *Analytical Chemistry*, *75*(18), 4895–4903.

Miwa, H. E., Song, Y., Alvarez, R., Cummings, R. D., & Stanley, P. (2012). The bisecting GlcNAc in cell growth control and tumor progression. *Glycoconjugate Journal*, *29*(8–9), 609–618.

Moremen, K. W., Tiemeyer, M., & Nairn, A. V. (2012). Vertebrate protein glycosylation: Diversity, synthesis and function. *Nature Reviews. Molecular Cell Biology, 13*(7), 448–462.

Mowry, R. W. (1956). Alcian blue technics for the histochemical study of acidic carbohydrates. *Journal of Histochemistry & Cytochemistry, 4*, 403–407.

Myers, R. B., Fredenburgh, J. L., & Grizzle, W. E. (2008). Carbohydrates. In J. D. Bancroft & M. Gamble (Eds.), *Theory and practice of histological techniques* (pp. 161–178). Philadelphia, PA: Elsevier.

Nairn, A. V., York, W. S., Harris, K., Hall, E. M., Pierce, J. M., & Moremen, K. W. (2008). Regulation of glycan structures in animal tissues: Transcript profiling of glycan-related genes. *The Journal of Biological Chemistry, 283*(25), 17298–17313.

Nonaka, M., Fukuda, M. N., Gao, C., Li, Z., Zhang, H., Greene, M. I., et al. (2014). Determination of carbohydrate structure recognized by prostate-specific F77 monoclonal antibody through expression analysis of glycosyltransferase genes. *The Journal of Biological Chemistry, 289*(23), 16478–16486.

Nyalwidhe, J. O., Betesh, L. R., Powers, T. W., Jones, E. E., White, K. Y., Burch, T. C., et al. (2013). Increased bisecting N-acetylglucosamine and decreased branched chain glycans of N-linked glycoproteins in expressed prostatic secretions associated with prostate cancer progression. *Proteomics. Clinical Applications, 7*, 677–689.

O'Connor, P. B., & Costello, C. E. (2001). A high pressure matrix-assisted laser desorption/ionization Fourier transform mass spectrometry ion source for thermal stabilization of labile biomolecules. *Rapid Communications in Mass Spectrometry, 15*, 1862–1868.

O'Connor, P. B., Mirgorordskaya, E., & Costello, C. E. (2002). High pressure matrix-assisted laser desorption/ionization Fourier transform mass spectrometry for minimization of ganglioside fragmentation. *Journal of the American Society of Mass Spectrometry, 13*, 402–407.

Ogrinc Potočnik, N., Porta, T., Becker, M., Heeren, R. M., & Ellis, S. R. (2015). Use of advantageous, volatile matrices enabled by next-generation high-speed matrix-assisted laser desorption/ionization time-of-flight imaging employing a scanning laser beam. *Rapid Communications in Mass Spectrometry, 29*(23), 2195–2203.

Pinho, S. S., & Reis, C. A. (2015). Glycosylation in cancer: Mechanisms and clinical implications. *Nature Reviews. Cancer, 15*(9), 540–555.

Powers, T. W., Holst, S., Wuhrer, M., Mehta, A. S., & Drake, R. R. (2015). Two-dimensional N-glycan distribution mapping of hepatocellular carcinoma tissues by MALDI-imaging mass spectrometry. *Biomolecules, 5*(4), 2554–2572.

Powers, T. W., Jones, E. E., Betesh, L. R., Romano, P. R., Gao, P., Copeland, J. A., et al. (2013). Matrix assisted laser desorption ionization imaging mass spectrometry workflow for spatial profiling analysis of N-linked glycan expression in tissues. *Analytical Chemistry, 85*, 9799–9806.

Powers, T. W., Neely, B. A., Shao, Y., Tang, H., Troyer, D. A., Mehta, A. S., et al. (2014). MALDI imaging mass spectrometry profiling of N-glycans in formalin-fixed-paraffin embedded clinical tissue blocks and tissue microarrays. *PloS One, 9*(9), e106255.

Ravn, V., & Dabelsteen, E. (2000). Tissue distribution of histo-blood group antigens. *Acta Pathologica, Microbiologica et Immunologica Scandinavica, 108*(1), 1–28.

Reid, P., Owen, D., Kruk, P., & Maitland, M. E. (1990). Histochemical classification based upon reaction types and its application to carbohydrate histochemistry. *The Histochemical Journal, 22*, 299–308.

Reiding, K. R., Blank, D., Kuijper, D. M., Deelder, A. M., & Wuhrer, M. (2014). High-throughput profiling of protein N-glycosylation by MALDI-TOF-MS employing linkage-specific sialic acid esterification. *Analytical Chemistry, 86*, 5784–5793.

Richter, S., & Makovitzky, J. (2006). Topo-optical visualization reactions of carbohydrate-containing amyloid deposits in the respiratory tract. *Acta Histochemica, 108*, 181–191.
Rini, J., Esko, J., & Varki, A. (2009). Glycosyltransferases and glycan-processing enzymes. In A. Varki, R. D. Cummings, J. D. Esko, H. H. Freeze, P. Stanley, & C. R. Bertozzi, et al. (Eds.), *Essentials of glycobiology* (2nd ed.). Cold Spring Harbor (NY): Cold Spring Harbor Laboratory Press. chapter 5.
Ruhaak, L. R., Miyamoto, S., & Lebrilla, C. B. (2013). Developments in the identification of glycan biomarkers for the detection of cancer. *Molecular and Cellular Proteomics, 12*(4), 846–855.
Scalia, C. R., Boi, G., Bolognesi, M. M., Riva, L., Manzoni, M., DeSmedt, L., et al. (2016). Antigen masking during fixation and embedding, dissected. *Journal of Histochemistry & Cytochemistry*. pii: 0022155416673995. [Epub ahead of print].
Schey, K. L., Anderson, D. M., & Rose, K. L. (2013). Spatially-directed protein identification from tissue sections by top-down LC-MS/MS with electron transfer dissociation. *Analytical Chemistry, 85*, 6767–6774.
Schuerenberg, M., & Deininger, S. O. (2010). Matrix application with ImagePrep. In M. Setou (Ed.), *Imaging mass spectrometry: Protocols for mass microscopy* (pp. 87–91). Tokyo, Japan: Springer. chapter 7.
Schultz, M. J., Swindall, A. F., & Bellis, S. L. (2012). Regulation of the metastatic cell phenotype by sialylated glycans. *Cancer Metastasis Reviews, 31*(3–4), 501–518.
Scott, J. E., Quintarelli, G., & Dellovo, M. C. (1964). The chemical and histochemical properties of Alcian blue. *Histochemie, 4*, 73–85.
Shao, C., Shi, X., Phillips, J. J., & Zaia, J. (2013). Mass spectral profiling of glycosaminoglycans from histological tissue surfaces. *Analytical Chemistry, 85*(22), 10984–10991.
Škrášková, K., Claude, E., Jones, E. A., Towers, M., Ellis, S. R., & Heeren, R. M. (2016). Enhanced capabilities for imaging gangliosides in murine brain with matrix-assisted laser desorption/ionization and desorption electrospray ionization mass spectrometry coupled to ion mobility separation. *Methods, 104*, 69–78.
Spicer, S. S. (1960). A correlative study of the histochemical properties of rodent acid mucopolysaccharides. *Journal of Histochemistry & Cytochemistry, 8*, 18–35.
Stanley, P., Schachter, H., & Taniguchi, N. (2009). N-glycans. In A. Varki, R. D. Cummings, J. D. Esko, H. H. Freeze, P. Stanley, & C. R. Bertozzi, et al. (Eds.), *Essentials of glycobiology* (2nd ed.). Cold Spring Harbor (NY): Cold Spring Harbor Laboratory Press. chapter 8.
Suga, A., Yamanishi, Y., Hashimoto, K., Goto, S., & Kanehisa, M. (2007). An improved scoring scheme for predicting glycan structures from gene expression data. *Genome Informatics, 18*, 237–246.
Taniguchi, N., & Kizuka, Y. (2015). Glycans and cancer: Role of N-glycans in cancer biomarker, progression and metastasis, and therapeutics. *Advances in Cancer Research, 126*, 11–51.
Thaysen-Andersen, M., & Packer, N. H. (2014). Advances in LC-MS/MS-based glycoproteomics: Getting closer to system-wide site-specific mapping of the N- and O-glycoproteome. *Biochimica et Biophysica Acta, 1844*(9), 1437–1452.
Toghi Eshghi, S., Yang, S., Wang, X., Shah, P., Li, X., & Zhang, H. (2014). Imaging of N-linked glycans from formalin-fixed paraffin-embedded tissue sections using MALDI mass spectrometry. *ACS Chemical Biology, 9*(9), 2149–2156.
Uemura, T., Shiozaki, K., Yamaguchi, K., Miyazaki, S., Satomi, S., Kato, K., et al. (2009). Contribution of sialidase NEU1 to suppression of metastasis of human colon cancer cells through desialylation of integrin beta4. *Oncogene, 28*(9), 1218–1229.
Varki, A. (2016). Biological roles of glycans. *Glycobiology*. http://dx.doi.org/10.1093/glycob/cww086. [Epub ahead of print].

Wang, D. (2012). N-glycan cryptic antigens as active immunological targets in prostate cancer patients. *Journal of Proteomics and Bioinformatics, 5,* 90–95.

Wang, D., Dafik, L., Nolley, R., Huang, W., Wolfinger, R. D., Wang, L. X., et al. (2013). Anti-oligomannose antibodies as potential serum biomarkers of aggressive prostate cancer. *Drug Development Research, 74,* 65–80.

Yang, J., & Caprioli, R. M. (2011). Matrix sublimation/recrystallization for imaging proteins by mass spectrometry at high spatial resolution. *Analytical Chemistry, 83*(14), 5728–5734.

Zoldoš, V., Novokmet, M., Bečeheli, I., & Lauc, G. (2013). Genomics and epigenomics of the human glycome. *Glycoconjugate Journal, 30*(1), 41–50.

CHAPTER FIVE

In Situ Metabolomics in Cancer by Mass Spectrometry Imaging

A. Buck[1], M. Aichler[1], K. Huber, A. Walch[2]

Research Unit Analytical Pathology, Helmholtz Zentrum München, Neuherberg, Germany
[2]Corresponding author: e-mail address: axel.walch@helmholtz-muenchen.de

Contents

1. Metabolomics in Cancer	117
2. In Situ Metabolomics by MALDI Imaging	119
3. Fresh-Frozen- vs Formalin-Fixed Paraffin-Embedded Tissue Samples	120
4. Tissue-Based Disease Classification—Diagnostic Markers and Metabolic Signatures	124
5. Therapy Response Prediction and Prognosis	125
6. Intra- and Intertumoral Heterogeneity	126
7. Conclusion	127
References	128

Abstract

Metabolomics is a rapidly evolving and a promising research field with the expectation to improve diagnosis, therapeutic treatment prediction, and prognosis of particular diseases. Among all techniques used to assess the metabolome in biological systems, mass spectrometry imaging is the method of choice to qualitatively and quantitatively analyze metabolite distribution in tissues with a high spatial resolution, thus providing molecular data in relation to cancer histopathology. The technique is ideally suited to study tissues molecular content and is able to provide molecular biomarkers or specific mass signatures which can be used in classification or the prognostic evaluation of tumors. Recently, it was shown that FFPE tissue samples are also suitable for metabolic analyses. This progress in methodology allows access to a highly valuable resource of tissues believed to widen and strengthen metabolic discovery-driven studies.

1. METABOLOMICS IN CANCER

Metabolomics can be broadly defined as the comprehensive analysis of all small molecule metabolites in a biological system. The metabolome is

---

[1] These authors contributed equally to this work.

built up from small molecular weight compounds that are present in cells, tissues, and/or whole organisms as influenced by multiple factors including genetics, diet, lifestyle, and pharmaceutical interventions (Ma, Zhang, Yang, Wang, & Qin, 2012; Wishart, 2007). These compounds are generally accepted to be ≤1500 Da and include small peptides, oligonucleotides, sugars, nucleosides, organic acids, ketones, aldehydes, amines, amino acids, lipids, steroids, and alkaloids (Aboud & Weiss, 2013; Wishart, 2007). Metabolites can directly or indirectly interact with molecular targets and thereby influence the risk and complications associated with various diseases, including cancer (Brenton, Carey, Ahmed, & Caldas, 2005; Goodacre, Vaidyanathan, Dunn, Harrigan, & Kell, 2004; Lindon, Holmes, & Nicholson, 2004; Ma et al., 2012).

In comparison to normal tissue, tumor cells exhibit dynamic metabolic profiles. The so-called "cancer metabolome" can be considered as the metabolites with relevance to oncologic processes and to the body's systemic responses to the tumor (Aboud & Weiss, 2013; Calvani et al., 2010; Romick-Rosendale et al., 2009). It is defined that there are six hallmarks of cancer-associated metabolic changes: (1) deregulated uptake of glucose and amino acids, (2) use of opportunistic modes of nutrient acquisition, (3) use of glycolysis/TCA cycle intermediates for biosynthesis and NADPH production, (4) increased demand for nitrogen, (5) alterations in metabolite-driven gene regulation, and (6) metabolic interactions with microenvironment (Pavlova & Thompson, 2016). Thereby alterations in tumor metabolism often include one or more of the listed hallmarks. The analysis of individual tumor metabolic state may contribute to improvements in tumors classification and allow prediction of disease outcome or therapy response.

Advances in the development and use of new imaging and analytical techniques to study cancer cell metabolism have been used to expand the knowledge about the mechanisms and functional consequences of tumor-associated metabolic alterations at various stages of tumorigenesis (Pavlova & Thompson, 2016). Classical biochemistry, for example, indicated that tumor cells exhibit increased glucose consumption relative to normal cells, known as the "Warburg effect" (Warburg, 1956). The "Warburg effect" later on has also been recognized as a diagnostic tool in the clinics, as it is used for in vivo imaging for tumor diagnosis and staging, as well as for monitoring responsiveness and treatment of tumors by ^{18}FDG-PET (2-18F-fluoro-2-deoxy-D-glucose—positron emission tomography) (Almuhaideb, Papathanasiou, & Bomanji, 2011; Gambhir, 2002). Beside in vivo metabolic analytical techniques, ex vivo analysis of endogenous metabolite profiles in tissues can lead to a deeper understanding of disease-related mechanisms.

The most frequently used strategies for analyzing the metabolome in cancer tissues include mass spectrometry (MS), such as matrix-assisted laser desorption/ionization mass spectrometry (MALDI MS), liquid chromatography–mass spectrometry (LC–MS), gas chromatography–mass spectrometry (GC–MS), and nuclear magnetic resonance spectroscopy (NMR), which represents a complementary technique to MS. Mass spectrometry and NMR are particularly suitable for the investigation of organic compounds and have been successfully used to profile metabolites from cells, bodily fluids, and tissue samples (Nordstrom & Lewensohn, 2010; Yuan, Breitkopf, Yang, & Asara, 2012). As in the last decade, the field of mass spectrometry has been quickly and permanently evolved in efficiency and sensitivity, it represents today the technique of choice for the global measurement of metabolites from biological samples. Among the great number of different mass analyzers available to laboratory analysts, high-resolution mass spectrometry instruments have improved mass measurements in sensitivity and accuracy with sub-ppm errors achieving optimal metabolome coverage (Junot, Fenaille, Colsch, & Becher, 2014). In untargeted metabolomics, the accurate measurement of a molecular mass is in particular need for the annotations of metabolic data based on a match of physicochemical properties and/or spectral similarity with available metabolite information in databases (e.g., HMDB, Metlin, LipidMaps, KEGG, or MassBank) (Vinaixa et al., 2016). More sophisticated analyses can be performed by the usage of tandem MS (MS/MS) approaches in mass spectrometry, allowing the identification of a molecule by comparing its fragmentation pattern against standards or reference MS/MS databases. The evaluation of data using MS/MS typically provides the most conclusive evidence in metabolites identification.

2. IN SITU METABOLOMICS BY MALDI IMAGING

Histochemical and immunohistochemical techniques are key tools for the molecular analysis of tissue samples. While immunohistochemistry requires an antibody to accurately determine an individual compound in tissue samples, histochemical stains can be used to detect specific molecular classes, e.g., periodic acid–Schiff stains the ensemble of polysaccharides including mucopolysaccharides, glycoproteins, and glycolipids present in tissues (Renshaw, 2007). In contrast to histopathological methods, mass spectrometry-based analyses of tissues allow the detection of a broad range of molecules in an untargeted fashion, but typically require the extraction of molecular data from homogenized samples to reveal metabolomic changes caused by a biological event. The disruption and homogenization of tissues destroy the anatomical context of data, which does not allow a comprehensive picture of the tissues biological organization. The maintenance of tissue integrity during molecular analyses is an issue as cancer is a heterogeneous disease with a complex cellular environment. The metabolic state of tumor cells can be affected by gene expression, cellular differentiation, as well as the cross-interaction occurring between tumor components, stromal cells, and recruited immune cells (Zhao et al., 2016). Therefore, the metabolome of individual cells and cell populations can vary considerably in cancer tissue. Thus bulk tissue analysis which uses the average signal for statistical analyses may distort biological effects. Although laser capture microdissection is a method allowing the selective isolation of individual cells and pieces from tissue sections for preparative procedures and analyses by molecular biological methods, its usage presents a tissue- and time-consuming procedure for mass spectrometry analyses (Liu et al., 2012; Umar, Luider, Foekens, & Pasa-Tolic, 2007). MALDI mass spectrometry imaging (MSI) is a label-free method for the multiplexed analysis of biomolecules in their tissue morphological context and therefore closes one of the last remaining open gaps in tissue-based research. The technique is ideally suited as a discovery tool allowing the detection of hundreds to thousands of biomolecules directly from tissue sections (Chaurand, Sanders, Jensen, & Caprioli, 2004). As MALDI MSI is a nondestructive technique, it maintains tissue integrity during biomolecular analysis. Tissue sections can be stained (e.g., hematoxylin and eosin; HE) and coregistered to the mass spectrometric data postmeasurement which leads to a new quality of molecular data owing to the incorporation of spatially resolved assignment of analytes to tissue histology. The assessment of molecular distributions using two-dimensional ion intensity maps in conjunction with histopathology enables the virtual microdissection of metabolic data from different tissue compartments. MALDI MSI enables therefore not only the separate investigation and comparison of tissue compartments but also allows combining metabolic information with patients' clinical data to augment clinical diagnostics and to stratify patients according to their clinical prognosis. Moreover, it provides a novel approach for the investigation of intra- and intertumoral heterogeneity.

3. FRESH-FROZEN- VS FORMALIN-FIXED PARAFFIN-EMBEDDED TISSUE SAMPLES

In tissue-based research there are typically two methods used to preserve tissue specimens after collection, either by freezing tissues in liquid nitrogen or by formalin-fixation and paraffin-embedding (FFPE). Both tissue types can be used to study metabolites, but with qualitative and quantitative differences in the yield of detected molecules. Until now, MALDI MS metabolite imaging has mainly carried out and demonstrated on frozen tissue sections (Fujimura & Miura, 2014; Miura, Fujimura, & Wariishi, 2012; Trim & Snel, 2016). This is explained by the fact, that snap-freezing of tissues results in samples that are ideally suited for the extraction and analysis of biomolecules with analytical techniques, especially for investigation of tissues' metabolism by MSI. The usage of frozen tissue specimen by MALDI MSI has determined a wide variety of metabolites and metabolic states in different human cancers, e.g., colorectal (Kurabe et al., 2013), gastric (Eberlin et al., 2014), prostate (Goto et al., 2015), thyroid (Dekker et al., 2015), breast (Calligaris et al., 2014; Dekker et al., 2015; Mao et al., 2016), and lung cancer (Li et al., 2015). Furthermore, using a high-mass-resolving MALDI imaging instrument, changes in metabolic pathways associated with the Warburg effect in cancer samples were demonstrated (Dekker et al., 2015). Nevertheless, fresh-frozen patient tissue samples constitute a rare material for molecular studies and are complex in terms of storage and handling. The majority of biopsied or surgically excised tissue specimens in clinics are formalin-fixed and paraffin-embedded. FFPE represents the gold standard for histopathological diagnosis providing an excellent insight in tissue morphology. The collection of tissue specimens and its investigation in routine diagnostics has led worldwide to the generation of archives with millions of FFPE biospecimens of various diseases including numerous cancer tissues. As such, FFPE tissue specimens are a golden mine for retrospective clinical studies investigating the altered metabolism of patients' cancer. Formalin fixation is known to stabilize tissues by the formation of methylene-bridges between certain amino acids resulting in a network of cross-linked proteins (Fox, Johnson, Whiting, & Roller, 1985), but also other formaldehyde-induced changes of biomolecules have been observed (Feldman, 1973). After the fixation process, tissues undergo dehydration steps followed by embedding in molten paraffin. Compared to frozen tissues, FFPE tissues can be stored at ambient temperature for decades with corresponding clinical data, allowing current researchers to obtain a more comprehensive understanding in cancer diseases and resistance mechanisms to common chemotherapeutic agents. However, the extraction of metabolites from archived material has been considered a challenging task, as tissue processing and the requirement of paraffin removal by solvents (e.g., xylene) could result to degradation or loss of molecular species. Recently, several groups have been actively working to overcome limitations in the analysis of FFPE tissue specimens providing unexpected achievements for clinical metabolomics. In a pioneering study Kelly et al. were able to discriminate sarcoma from normal FFPE tissue samples based on 106 detected metabolites using targeted LC-MS/MS (Kelly et al., 2011). By developing a suitable high-mass resolution MALDI MSI protocol for metabolite imaging from FFPE tissue samples, we have recently compared fresh-frozen with FFPE tissues and found an overlap of 72% of m/z species, while 1.700 m/z species were detected (Buck et al., 2015). The percentage is similar to a study of Wojakowska, Marczak, et al. (2015) using GC-MS which found that 75% of compounds detected in frozen tissue overlapped in metabolite profiles from FFPE sample. In our MALDI MSI study, the highest differences in metabolite composition between fresh-frozen and FFPE tissues were observed in the mass range m/z 600–1000 which mainly represents lipid ions, while metabolite peaks in the low mass range (m/z 50–400) were comparable between tissue types (Buck et al., 2015). The effects of sample preparation on metabolite and lipid composition was examined in two separate studies comparing fresh-frozen, formalin-fixed/frozen, and FFPE tissues (Pietrowska, Gawin, Polanska, & Widlak, 2016; Wojakowska, Marczak, et al., 2015). It has been shown that mainly lipid species leach away during the processing steps of tissue embedding and the removal of paraffin with solvents, while solely tissue fixation in formalin does not in general affect tissue molecular content (Pietrowska et al., 2016; Wojakowska, Marczak, et al., 2015). Nevertheless, several solvent-resistant membrane lipids, e.g., phosphatic acids and phosphatidylinositols, remain in deparaffinized FFPE tissues (Buck et al., 2015; Hughes, Gaunt, Brown, Clarke, & Gardner, 2014). The recovery of various metabolites in FFPE tissue samples is not unexpected, as it is generally known, that cellular components such as nucleic acids, carbohydrates, and lipids are not directly fixed by formaldehyde, but remain trapped in the network of insoluble protein cross-links (Renshaw, 2007). Thus the preservation of different classes of small molecules in deparaffinized FFPE tissue samples allows reliable targeted and nontargeted metabolic analyses by MSI (Fig. 1). For example, the metabolite content from FFPE tissue samples of thyroid cancers has been examined using a GC-MS-based method (Wojakowska, Chekan, et al., 2015). Five different thyroid malignancies (follicular, papillary/classical variant, papillary/follicular variant, medullary, and anaplastic cancers), benign follicular adenoma, and normal thyroid could be distinguished in a classification approach using metabolic signatures. Metabolites significantly discriminated the different thyroid lesions including lipids, carboxylic acids, and saccharides (Wojakowska, Chekan, et al., 2015). Bruinen et al. (2016) used

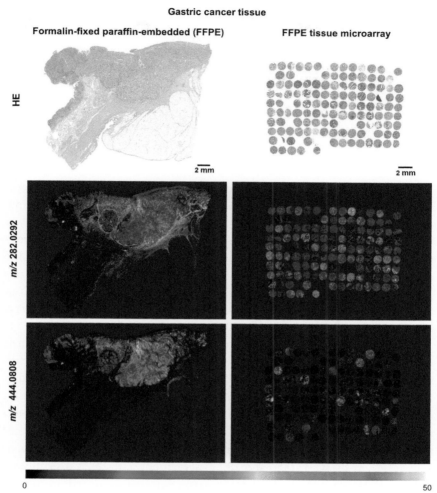

Fig. 1 Mass spectrometry metabolite imaging of a tissue section derived from surgical resection (FFPE gastric cancer patient) (*left*) and a multipatient FFPE gastric cancer tissue microarray (*right*). The tissue microarray is composed of 102 patient cores. Localization of metabolites such as *m/z* 282.0292 coincide with desmoplastic stroma and *m/z* 444.0808 reveal tumor regions in the resected patient sample and in individual tissue microarray cores. *HE*, hematoxylin and eosin-stained tissue samples after MALDI MSI measurement. *Image reproduced with permission from Buck, A., Ly, A., Balluff, B., Sun, N., Gorzolka, K., Feuchtinger, A.,..., Walch, A. (2015). High-resolution MALDI-FT-ICR MS imaging for the analysis of metabolites from formalin-fixed, paraffin-embedded clinical tissue samples.* Journal of Pathology, 237(1), 123–132. *Copyright 2015 Pathological Society of Great Britain and Ireland. Published by John Wiley & Sons, Ltd.*

targeted MSI to image drug-related crystal-like structures in formalin-fixed frozen and FFPE rabbit kidney tissue sections. The advantages of MALDI MSI as a potential diagnostic, predictive, and also a prognostic tool using FFPE tissue specimens have been demonstrated in a recently published study on the basis of low-molecular-weight molecules (Buck et al., 2015).

4. TISSUE-BASED DISEASE CLASSIFICATION— DIAGNOSTIC MARKERS AND METABOLIC SIGNATURES

MALDI imaging can be used to differentiate among tumors of varying subtype or tumor stage. Furthermore, the correct identification of the tumor origin is crucial for a personalized, individually tailored treatment regimen. Clinical diagnosis typically relies on histological and often extensive immunohistochemical analyses of tumor tissues (Bugat et al., 2003). By using MSI, molecular tissue classification can be generated with a set of molecular parameters that allow differentiation among different tissues. It was shown that ambient MSI can be used for the intraoperative molecular diagnosis of human brain tumors. A classifier was built using lipidomic signatures to enable the discrimination between gliomas and meningiomas (Eberlin et al., 2013). A recently published study distinguished between breast invasive ductal carcinoma and breast ductal carcinoma in situ (Mao et al., 2016). It was demonstrated that phospholipids were more abundant in invasive ductal carcinoma than in ductal carcinoma in situ, whereas fatty acids were more abundant in ductal carcinoma in situ than in invasive ductal carcinoma (Mao et al., 2016). The classification of specimens in the subtype and grade validation sets showed 100% and 78.6% agreement with the histopathological diagnosis, respectively (Mao et al., 2016). Another study investigated the overall level of phosphatidylcholine (PC) as elevated in colorectal cancer (Kurabe et al., 2013). To investigate which species of PC is overexpressed in colorectal cancer, MSI was performed using a panel of nonneoplastic mucosal and colorectal cancer tissues (Kurabe et al., 2013). In the study, the authors identified a novel biomarker, PC(16:0/16:1), in colorectal cancer (Kurabe et al., 2013). Specifically, elevated levels of PC(16:0/16:1) expression were observed in the more advanced stage of colorectal cancer (Kurabe et al., 2013). Their data further showed that PC(16:0/16:1) were specifically localized in the cancer region when examined using imaging mass spectrometry (Kurabe et al., 2013). In another application, metabolic signatures were used as classifiers to evaluate tumor surgical resection margin for malignant lesions. By examining breast and gastric cancer tissue samples metabolic signatures were found able to distinct cancerous from noncancerous tissue with the potential to delineate tumor boundaries (Calligaris et al., 2014; Eberlin et al., 2014). For example, several fatty acids, including oleic acid, were more abundant in breast cancer tissue than in normal tissues (Calligaris et al., 2014). In a recently published work, the distinction between normal and tumor colon tissue by in situ metabolomics has been demonstrated (Buck et al., 2015). But even more impressive in the same work a metabolomic signature discriminating between chromophobe renal cell carcinoma and renal oncocytoma was presented (Buck et al., 2015).

5. THERAPY RESPONSE PREDICTION AND PROGNOSIS

Predicting the functional relevance and efficacy of a therapy for individual patients is an important emerging topic in oncologic care and presents an essential part of modern predictive diagnosis. Tumors are usually sampled at the time of initial diagnosis and obtained during surgical biopsy or resection. As classical histopathology is only partly able to mirror and predict clinical behavior of individual tumors, new molecular tests and methods will have to be added into the morphology-based diagnostic procedure to read a patient's tissue as "deeply" as possible and to obtain combined information on morphological, genetic, epigenetic, proteomic as well as on metabolomic information (Dietel et al., 2013, 2015). Metabolic profiles containing mass signatures represent the possibility to enable response prediction and thus specific mass signatures can serve as potential biomarkers in diagnostics and therapy. In cancer research, metabolomic studies employing classical techniques have been shown to be valuable tools for discovery of predictive and prognostic markers. In the following we refer only a few out of many classical metabolomic studies. In a recently published study on the utility of serum metabolomic analysis in neuroblastoma patients, Beaudry et al. found differences in nitrogen, amino acids, carbohydrate metabolism, and ketosis, which allowed the differentiation of high- and low-risk patients as well as active disease from complete response/very good response (Beaudry et al., 2016). A further study showed the metabolomic response of myasthenia gravis patients to chronic prednisone treatment. Metabolic fingerprints were discovered that can now be validated as prednisone-responsive biomarkers for the improvement in diagnostic accuracy and prediction of therapeutic outcome (Sengupta, Cheema, Kaminski, & Kusner, 2014). In a population of cervical cancer patients, plasma metabolite profiling was performed in order to identify biomarkers for the prediction of response to neoadjuvant chemotherapy (Hou et al., 2014).

Until now there are several MALDI imaging studies which also provided valuable molecular signatures for therapy response prediction and prognosis. A molecular signature based on seven proteins serves, for example, as a new independent indicator of unfavorable overall survival after surgical resection of gastric cancer (Balluff et al., 2011). One of the first biomarker discovery studies using MALDI imaging was carried out in a patient collection in which protein markers in tumor tissue were identified that were associated with the response to a neoadjuvant therapy with paclitaxel in breast cancer (Bauer et al., 2010). In another study, the use of MALDI imaging facilitated the detection of previously unrecognized defects in the mitochondrial respiratory chain, which lead to individual patient response to cisplatin-based chemotherapy in advanced adenocarcinoma of the esophagus (Aichler et al., 2013). However, even as these are examples from in situ proteomic studies, the studies demonstrate the high relevance of MSI in disease prognosis and to predict therapeutic responses. Complementary, metabolic signatures, and biomarkers can enrich and support the prognostic and/or predictive value. Recently, a decreased expression of lysophosphatidylcholine (16:0/OH) independently predicted biochemical recurrence after surgical treatment for prostate cancer (Goto et al., 2015). High-mass resolution MALDI MSI showed that the expression of LPC(16:0/OH) and SM(d18:1/16:0) was lower in prostate cancer than in benign prostate epithelium (Goto et al., 2015). Another study using FFPE tissue samples demonstrated further the ability of the discovery of prognostic markers by in situ metabolomic MALDI imaging (Buck et al., 2015). In this study, a survival analysis revealed an independent prognostic marker in esophageal adenocarcinoma patients (Buck et al., 2015).

6. INTRA- AND INTERTUMORAL HETEROGENEITY

Understanding tumor heterogeneity presents an important task in cancer research toward improving both diagnosis and treatment of patients (Dalerba et al., 2011; Gerlinger et al., 2012). Heterogeneity of tumors contributes to treatment failure and disease recurrence, whereas the interaction between tumor cells and the associated stroma poses novel therapeutic opportunities. Intratumoral heterogeneity refers to the fact that cells within a tumor mass can be highly diverse due to an evolving process in cancer development driven by the stepwise accumulation of molecular changes and clonal selection, while intertumoral heterogeneity describes molecular alterations in several (metastatic) tumors present within or between patients (Katona et al., 2007; Loeb, 2011; Merlo, Pepper, Reid, & Maley, 2006). In malignant cells alteration in molecular features and clonal selection constantly takes place and provides either a selective advantage in function such as proliferation and survival, proceed neutral without any effects, or has negative cell damaging effects. Thereby, the fitness of a neoplastic cell is affected by the interactions with itself and other cells in its microenvironment competing with one another for the available resources (Merlo et al., 2006). Tumor heterogeneity has mainly been described at a genetic, chromosomal, or transcriptomal level (Swanton, 2012). The approaches used to analyze tumor heterogeneity are mostly targeted either focusing on the distribution of a single molecule in tissues such as proteins by using immunohistochemistry, or provide detailed molecular insight in tumor subpopulations by the targeted selection of tumor subpopulations. However, selection of representative tumor areas leads only to an average picture which describes the most dominant clone and therefore underestimates the extent and pattern of clonal heterogeneity. MALDI MSI has demonstrated its suitability to study tumor heterogeneity in its native microenvironment (Cimino et al., 2013; Jones et al., 2011; Willems et al., 2010). Imaging protein, peptide, and lipid levels of myxofibrosarcoma and myxoid liposarcoma allowed the classification of tumors according to their grade-specific phenotype and indicated that the technique reveals the inter- and intratumoral biomolecular heterogeneity of histologically contiguous tumors (Willems et al., 2010). In particular, lipid changes observed in myxoid liposarcomas could be related to pathways known to be affected during tumor progression (Willems et al., 2010). Bao et al. (2013) examined the energy management in tumor-bearing livers and demonstrated that tumor cells in the G1 phase exhibited higher concentrations of ATP, NADH, and UDP-N-acetylglucosamine than those in S and G2–M phases, suggesting accelerated glycolysis in G1 phase cells. In a recent study, the molecular heterogeneity within tumor areas was successfully linked with clinical data demonstrating that patients' overall survival is associated with respective tumor subpopulations (Balluff et al., 2015).

7. CONCLUSION

In biomedical research, working with tissue samples is indispensable because it permits direct insights into the biological reality of patients. In recent years, MSI metabolomic analyses have achieved encouraging results in tissue characterization, therapeutic treatment prediction, and cancer prognosis becoming a promising clinical tool. With technical and methodological progresses, in MSI, it can be expected that following metabolomic studies will generate metabolic signatures and markers with a high clinical impact.

REFERENCES

Aboud, O. A., & Weiss, R. H. (2013). New opportunities from the cancer metabolome. *Clinical Chemistry*, *59*(1), 138–146. http://dx.doi.org/10.1373/clinchem.2012.184598.

Aichler, M., Elsner, M., Ludyga, N., Feuchtinger, A., Zangen, V., Maier, S. K., ... Walch, A. K. (2013). Clinical response to chemotherapy in oesophageal adenocarcinoma patients is linked to defects in mitochondria. *The Journal of Pathology*, *230*(4), 410–419. http://dx.doi.org/10.1002/path.4199.

Almuhaideb, A., Papathanasiou, N., & Bomanji, J. (2011). 18F-FDG PET/CT imaging in oncology. *Annals of Saudi Medicine*, *31*(1), 3–13. http://dx.doi.org/10.4103/0256-4947.75771.

Balluff, B., Frese, C. K., Maier, S. K., Schone, C., Kuster, B., Schmitt, M., ... McDonnell, L. A. (2015). De novo discovery of phenotypic intratumour heterogeneity using imaging mass spectrometry. *The Journal of Pathology*, *235*(1), 3–13. http://dx.doi.org/10.1002/path.4436.

Balluff, B., Rauser, S., Meding, S., Elsner, M., Schone, C., Feuchtinger, A., ... Walch, A. (2011). MALDI imaging identifies prognostic seven-protein signature of novel tissue markers in intestinal-type gastric cancer. *The American Journal of Pathology*, *179*(6), 2720–2729. http://dx.doi.org/10.1016/j.ajpath.2011.08.032.

Bao, Y., Mukai, K., Hishiki, T., Kubo, A., Ohmura, M., Sugiura, Y., ... Minamishima, Y. A. (2013). Energy management by enhanced glycolysis in G1-phase in human colon cancer cells in vitro and in vivo. *Molecular Cancer Research*, *11*(9), 973–985. http://dx.doi.org/10.1158/1541-7786.MCR-12-0669-T.

Bauer, J. A., Chakravarthy, A. B., Rosenbluth, J. M., Mi, D., Seeley, E. H., De Matos Granja-Ingram, N., ... Pietenpol, J. A. (2010). Identification of markers of taxane sensitivity using proteomic and genomic analyses of breast tumors from patients receiving neoadjuvant paclitaxel and radiation. *Clinical Cancer Research*, *16*(2), 681–690. http://dx.doi.org/10.1158/1078-0432.ccr-09-1091.

Beaudry, P., Campbell, M., Dang, N. H., Wen, J., Blote, K., & Weljie, A. M. (2016). A pilot study on the utility of serum metabolomics in neuroblastoma patients and xenograft models. *Pediatric Blood & Cancer*, *63*(2), 214–220. http://dx.doi.org/10.1002/pbc.25784.

Brenton, J. D., Carey, L. A., Ahmed, A. A., & Caldas, C. (2005). Molecular classification and molecular forecasting of breast cancer: Ready for clinical application? *Journal of Clinical Oncology*, *23*(29), 7350–7360. http://dx.doi.org/10.1200/JCO.2005.03.3845.

Bruinen, A. L., van Oevelen, C., Eijkel, G. B., Van Heerden, M., Cuyckens, F., & Heeren, R. M. (2016). Mass spectrometry imaging of drug related crystal-like structures in formalin-fixed frozen and paraffin-embedded rabbit kidney tissue sections. *Journal of the American Society for Mass Spectrometry*, *27*(1), 117–123. http://dx.doi.org/10.1007/s13361-015-1254-3.

Buck, A., Ly, A., Balluff, B., Sun, N., Gorzolka, K., Feuchtinger, A., ... Walch, A. (2015). High-resolution MALDI-FT-ICR MS imaging for the analysis of metabolites from formalin-fixed, paraffin-embedded clinical tissue samples. *The Journal of Pathology*, *237*(1), 123–132. http://dx.doi.org/10.1002/path.4560.

Bugat, R., Bataillard, A., Lesimple, T., Voigt, J. J., Culine, S., Lortholary, A., ... Fizazi, K. (2003). Summary of the standards, options and recommendations for the management of patients with carcinoma of unknown primary site (2002). *British Journal of Cancer*, *89*(Suppl. 1), S59–S66. http://dx.doi.org/10.1038/sj.bjc.6601085.

Calligaris, D., Caragacianu, D., Liu, X., Norton, I., Thompson, C. J., Richardson, A. L., ... Agar, N. Y. (2014). Application of desorption electrospray ionization mass spectrometry imaging in breast cancer margin analysis. *Proceedings of the National Academy of Sciences of the United States of America*, *111*(42), 15184–15189. http://dx.doi.org/10.1073/pnas.1408129111.

Calvani, R., Miccheli, A., Capuani, G., Tomassini Miccheli, A., Puccetti, C., Delfini, M., ... Mingrone, G. (2010). Gut microbiome-derived metabolites characterize a peculiar obese urinary metabotype. *International Journal of Obesity*, *34*(6), 1095–1098. http://dx.doi.org/10.1038/ijo.2010.44.

Chaurand, P., Sanders, M. E., Jensen, R. A., & Caprioli, R. M. (2004). Proteomics in diagnostic pathology: Profiling and imaging proteins directly in tissue sections. *The American Journal of Pathology*, *165*(4), 1057–1068. http://dx.doi.org/10.1016/s0002-9440(10)63367-6.

Cimino, J., Calligaris, D., Far, J., Debois, D., Blacher, S., Sounni, N. E., ... De Pauw, E. (2013). Towards lipidomics of low-abundant species for exploring tumor heterogeneity guided by high-resolution mass spectrometry imaging. *International Journal of Molecular Sciences*, *14*(12), 24560–24580. http://dx.doi.org/10.3390/ijms141224560.

Dalerba, P., Kalisky, T., Sahoo, D., Rajendran, P. S., Rothenberg, M. E., Leyrat, A. A., ... Quake, S. R. (2011). Single-cell dissection of transcriptional heterogeneity in human colon tumors. *Nature Biotechnology*, *29*(12), 1120–1127. http://dx.doi.org/10.1038/nbt.2038.

Dekker, T. J., Jones, E. A., Corver, W. E., van Zeijl, R. J., Deelder, A. M., Tollenaar, R. A., ... McDonnell, L. A. (2015). Towards imaging metabolic pathways in tissues. *Analytical and Bioanalytical Chemistry*, *407*(8), 2167–2176. http://dx.doi.org/10.1007/s00216-014-8305-7.

Dietel, M., Johrens, K., Laffert, M., Hummel, M., Blaker, H., Muller, B. M., ... Anagnostopoulos, I. (2013). Predictive molecular pathology and its role in targeted cancer therapy: A review focussing on clinical relevance. *Cancer Gene Therapy*, *20*(4), 211–221. http://dx.doi.org/10.1038/cgt.2013.13.

Dietel, M., Johrens, K., Laffert, M. V., Hummel, M., Blaker, H., Pfitzner, B. M., ... Anagnostopoulos, I. (2015). A 2015 update on predictive molecular pathology and its role in targeted cancer therapy: A review focussing on clinical relevance. *Cancer Gene Therapy*, *22*(9), 417–430. http://dx.doi.org/10.1038/cgt.2015.39.

Eberlin, L. S., Norton, I., Orringer, D., Dunn, I. F., Liu, X., Ide, J. L., ... Cooks, R. G. (2013). Ambient mass spectrometry for the intraoperative molecular diagnosis of human brain tumors. *Proceedings of the National Academy of Sciences of the United States of America*, *110*(5), 1611–1616. http://dx.doi.org/10.1073/pnas.1215687110.

Eberlin, L. S., Tibshirani, R. J., Zhang, J., Longacre, T. A., Berry, G. J., Bingham, D. B., ... Poultsides, G. A. (2014). Molecular assessment of surgical-resection margins of gastric cancer by mass-spectrometric imaging. *Proceedings of the National Academy of Sciences of the United States of America*, *111*(7), 2436–2441. http://dx.doi.org/10.1073/pnas.1400274111.

Feldman, M. Y. (1973). Reactions of nucleic acids and nucleoproteins with formaldehyde. *Progress in Nucleic Acid Research and Molecular Biology*, *13*, 1–49.

Fox, C. H., Johnson, F. B., Whiting, J., & Roller, P. P. (1985). Formaldehyde fixation. *The Journal of Histochemistry and Cytochemistry*, *33*(8), 845–853.

Fujimura, Y., & Miura, D. (2014). MALDI mass spectrometry imaging for visualizing in situ metabolism of endogenous metabolites and dietary phytochemicals. *Metabolites*, *4*(2), 319–346. http://dx.doi.org/10.3390/metabo4020319.

Gambhir, S. S. (2002). Molecular imaging of cancer with positron emission tomography. *Nature Reviews. Cancer*, 2(9), 683–693. http://dx.doi.org/10.1038/nrc882.

Gerlinger, M., Rowan, A. J., Horswell, S., Larkin, J., Endesfelder, D., Gronroos, E., ... Swanton, C. (2012). Intratumor heterogeneity and branched evolution revealed by multiregion sequencing. *The New England Journal of Medicine*, 366(10), 883–892. http://dx.doi.org/10.1056/NEJMoa1113205.

Goodacre, R., Vaidyanathan, S., Dunn, W. B., Harrigan, G. G., & Kell, D. B. (2004). Metabolomics by numbers: Acquiring and understanding global metabolite data. *Trends in Biotechnology*, 22(5), 245–252. http://dx.doi.org/10.1016/j.tibtech.2004.03.007.

Goto, T., Terada, N., Inoue, T., Kobayashi, T., Nakayama, K., Okada, Y., ... Ogawa, O. (2015). Decreased expression of lysophosphatidylcholine (16:0/OH) in high resolution imaging mass spectrometry independently predicts biochemical recurrence after surgical treatment for prostate cancer. *Prostate*, 75(16), 1821–1830. http://dx.doi.org/10.1002/pros.23088.

Hou, Y., Yin, M., Sun, F., Zhang, T., Zhou, X., Li, H., ... Li, K. (2014). A metabolomics approach for predicting the response to neoadjuvant chemotherapy in cervical cancer patients. *Molecular BioSystems*, 10(8), 2126–2133. http://dx.doi.org/10.1039/c4mb00054d.

Hughes, C., Gaunt, L., Brown, M., Clarke, N. W., & Gardner, P. (2014). Assessment of paraffin removal from prostate FFPE sections using transmission mode FTIR-FPA imaging. *Analytical Methods*, 6(4), 1028–1035. http://dx.doi.org/10.1039/C3AY41308J.

Jones, E. A., van Remoortere, A., van Zeijl, R. J., Hogendoorn, P. C., Bovee, J. V., Deelder, A. M., & McDonnell, L. A. (2011). Multiple statistical analysis techniques corroborate intratumor heterogeneity in imaging mass spectrometry datasets of myxofibrosarcoma. *PloS One*, 6(9), e24913. http://dx.doi.org/10.1371/journal.pone.0024913.

Junot, C., Fenaille, F., Colsch, B., & Becher, F. (2014). High resolution mass spectrometry based techniques at the crossroads of metabolic pathways. *Mass Spectrometry Reviews*, 33(6), 471–500. http://dx.doi.org/10.1002/mas.21401.

Katona, T. M., Jones, T. D., Wang, M., Eble, J. N., Billings, S. D., & Cheng, L. (2007). Genetically heterogeneous and clonally unrelated metastases may arise in patients with cutaneous melanoma. *The American Journal of Surgical Pathology*, 31(7), 1029–1037. http://dx.doi.org/10.1097/PAS.0b013e31802b3488.

Kelly, A. D., Breitkopf, S. B., Yuan, M., Goldsmith, J., Spentzos, D., & Asara, J. M. (2011). Metabolomic profiling from formalin-fixed, paraffin-embedded tumor tissue using targeted LC/MS/MS: Application in sarcoma. *PloS One*, 6(10), e25357. http://dx.doi.org/10.1371/journal.pone.0025357.

Kurabe, N., Hayasaka, T., Ogawa, M., Masaki, N., Ide, Y., Waki, M., ... Sugimura, H. (2013). Accumulated phosphatidylcholine (16:0/16:1) in human colorectal cancer; possible involvement of LPCAT4. *Cancer Science*, 104(10), 1295–1302. http://dx.doi.org/10.1111/cas.12221.

Li, T., He, J., Mao, X., Bi, Y., Luo, Z., Guo, C., ... Abliz, Z. (2015). In situ biomarker discovery and label-free molecular histopathological diagnosis of lung cancer by ambient mass spectrometry imaging. *Scientific Reports*, 5, 14089. http://dx.doi.org/10.1038/srep14089.

Lindon, J. C., Holmes, E., & Nicholson, J. K. (2004). Metabonomics and its role in drug development and disease diagnosis. *Expert Review of Molecular Diagnostics*, 4(2), 189–199. http://dx.doi.org/10.1586/14737159.4.2.189.

Liu, N. Q., Braakman, R. B., Stingl, C., Luider, T. M., Martens, J. W., Foekens, J. A., & Umar, A. (2012). Proteomics pipeline for biomarker discovery of laser capture microdissected breast cancer tissue. *Journal of Mammary Gland Biology and Neoplasia*, 17(2), 155–164. http://dx.doi.org/10.1007/s10911-012-9252-6.

Loeb, L. A. (2011). Human cancers express mutator phenotypes: Origin, consequences and targeting. *Nature Reviews. Cancer, 11*(6), 450–457. http://dx.doi.org/10.1038/nrc3063.

Ma, Y., Zhang, P., Yang, Y., Wang, F., & Qin, H. (2012). Metabolomics in the fields of oncology: A review of recent research. *Molecular Biology Reports, 39*(7), 7505–7511. http://dx.doi.org/10.1007/s11033-012-1584-1.

Mao, X., He, J., Li, T., Lu, Z., Sun, J., Meng, Y., … Chen, J. (2016). Application of imaging mass spectrometry for the molecular diagnosis of human breast tumors. *Scientific Reports, 6*, 21043. http://dx.doi.org/10.1038/srep21043.

Merlo, L. M., Pepper, J. W., Reid, B. J., & Maley, C. C. (2006). Cancer as an evolutionary and ecological process. *Nature Reviews. Cancer, 6*(12), 924–935. http://dx.doi.org/10.1038/nrc2013.

Miura, D., Fujimura, Y., & Wariishi, H. (2012). In situ metabolomic mass spectrometry imaging: Recent advances and difficulties. *Journal of Proteomics, 75*(16), 5052–5060. http://dx.doi.org/10.1016/j.jprot.2012.02.011.

Nordstrom, A., & Lewensohn, R. (2010). Metabolomics: Moving to the clinic. *Journal of Neuroimmune Pharmacology, 5*(1), 4–17. http://dx.doi.org/10.1007/s11481-009-9156-4.

Pavlova, N. N., & Thompson, C. B. (2016). The emerging hallmarks of cancer metabolism. *Cell Metabolism, 23*(1), 27–47. http://dx.doi.org/10.1016/j.cmet.2015.12.006.

Pietrowska, M., Gawin, M., Polanska, J., & Widlak, P. (2016). Tissue fixed with formalin and processed without paraffin embedding is suitable for imaging of both peptides and lipids by MALDI-IMS. *Proteomics, 16*(11–12), 1670–1677. http://dx.doi.org/10.1002/pmic.201500424.

Renshaw, S. (2007). Immunochemical staining techniques. In S. Renshaw (Ed.), *Immunohistochemistry: Methods express*. Bloxham, UK: Scion Publishing Ltd.

Romick-Rosendale, L. E., Goodpaster, A. M., Hanwright, P. J., Patel, N. B., Wheeler, E. T., Chona, D. L., & Kennedy, M. A. (2009). NMR-based metabonomics analysis of mouse urine and fecal extracts following oral treatment with the broad-spectrum antibiotic enrofloxacin (Baytril). *Magnetic Resonance in Chemistry, 47*(Suppl. 1), S36–S46. http://dx.doi.org/10.1002/mrc.2511.

Sengupta, M., Cheema, A., Kaminski, H. J., & Kusner, L. L. (2014). Serum metabolomic response of myasthenia gravis patients to chronic prednisone treatment. *PloS One, 9*(7), e102635. http://dx.doi.org/10.1371/journal.pone.0102635.

Swanton, C. (2012). Intratumor heterogeneity: Evolution through space and time. *Cancer Research, 72*(19), 4875–4882. http://dx.doi.org/10.1158/0008-5472.can-12-2217.

Trim, P. J., & Snel, M. F. (2016). Small molecule MALDI MS imaging: Current technologies and future challenges. *Methods, 104*, 127–141. http://dx.doi.org/10.1016/j.ymeth.2016.01.011.

Umar, A., Luider, T. M., Foekens, J. A., & Pasa-Tolic, L. (2007). NanoLC-FT-ICR MS improves proteome coverage attainable for approximately 3000 laser-microdissected breast carcinoma cells. *Proteomics, 7*(2), 323–329. http://dx.doi.org/10.1002/pmic.200600293.

Vinaixa, M., Schymanski, E. L., Neumann, S., Navarro, M., Salek, R. M., & Yanes, O. (2016). Mass spectral databases for LC/MS- and GC/MS-based metabolomics: State of the field and future prospects. *TrAC Trends in Analytical Chemistry, 78*, 23–35. http://dx.doi.org/10.1016/j.trac.2015.09.005.

Warburg, O. (1956). On the origin of cancer cells. *Science, 123*(3191), 309–314.

Willems, S. M., van Remoortere, A., van Zeijl, R., Deelder, A. M., McDonnell, L. A., & Hogendoorn, P. C. (2010). Imaging mass spectrometry of myxoid sarcomas identifies proteins and lipids specific to tumour type and grade, and reveals biochemical intra-tumour heterogeneity. *The Journal of Pathology, 222*(4), 400–409. http://dx.doi.org/10.1002/path.2771.

Wishart, D. S. (2007). Proteomics and the human metabolome project. *Expert Review of Proteomics, 4*(3), 333–335. http://dx.doi.org/10.1586/14789450.4.3.333.

Wojakowska, A., Chekan, M., Marczak, L., Polanski, K., Lange, D., Pietrowska, M., & Widlak, P. (2015). Detection of metabolites discriminating subtypes of thyroid cancer: Molecular profiling of FFPE samples using the GC/MS approach. *Molecular and Cellular Endocrinology*, *417*, 149–157. http://dx.doi.org/10.1016/j.mce.2015.09.021.

Wojakowska, A., Marczak, L., Jelonek, K., Polanski, K., Widlak, P., & Pietrowska, M. (2015). An optimized method of metabolite extraction from formalin-fixed paraffin-embedded tissue for GC/MS analysis. *PloS One*, *10*(9), e0136902. http://dx.doi.org/10.1371/journal.pone.0136902.

Yuan, M., Breitkopf, S. B., Yang, X., & Asara, J. M. (2012). A positive/negative ion-switching, targeted mass spectrometry-based metabolomics platform for bodily fluids, cells, and fresh and fixed tissue. *Nature Protocols*, *7*(5), 872–881. http://dx.doi.org/10.1038/nprot.2012.024.

Zhao, H., Yang, L., Baddour, J., Achreja, A., Bernard, V., Moss, T., ... Nagrath, D. (2016). Tumor microenvironment derived exosomes pleiotropically modulate cancer cell metabolism. *eLife*, *5*, 1–27. http://dx.doi.org/10.7554/eLife.10250.

CHAPTER SIX

Mass Spectrometry Imaging in Oncology Drug Discovery

R.J.A. Goodwin[*,1], J. Bunch[†,‡], D.F. McGinnity[§]

[*]Mass Spectrometry Imaging Group, Pathology Sciences, Drug Safety & Metabolism, Innovative Medicines and Early Development, AstraZeneca, Cambridge, United Kingdom
[†]National Physical Laboratory, Teddington, United Kingdom
[‡]School of Pharmacy, University of Nottingham, Nottingham, United Kingdom
[§]DMPK, Oncology IMED, Innovative Medicines and Early Development, AstraZeneca, Cambridge, United Kingdom
[1]Corresponding author: e-mail address: Richard.Goodwin@astrazeneca.com

Contents

1. Introduction — 134
2. How MSI Can Inform Our Understanding of Pharmacokinetic–Pharmacodynamic Relationships — 135
3. Biodistribution — 137
4. Tumor Metabolism: MSI Analysis for More Than Just Drug Distribution — 138
5. Sample Preparation — 139
6. Quantitation — 141
7. Toxicity and Safety Assessment — 142
8. Biomarkers for Efficacy — 144
9. Drug Delivery — 145
10. Tumor Microenvironment — 147
11. Assessing Hypoxia — 150
12. BBB Penetration — 150
13. Beyond Small Molecules — 153
14. Clinical Translation — 156
15. Emerging Applications: Spheroids — 157
16. Increased Spatial Resolution — 159
17. Metrology for MS Imaging — 160
18. Conclusion — 162
References — 163

Abstract

Over the last decade mass spectrometry imaging (MSI) has been integrated in to many areas of drug discovery and development. It can have significant impact in oncology drug discovery as it allows efficacy and safety of compounds to be assessed against the backdrop of the complex tumour microenvironment. We will discuss the roles of MSI in investigating compound and metabolite biodistribution and defining pharmacokinetic -pharmacodynamic relationships, analysis that is applicable to all drug discovery projects. We will then look more specifically at how MSI can be used to understand tumour metabolism and other applications specific to oncology research. This will all be described alongside the challenges of applying MSI to industry research with increased use of metrology for MSI.

1. INTRODUCTION

Modern drug discovery and development is a lengthy, high-risk, and competitive business. It may take a decade and costs billions of dollars to get new medicines to market. High attrition rates, together with rising R&D and clinical trial costs, have presented significant challenges to the pharmaceutical industry over the last two decades. While the financial rewards for new blockbuster drugs are substantial, it is worth remember that a key driving force behind the scientists developing new therapeutics is the desire to get new medicines to patients to save or improve lives. This is particularly true for oncology drug discovery research teams, where there continues to be significant unmet medical need for treating this most devastating disease.

The primary reasons for drug attrition have consistently been lack of efficacy and toxicological or clinical safety risk (Kubinyi, 2003; Schuster, Laggner, & Langer, 2005). Oncology as a therapeutic area has suffered similarly high attrition compared to other diseases, but there are distinct trends that have shaped the modern drug discovery environment. The demand for effective new treatments is being met by oncology researchers developing ever more complex therapeutic regimes including use of new modalities, drug combinations, and delivery approaches. Advanced drug delivery technologies, such as nanomedicines, are being designed to increase therapeutic index by improving the delivery of drug to tumors relative to normal tissue, albeit with limited clinical success up to now (Hare et al., 2016). Historically, treatment of cancer relied predominantly on DNA-damaging radiotherapy and chemotherapy. In recent years, more specific target-based approaches have come to the fore. There has also been the tailoring of therapies to those specific patient populations most likely to benefit and increased use of oral administration. These developments, together with therapeutic and technological advances, have led to a greater emphasis on applying DMPK insight and know-how to develop oncology drugs with an acceptable pharmacokinetic profile while maximizing the therapeutic index and minimizing the drug–drug interaction potential.

The complexity of modern drug discovery and development, utilizing novel therapeutics, delivery systems, and innovative schedule design, provides certain challenges for the traditional bioanalytical methods employed across the value chain. While not replacing any established method for measuring and monitoring of drug exposure, metabolism, and disposition, mass spectrometry imaging (MSI) is increasingly being applied to complement these established bioanalysis approaches and is providing additional insights in both a timely and cost-effective way. MSI has been employed by both academic and industrial researchers for over a decade now, with increasing utility and breadth of applications reported in the literature. However, by the nature of commercial pharmaceutical research, the number of publications does not truly reflect the uptake and effectiveness of MSI in drug development. We will now outline the current and emerging applications of MSI within the field of oncology drug discovery, drug delivery, and clinical applications. Where no current examples from oncology have been reported, we will reference wider drug discovery examples. There are a number of comprehensive reviews on the range of MSI technologies, detailing their relative merits and utility to a wide range of applications, often with a focus on drug discovery (Cobice et al., 2015; Goodwin & Webborn, 2015; Nilsson et al., 2015). We will not therefore consider in detail the respective data acquisition speeds, mass analyzer sensitivities, and spatial and spectral resolutions of the increasing number of ionization systems that form the wide array of MSI platforms open to researchers. We will, however, focus on how MSI can impact oncology drug discovery. In recent times, the pharmaceutical industry has increased the diversity of approaches to therapeutic intervention, from traditional "small molecule" new chemical entities to "large molecule" biologic drug candidates, including peptides and monoclonal antibodies due to advances in biotechnology and apparent lower attrition rates relative to small molecules (Jiunn, 2009). Within this chapter, we will pay particular attention to the larger field of small molecule drug discovery, but we will also consider the analysis of new and emerging large molecule therapeutics.

2. HOW MSI CAN INFORM OUR UNDERSTANDING OF PHARMACOKINETIC–PHARMACODYNAMIC RELATIONSHIPS

The consideration and application of pharmacokinetic principles in drug discovery are now ubiquitous due to the recognition of the role of free drug concentrations as a surrogate for measuring pharmacological effects. A basic tenet of pharmacology is that the magnitude of a pharmacological response is a function of the drug concentration at the site of action. Thus the objective of therapy may be achieved by maintaining sufficient concentration of drug at the site of action for the necessary duration, yet not so high as to elicit a deleterious effect. The concentration of drug has been typically measured in blood or plasma as the direct measurement of the available compound at the site of action, including tumors, can be challenging. However, the relevance of this surrogate site, and the drivers for the observed changes over time, will depend upon a number of factors and assumptions. Our understanding of such factors has been via the science of pharmacokinetics (PK), which is defined as the study of change of drug concentration over time and describes a systematic approach to relating dose to amount of drug in the body. Pharmacodynamics (PD) is the study of how drug concentration relates to effect. By modeling the mechanism of drug action, quantitative pharmacology or pharmacokinetics–pharmacodynamics (PKPD) describes the relationship between dose, concentration, and intensity and duration of response. The relatively recent use of new analytical tools including MSI in drug discovery and development now can provide insight into not just the identity and spatial distributions of therapeutic modalities in both normal tissues and tumors but also endogenous substrates, ligands, target proteins, and downstream biomarkers of target engagement and drug effect. Factors influencing the distribution of candidate drugs to the site(s) of pharmacological effect include intrinsic membrane permeability, the relative affinity for components in tissue vs blood, and whether or not the candidate drugs are substrates for drug transporters. Candidate drug or metabolite (which may be active or inactive against the same or different pharmacologically relevant targets) levels measured in plasma may not represent those present at the site of action and therefore cannot explain any observed efficacy or toxicological sequelae. It is important to stress that even in the oncology therapeutic area, for solid tumors, it remains the goal preclinically to build a PKPD relationship from the plasma compartment. This facilitates translatability to the clinical situation where sample availability and analysis from the plasma will be more achievable than from tumor biopsies. Understanding tumor PK and PD supports the building of this pharmacokinetic–pharmacodynamic–efficacy (PK/PD/E) relationship. Moreover, tumor PK can bridge the gap between plasma PK and tumor PD kinetics. A key question in oncology drug discovery that needs to be asked includes whether in vivo pharmacodynamics (and efficacy) can be robustly modeled using plasma PK? Of course, this assumes that unbound concentrations of drug candidates are determined relative to PD endpoints via a time course of total plasma level measurements. This requires applying a correction for plasma protein binding. Preferably all measurements are determined from experiments at a range of doses via the relevant administration route.

3. BIODISTRIBUTION

To date, the core role for MSI in pharma R&D has been to provide projects with data on the biodistribution of compounds in situations when plasma measurements are thought to poorly represent tissue concentration. For oncology, this is not limited to abundance and distribution within a tumor but across all tissues. The MSI data generated also should not be considered in isolation but related to the range of complementary in vivo, in vitro, and in silico assays. However, when successfully performed, MSI can generate insights in a timely and cost-effective way. Determining the biodistribution of a compound can require several iterative MSI experiments to be performed. The first that a researcher could consider performing, to obtain an overall view on distribution of target analyte, would be whole body tissue section MSI (Shahidi-Latham, Dutta, Prieto Conaway, & Rudewicz, 2012). While this can generate images that appear initially impressive and be interpreted in a similar way as the radiolabeled mainstay in DMPK of quantitative whole body autoradiography (QWBA), the collected data come with many complications that need to be interpreted (Drexler, Tannehill-Gregg, Wang, & Brock, 2011). These have been discussed in previous reviews (Amstalden van Hove, Smith, & Heeren, 2010; Goodwin, 2012), but in brief the main issues are localized ionization suppression requiring every tissue to need a correction factor to allow interorgan comparison (Hamm et al., 2012; Källback, Shariatgorji, Nilsson, & Andrén, 2012) and requirement to compromise on spatial resolution. Therefore, an alternative approach for preliminary analysis is the collection of the target tissues and analysis at higher spatial resolution. Regardless of size of the sample, subsequent MS/MS analysis is usually performed to provide confirmation and validation of target, albeit this means that no additional metabolite or biomarker distributions can be simultaneously detected when acquiring MS/MS data. Therefore, analysis often involves iterative qualitative and quantitative analysis at a range of spatial resolutions, as required to address specific project needs. Quantitative MSI can also be performed in either full spectrum or targeted MS/MS mode and allows inferences beyond relative distribution to be made and is discussed in more detail later.

As regularly asserted, the power of MSI is that, unlike other probe or labeling assays, thousands of endogenous molecular species are simultaneously detected at the same time as any exogenous compound or metabolite. This allows MSI experiments to relate the distribution of the drug to the distribution of endogenous metabolites. The endogenous molecules also provide molecular maps of the target tissue or the tumor architecture (Mascini et al., 2016) and can be biomarkers for compound efficacy or safety and these applications will be explored later.

4. TUMOR METABOLISM: MSI ANALYSIS FOR MORE THAN JUST DRUG DISTRIBUTION

Before identifying new therapeutic targets, a key component is to understand the target tumor metabolism. Increased knowledge of cancer metabolism shows that the metabolic processes are extremely heterogeneous, derived from factors including genetic diversity, multiple and redundant metabolic pathways, and the tumor microenvironment. Metabolomic analysis has been extensively studied since the discovery that cancer cells rely on glycolysis followed by lactic acid fermentation for energy production, even in the presence of oxygen (Vander Heiden, Cantley, & Thompson, 2009; Warburg, 1956). While the Warburg effect remains debated, the consensus is that cancer cells can salvage organic carbon for the synthesis of large quantities of the biomolecules required for cell proliferation. Therefore, understanding tumor metabolism, crucial in identifying new druggable targets, requires technologies that are able to measure the level and flux of metabolism within the tumor microenvironment (Zhou & Lu, 2016). Examples of metabolic changes, often linked to specific genetic alterations, include the following: (1) reductive glutamine metabolism involving cytoplasmic IDH1 or mitochondrial IDH2 (Maddocks et al., 2013; Metallo et al., 2012; Wise et al., 2011); (2) oncogenic KRAS that promotes nonoxidative pentose phosphate pathway metabolism in pancreatic tumors (Ying et al., 2012); (3) oncogenic BRAF V600E that upregulates a ketogenic enzyme 3-hydroxy-3-methylglutaryl-CoA lyase, leading in turn to the increased induction of the ketone body acetoacetate that selectively enhances BRAF V600E to MEK1 binding to promote MEK–ERK signaling (Kang et al., 2015); (4) inappropriate activation of PI3K/Akt also promotes glycolysis (Deprez, Vertommen, Alessi, Hue, & Rider, 1997; Gottlob et al., 2001; Kohn, Summers, Birnbaum, & Roth, 1996; Rathmell et al., 2003) and stimulates de novo lipogenesis through the direct phosphorylation and activation of ACL (Berwick, Hers,

Heesom, Moule, & Tavare, 2002); and (5) mTORC1-mediated activation of SREBF (Porstmann et al., 2008). Metabolic changes are also observed for tumor cells with loss or mutation of well-established tumor suppressors, such as for p53, where p53-null cells fail to complete the response to serine deprivation, resulting in oxidative stress-induced inhibition of cell viability and proliferation (Maddocks et al., 2013). To further exacerbate the situation, cancers also elicit extremely high metabolic flexibility, making some metabolite synthesis-targeted therapeutic strategies fail. For example, fatty acid synthase (FASN) inhibitors, targeting a key enzyme involved in neoplastic lipogenesis, can block de novo fatty acid synthesis. However, this leads to reliance on the circulating lipid pool for the synthesis of new membranes, resulting in slower tumor growth but not complete remission (Flavin, Peluso, Nguyen, & Loda, 2010). Therefore, the MSI analysis of metabolomic profiles in tumor samples, at both baseline and following therapeutic interventions, will be of significance in both understanding and characterizing the tumor microenvironment in the preclinical and clinical setting.

Steroid concentrations within tissues, modulated by intracellular enzymes, can influence hormone-dependent cancers and have proven elusive endogenous targets for bioanalysis. However, mapping and quantifying the distribution of steroids within tissue sections has been possible by simple adaptation to standard MSI protocols. Effective detection of previously poorly detected steroid targets has been accomplished using on-tissue derivatization with Girard T reagent. This modified the targets to generate increased ionization, so quantification of substrate and product (11-dehydrocorticosterone and corticosterone) of the glucocorticoid-amplifying enzyme 11β-HSD1 could be made (Cobice et al., 2013). The use of on-tissue derivatization is proving to be an effective strategy for poorly detected targets (Shariatgorji et al., 2014). While not simple or quick, it does mean whole classes of endogenous and exogenous targets can be analyzed by MSI. There are a number of sample processing and optimization strategies that can be used to increase the success of MSI in drug development, and the importance of sample preparation will now be discussed.

5. SAMPLE PREPARATION

Having considered the endogenous molecular targets, we will now consider aspects of sample collection and processing. Sample preparation is crucial in performing robust and accurate MSI experiments. Any failings early in the MSI analysis can render subsequent interpretation void. Specifics of the sample preparation for the different MSI technologies have been extensively reviewed (Goodwin, 2012) so will not be described here. However, before starting a study, adequate consideration must be made into how each tissue, biopsy, or tumor sample will be analyzed to allow effective interassay comparisons. Each preclinical or patient sample may be analyzed by a range of imaging and molecular assays; therefore, tissue collection is the first point where protocol optimization and standardization is required for a multiassay tissue study. As discussed, MSI can be performed on whole rodent body sections, dissected tissues, or biopsies, each with advantages and challenges in the subsequent analyses. Dissected tissue analysis is probably the most common approach for studying PK groups, and using such tissues means that control and dosed samples can be readily analyzed within a single experiment if tissue sections are placed on the same slide. Typically, six rat organ sections can be placed on the same MS compatible slide. To allow a full study means running interslide standards, utilizing effective quantification strategies, and mixing groups so no bias is introduced due to the order of analysis. It is important to note that time taken to section can affect the detected abundance of endogenous (Goodwin, Dungworth, Cobb, & Pitt, 2008; Goodwin, Pennington, & Pitt, 2008) or exogenous compounds (Goodwin, Iverson, & Andren, 2012). This can be mitigated by reducing sample preparation time through the embedding of multiple samples into a support media, allowing all samples to be sectioned using a cryomicrotome at the same time rather than sequentially. This means that all samples spend the same period in cryostat and all sections on a given slide will undergo identical treatment for the remainder of the experiment. Common embedding media are carboxymethyl cellulose and gelatin as they do not cause detectable sample contamination yet provide sufficient support for sectioning thin (<5 μm) tissue sections required for optimal histology (immunohistochemistry—IHC and hematoxylin and eosin—H&E) of matched MSI samples. New MSI embedding support media (Strohalm et al., 2011) and even blends of embedding media (Nelson, Daniels, Fournie, & Hemmer, 2013) are used for optimal tissue support. For analysis of small tumors, such as those from efficacy studies, embedding is the only practicable option. This is also the case for the analysis of tissue needle biopsies from clinical studies. Larger biopsies and resection material can be treated in a similar way to dissected rodent organs and sectioned either embedded in media or with minimal mounting media.

Postsample sectioning, tissue washing (briefly bathing tissue sections placed on slides into water and ethanol solutions) can be performed to remove low molecular weight molecules and salts that limit the ionization and detection of larger peptides and proteins. In our experience, this has limited usefulness if attempting analysis of either lower molecular weight endogenous biomarkers or therapeutics as they are also removed or delocalized. Researchers have however demonstrated that by modifying the pH of the wash solution to a level where the target compound is poorly soluble, tissues can be washed and efficiency of detection of the target can be improved by removal of ionization suppressing material (Shariatgorji et al., 2012). In a similar way to derivatization, this is not a strategy that is widely used in the MSI analysis due to the extra optimization and validation time required. However, it is particularly effective within a pharma environment when a single target is often pursued and teams are able to make more in-depth investments in analysis time for each target.

6. QUANTITATION

Drug quantitation measured directly from tissue sections is now routinely performed by MSI, though there remain many complications and caveats that need to be considered (Nilsson et al., 2010). The standard approach is to spot an array of known concentration droplets onto the surface of adjacent control tissue sections and analyze in parallel to the target tissue. This allows a simple calibration curve to be generated and subsequent calculation of concentration of target analyte in the sample. However, this method fails to take into account any localized suppression or enhancement of analyte ionization caused by the architecture of the sample. This can be overcome by correcting for ionization efficiency using a stable-labeled analogue of the drug. The compound is sprayed over the tissue at an appropriate concentration to give moderated detection by subsequent MSI analysis (Källback et al., 2012). Note that too strong a detected response may cause suppression of the target analyte. The application of the labeled standard should be homogeneous and in minimal solvent to prevent delocalization of analytes within the tissue section. This is typically performed using similar systems for applying the even matrix coating required for MALDI MSI. Correction for ionization suppression can then be made pixel by pixel and has been shown to be an improvement over other normalization correction factors such as total ion count. However, a stable-labeled analogue of the compound is often not available for researchers working in drug discovery when multiple compounds in a chemical series are being evaluated. In such situations, two alternative approaches can be attempted. Control tissue can be coated with the drug, as just described, and regions of ionization suppression and enhancement identified and excluded from subsequent quantification. Alternatively a compound with similar structure, physiochemical properties, and ionization efficiency can be used as a substitute for the labeled analogue. Neither method is ideal; however, the analysis is often performed to differentiate and identify major differences between candidates that will be more fully characterized and assessed as they are progressed. So, for example, precoating a tissue sample with the candidate drug may show that a calibration curve needs to be applied to multiple tissue regions or over entire samples for more accurate quantification calculations (Chumbley et al., 2016).

Safety and efficacy biomarkers, discussed later, can also be quantified directly from tissue sections. Such analysis is also routinely performed, but the main complication is the fact that control tissue will also contain the endogenous molecular target, either at higher or at lower abundance depending on the effect of compound or disease state. This means that the calibration spots need to contain a labeled target. Labeled endogenous metabolites are readily available for purchase with various degrees and positioning of the labeling. It should be checked that the selected labeled standard is sufficiently different in mass to the unlabeled version to avoid overlap of the isotope peaks of the endogenous molecule. A further complication for many efficacy biomarkers for oncology projects is endogenous metabolites, which may be readily turned over and unstable during standard bioanalysis (homogenization and LC-MS quantification). Care must therefore be taken to evaluate the stability of the calibration stock solutions as well as the stability of the endogenous metabolites within the tissue section. While this can be problematic, the MSI analysis can actually prove to be superior to homogenization in this regard. The sample preparation for MSI, performed on frozen samples, does not cause mixing of the endogenous metabolites, enzymes, and reactive molecules in an extraction solution such as used for homogenization. So, MSI can provide means to relate biomarker distribution to drug distribution and tissue microenvironment. Arguably, MSI provides a simple, more accurate means to quantify the absolute abundance of unstable endogenous metabolites (Fujimura & Miura, 2014). In summary, quantitation by MSI has many challenges to be overcome, but it does provide multiple opportunities to add value to imaging analysis of tissues and tumors. Though it is imperative, given the number of possible complications, all MSI quantitative data should be carefully reviewed and where possible related to complementary data.

7. TOXICITY AND SAFETY ASSESSMENT

While oncology researchers may typically countenance lower therapeutic safety/toxicity margins, due to the serious and life-threatening nature of the disease, relative to other more chronic diseases, the primary objective of oncology animal toxicology studies remains evaluating the safety profile of drug candidates. Preclinical analysis and effective understanding and translation to man of any toxicological finding in drug discovery can lead to the termination of a compound. Ideally, information will feed into the subsequent optimization of the chemical series to avoid any off-target pharmacology (adverse pharmacologic effects at proteins other than the primary therapeutic target) or the termination of the project against a target with the elucidation of an unacceptable on-target effect. While it is customary for a compound's efficacy in tumors to be investigated in detail during drug discovery, possible adverse sequelae in the range of target organs from toxicological studies can be extremely challenging to assess (Pellegatti & Pagliarusco, 2011). In our experience, MSI can and will increasingly provide significant insight into the elucidation of safety and toxicity signals. Again, this technique is most powerful when performing in combination with other technologies such as histopathological analysis and LC-MS of tissue homogenates and blood/plasma bioanalysis. Multiple examples of investigatory MSI analysis in drug discovery have been reported to date, but few have specifically related to oncology drug development and/or drug safety. The following examples may have been derived from oncology candidate drugs, demonstrating either on- or off-target toxicity. Their inclusion is to exemplify how MSI can be used for investigatory toxicity studies. Early reported toxicity studies reported the characterization of crystal deposits within rat kidney or spleen following administration of compounds. MSI has able to identify the crystal deposits as metabolites of the compound compounds, in situ and with sufficient spatial resolution to map the deposition of the crystals (Drexler et al., 2007; Kim et al., 2010; Nilsson et al., 2012).

The effectiveness of MSI to measure, monitor, and investigate a compound's ability to penetrate the blood–brain barrier (BBB) and subsequent accumulation within the central nervous system (CNS) will be considered in detail later. Sample homogenization can be sufficient for certain tissues, when an averaged abundance is an acceptable comprise for increased sensitivity (using sample cleanup and separation by liquid chromatography). However, it is a poor substitute when looking to study subtle or slight concentration changes within the brain architecture. MSI toxicity studies have therefore been readily applied to the study of the CNS, such as reported by researchers at GlaxoSmithKline who assessed the disposition and metabolism of fosdevirine, a nonnucleoside reverse transcriptase inhibitor, on the CNS (Castellino et al., 2012). Using MSI detection and quantitation of multiple neurotransmitters, and monitoring effects of xenobiotics can be a crucial component of drug safety programs. Recent improvement in the detection of neurotransmitters has been reported, and the utility is exemplified by following the example including drug treatments using MALDI (Shariatgorji et al., 2014) and DESI analysis (Shariatgorji et al., 2016). Methods used for on-tissue derivatization, for example, targeting primary amines in neurotransmitters such as GABA, dopamine, serotonin, and glutamate, are increasingly being employed. Chemical modification increases the ionization efficiency and hence detection and mapping in target tissue samples. Such endogenous metabolites may be measured to assess off-target effects that cause toxicity or safety concerns. The same endogenous markers may also be used as efficacy biomarkers. Oncology MSI drug safety studies have also reported that a compound's toxicity can be mitigated through the use of nanoparticle delivery systems. The MSI analysis has shown that delivery of paclitaxel into tumors in a nanoparticle formulation manipulated exposure. This was also associated with less neurotoxicity in mice (Yasunaga et al., 2013). Moreover, in our laboratory, the ability to monitor in situ drug concentrations and rate of active pharmaceutical ingredient (API) release, using novel delivery systems, in relation to the target tissue architecture offers powerful insights to bioscientists and medicine formulation groups.

8. BIOMARKERS FOR EFFICACY

Rapid feedback to medicinal chemistry-led design teams providing estimates of potency, selectivity, predictions of human PK, and therapeutic dose is pivotal to the efficient optimization of lead compounds into candidate drugs. The design-make-test-analyze (DMTA) cycle is the multidisciplinary engine of lead optimization and drug discovery screening cascades. The most efficient DMTA cycles are based on screening cascades appropriate for projects typically starting with a panel of assays in "Wave I" determining physicochemical properties and basic in vitro data such as potency against the target, key selectivity screens, metabolic stability, and perhaps cytochrome P450 inhibition assessment. Subsequent studies further profile lead compounds, and these include establishing in vivo PK, PD, and efficacy relationships only with quality compounds. There are many available PK optimization approaches to input into compound design and to predict key human PK properties with differing power, cost, and complexity, ranging from purely in silico methods via a multitude of in vitro assays and in vivo models, to physiologically based pharmacokinetic techniques, which characterize drug distribution to specific organs in terms of physiologically relevant variables (Grime, Barton, & McGinnity, 2013). MSI is a powerful tool for analyses of known targets and is valuable in PK tissue analysis. However, the ability to also monitor endogenous as well as exogenous compounds makes it suitable for mapping and quantifying the abundance of biomarkers of compound efficacy in PD studies. Over a decade ago, researchers described solid tumor distribution of drug and active metabolite and related distribution simultaneously to intratumor distribution of ATP, indicating that the cytotoxic metabolite was confined to hypoxic tumor regions (Atkinson, Loadman, Sutton, Patterson, & Clench, 2007). Analysis of endogenous molecular changes can also be expanded to include identification of new biomarkers, and their characterization and distribution within the sample, to evaluate disease or toxicity in preclinical and clinical samples (Balluff et al., 2011; Elsner et al., 2012; Ye et al., 2014). Such MSI analysis can also monitor disease progression or regression in response to therapeutic intervention. Endogenous masses can be matched with metabolite databases to identify unknown markers within the samples. This typically requires high mass resolving mass analyzers or a hypothesis-driven experiment where MS/MS data from the tissue can be compared to the theorized molecular target. It is worth noting that many reported biomarkers are proteomic (Bauer et al., 2010; Cole et al., 2011; Meistermann et al., 2006; Reyzer et al., 2004) or for endogenous metabolites that related to energy metabolism (Bao et al., 2013; Miura, Fujimura, & Wariishi, 2012; Miura et al., 2010).

9. DRUG DELIVERY

To mitigate toxicity and to allow for optimal exposure profiles, modern therapeutics can be delivered by a number of methods. While oral administration is preferred, intravenous chemotherapy is a common route and other more advanced methods also exist, such as slow release depot implants. MSI has been used to assess the release of theophylline and propranolol hydrochloride from such implants (Kreye et al., 2012). The researchers used a cylindrical implant embedded in gelatin, frozen and sectioned as standard for tissues. The MSI analysis showed the change in the structure of the implant over a range of time points and release into external medium. A number of oncology researchers have attempted to incorporate tumor targeting by enhanced permeability retention effect (EPR) at target site. The aim is increased efficacy through higher localized exposure at the target site (e.g., tumor) and lower systemic exposure, thus increasing the therapeutic index and reducing the dose-limiting toxicities. In oncology-targeted approaches, the premise is that by using larger molecules, delivery vesicles, or particles, therapeutics accumulate more in tumors than in the surrounding normal tissues (Gabizon & Papahadjopoulos, 1988; Golan et al., 2015; Maeda, Fang, Inutsuka, & Kitamoto, 2003). The encapsulation of the drug in a delivery particle can also protect it from early degradation or metabolism. The relevance of the EPR phenomenon in increasing therapeutic index is still hotly contested, in part because the clinical translation of a specific EPR effect has yet to be unequivocally demonstrated and to date most anticancer nanoparticle medicines on the market take form of a small molecule drug within a liposome with diameters of 30–150 μm.

Some feel that the full impact of carrier-mediated drugs remains to be fully utilized, as exemplified in recent meta-analysis of clinical and preclinical studies comparing the anticancer efficacy of liposomal vs conventional nonliposomal doxorubicin (Petersen, Alzghari, Chee, Sankari, & La-Beck, 2016) and summarized in Table 1. The researchers concluded that the optimal dosing regimen for carrier-mediated agents has not been thoroughly investigated, and it is possible that this will differ from that of the conventional formulation. Moreover, the contribution of the EPR effect and the tumor microenvironment to clinical efficacy remains to be fully elucidated. These are areas where MSI could make a significant impact. As MSI is able to detect

Table 1 Anticancer Nanoparticles Approved for Clinical Use

Trade Name	Carrier	Drug Cargo	Size	Approved Indications
Doxil/Caelyx	PEGylated liposome	Doxorubicin	90 nm	AIDS-related Kaposi sarcoma, multiple myeloma, ovarian cancer
Myocet	Liposome	Doxorubicin	150 nm	Metastatic breast cancer
DaunoXome	Liposome	Daunorubicin	45 nm	AIDS-related Kaposi sarcoma
Marqibo	Liposome	Vincristine	100 nm	Acute lymphoblastic leukemia
Onivyde	PEGylated liposome	Irinotecan	110 nm	Pancreatic adenocarcinoma
DepoCyt	Liposome	Cytarabine	3–30 μm	Lymphomatous meningitis

Reproduced with permission from Petersen, G. H., Alzghari, S. K., Chee, W., Sankari, S. S., & La-Beck, N. M. (2016). Meta-analysis of clinical and preclinical studies comparing the anticancer efficacy of liposomal versus conventional non-liposomal doxorubicin. *Journal of Controlled Release, 232*, 255–264.

multiple molecular targets in every mass spectrum, researchers are able to determine the abundance of the API as well as constituents of the nanoparticle. This can offer insights into the total abundance of drug, released API, and possible carrier nanoparticle accumulation within tumors. If a labeled or tagged method was to be used, information would only be obtained on one component. For example, a radiolabeled drug could show the distribution with the tumor but could not differentiate if API was still within the nanoparticle or released. This is further complicated if there is localized metabolism and radiolabel remains on the metabolite. The inverse scenario is also true. A fluorescent nanoparticle could be formulated, and the distribution measured within tumor sections. However, no information on whether API was still encapsulated, released or if the nanoparticle had degraded and the label itself was accumulated at target site. MSI has the potential to assess such situations in a pixel-by-pixel and spectra-by-spectra fashion (Ashton et al., 2016).

Looking forward, many pharmaceutical companies have aspirations of combining small molecule treatments with a range of new modalities including macromolecules and gene therapy therapeutics. Recent reviews of advances on nanoparticles have highlighted that the use of delivery systems can circumvent the induction of immune response through the use of degradable and biocompatible polymeric particles. Again, delivery of such systems provides biocompatibility and prolonged systemic exposure, prevents payload degradation, and possibly benefits tumor targeting via the EPR effect. However, such delivery systems still have limitations to be overcome, including suitable stability and optimized releases (Mokhtarzadeh et al., 2016). Future oncology treatments may employ even more advanced drug delivery systems such as carbon-based nanomaterials. These include a range of structures such as carbon nanotubes, graphene, and carbon nanodots that have been proposed as drug delivery vehicles (Cai et al., 2005; Miyako et al., 2012). Directly assessing the biodistribution of such carbon carriers by MSI is complicated due to the larger molecular weights of the materials. However, researchers recently demonstrated the suborgan distribution of carbon nanomaterials using laser desorption/ionization MSI to map the intrinsic carbon cluster fingerprint signal of the nanomaterials (Chen et al., 2015).

10. TUMOR MICROENVIRONMENT

As an oncology compound progresses from in vitro to in vivo studies, researchers must use a range of preclinical models that are derived from human or rodent cell lines. The four predominant models are (1) cell line xenograft models, (2) patient-derived xenograft models, (3) syngeneic mouse models, and (4) genetically engineered mouse model. Each model system has advantages and challenges as summarized in Table 2 reproduced from the comprehensive review by Gould, Junttila, and de Sauvage (2015).

Table 2 Preclinical In Vivo Efficacy Models Used for Oncology Drug Discovery: Strengths and Caveats

	Principal Components	Benefits	Caveats
Cell line xenograft models	Established human tumor cell lines transplanted into immune-deficient host	(i) Numerous established and well-annotated cell lines (ii) Representation from various human tumor types (iii) Features of the tumor microenvironment, including vascular and stromal cells incorporated within the tumor (iv) Tumors are easily and precisely measured	(i) Immune deficient (ii) Subcutaneous location may not foster important tissue-specific stromal infiltrate (iii) Cross-species disconnect, stromal components are mouse, whereas tumor cells are human (iv) Limited or no genetic heterogeneity is present within the tumor
Patient-derived xenograft models	Human tumor explants grown in immune-deficient host	(i) Genetic diversity and heterogeneity within tumors (ii) Representation from various human tumor types (iii) Features of the tumor microenvironment, including vascular and stromal cells incorporated within the tumor (iv) Tumors are easily and precisely measured	(i) Immune deficient (ii) Subcutaneous location may not foster important tissue-specific stromal infiltrate (iii) Surgical implantation required (iv) Cross-species disconnect, stromal components are mouse, whereas tumor cells are human (v) Genetic and phenotypic drift with passage

Table 2 Preclinical In Vivo Efficacy Models Used for Oncology Drug Discovery: Strengths and Caveats—cont'd

	Principal Components	Benefits	Caveats
Syngeneic models	Established mouse tumor cell lines transplanted into immune-competent host	(i) Presence of an intact immune system (ii) Features of the tumor microenvironment, including vascular and stromal cells incorporated within the tumor (iii) All cell types within the tumor are of mouse origin (iv) Tumors easily and precisely measured	(i) Limited number of established cell lines, poorly annotated (ii) Strong immunogenicity of some lines promotes spontaneous regression (iii) Rapid growth rate of many lines limits use in longer-term studies
Genetically engineered mouse models	Genetic modification that permits spontaneous or induced tumor development	(i) Tumors develop in the tissue of origin (ii) Presence of an intact immune system (iii) All cell types within the tumor are of mouse origin (iv) Incorporates features of the tumor microenvironment, including vascular and stromal cells and immune components	(i) Limited genetic mosaicism and heterogeneity of tumors (ii) Technical hurdles for monitoring tumor response when on internal organs (iii) Low throughput and high investment

Reproduced with permission from Gould, S. E., Junttila, M. R., & de Sauvage, F. J. (2015). Translational value of mouse models in oncology drug development. *Nature Medicine, 21*(5), 431–439.

A common challenge for all preclinical models is the heterogeneity within the tumor. This complex microenvironment is composed of multiple cell types, with oxygenation ranging from normoxia to hypoxia, and with huge interplay between viable and necrotic regions. This provides a challenge for pharmacokinetic and biomarker studies using tissue homogenization and analysis. However, this challenge is increasingly being met by MSI with the suite of technologies allowing researchers to map the architecture of the microenvironment in preclinical models and patient samples using endogenous metabolites and lipids (Calligaris et al., 2014; Eberlin et al., 2012, 2014). However, pharmaceutical research assessment of the tumor microenvironment can go beyond simple comparison of endogenous and exogenous compounds in ex vivo tumor sections. Recent work has demonstrated an approach whereby in vivo imaging can be combined with ex vivo MSI analysis, using MRI techniques. Researchers described how following administration of a contrast reagent, tumors were excised and analyzed by DESI MS imaging. Researchers were able to map the distribution of the contrast reagent gadoteridol as well as endogenous lipids (Tata et al., 2015).

As the tumor microenvironment is thought to play a role in solid organ cancer development, it is also worth noting that microenvironment characterization is not restricted to preclinical models but has also been applied to patient samples. Such work has exemplified by groups investigating lipidomic differences between cancerous and healthy colorectal tissue (Mirnezami et al., 2014). Such data can feed back to drug discovery environment to generate better preclinical models as well as offer insights into the molecular events in tumor environment that can help researchers understand the lack of efficacy following transfer into the clinic.

11. ASSESSING HYPOXIA

Recent refinement of the lipid microenvironment mapping allows characterization of the regions of hypoxia. This has been achieved by relating changes in lipid profiles to IHC for hypoxia (Chughtai, Jiang, Greenwood, Glunde, & Heeren, 2013), thereby allowing co-registered MSI, IHC, and H&E data for multivariate image analysis. This allows statistical interrogation of the lipid abundances, which when validated can be used in subsequent MSI experiments for the same cell line to define tumor regions. The process can be refined by addition of specific IHC markers of tissue hypoxia, injected prior to tissue extraction, which can be directly detected by MSI. Such has been reported for pimonidazole, an exogenous hypoxia marker detected in tumor sections by MALDI MSI (Mascini et al., 2016).

12. BBB PENETRATION

As previously alluded to, there are scenarios in which there is a discrepancy between the measured circulating plasma levels of a compound and the abundance in a target tissue, and in such situations, MSI analysis can prove particularly beneficial. Such may be the case for tumors or when tissue-targeting delivery strategies are used. However, during drug discovery a common such situation occurs when trying to understand the exposure to the brain. There will be situations where researchers want to make sure that compounds do not penetrate the CNS, risking off-target deleterious toxicity, as well as situations when brain penetration is highly desirable for a compound in order to have required neuropharmacological efficacy or to reach the site of brain tumors. The BBB tight junctions present between the brain microvessel endothelial cells, in combination with efflux transporters such as P-glycoprotein (P-gp), multidrug resistance-associated proteins, or organic anion-transporting polypeptides, prevent endogenous and exogenous compounds from affecting neurological activity. Therefore, it is important to determine BBB permeability of drug candidates during drug discovery both in animal models and in human. A range of alternative analytical strategies can be used to predict the impact of the BBB on new therapeutics (Löscher & Potschka, 2005), but clearly using the spatially resolving MSI can be a rapid and cost-effective tactic. Analysis can be further complicated if researchers are trying to make assessment of the exposure of compound not just to the brain but to brain tumors, with associated damage and leakage to BBB (Agarwal, Manchanda, Vogelbaum, Ohlfest, & Elmquist, 2013). Therefore, imaging methods that move analysis away from brain homogenization and calculation for peripheral blood contamination are an improvement. The simplest approach to detect if the compound has crossed the BBB is to use a low spatial resolution technology like liquid extraction surface analysis (LESA). This allows researchers to profile or probe tissue sections rather than higher resolution imaging. With spatial resolution of approximately 1 mm, such methods are suitable for analysis of rat brain sections but may not be suitable for smaller mouse brains or for analysis of brain tumors. However, the large sampling area means that such an approach can be very sensitive. By measuring for Heme β (m/z 616) as a marker for blood, any compound detected can be shown to either correlate to blood distribution or show BBB penetration (Swales et al., 2015). The validation of heme as a biomarker of vasculature in the brain has been performed at higher spatial resolution using MALDI MSI analysis and relating the MSI detected distribution of the blood to the fluorescence image distribution of fluorescein as reproduced in Fig. 1 (Liu, Ide, et al., 2013). The authors went on to show detection and distribution of RAF265, a small molecule inhibitor of the RAF serine/threonine protein kinases by MALDI MSI and the impact of aberrant tumor vasculature on compound distribution.

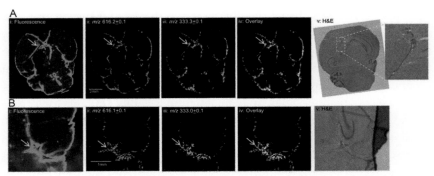

Fig. 1 Comparison of heme and fluorescein images from MALDI TOF MSI at 50 μm resolution with fluorescence image in the same mouse brain section (10 μm thickness) with preinjected fluorescein. (i) Fluorescence image of blood vessels from fluorescein (Ex 490 nm, Em 520 nm); (ii) heme image (*red, m/z* 616.2±0.1) from MALDI MSI; (iii) fluorescein image (*blue, m/z* 333.3±0.1) from MALDI MSI; (iv) overlay of heme (*red*) and fluorescein (*blue*) from MALDI MSI; (v) H&E staining of a sister section from (A) with the expanded view showing the lateral ventricle. The *yellow arrow* indicates the lateral ventricle delineated by fluorescein with the absence of heme. The *red arrow* shows blood in the H&E staining image. (B) Selected view of heme and fluorescein images from MALDI MSI under 25 μm resolution and fluorescence image in the same mouse brain section. (i) Fluorescence image of blood vessels from fluorescein (Ex 490 nm, Em 520 nm); (ii) heme image (*red, m/z* 616.1±0.1) from MALDI MSI; (iii) fluorescein image (*blue, m/z* 333.0±0.1) from MALDI MSI; (iv) overlay of heme (*red*) and fluorescein (*blue*) from MALDI MSI; (v) H&E staining of a sister section. The *arrow* shows the region of blood. *Reproduced with permission from Liu, X., Ide, J. L., Norton, I., Marchionni, M. A., Ebling, M. C., Wang, L. Y., . . . Agar, N. Y. R. (2013). Molecular imaging of drug transit through the blood-brain barrier with MALDI mass spectrometry imaging.* Scientific Reports, 3, 2859.

MSI can also be used to study effects of efflux transporters for poorly brain penetrative compounds. The recent elegant study by Genentech highlighted the utility of MSI in a study investigating the role of P-gp on the brain penetration and brain PD activity of the MEK inhibitor cobimetinib (Choo et al., 2014). The researchers stated that the objective of the study was to determine—in vitro and in vivo—whether cobimetinib is a substrate of P-gp and/or breast cancer resistance protein (Bcrp1). The aim was also to assess the implications of efflux on cobimetinib PK, brain penetration, and target modulation. The data showed that for the preclinical study, the ability of cobimetinib to elicit pathway modulation in the brain is prevented by P-gp efflux. The study exemplified how MSI data can be effectively integrated into a wider PK/PD study. A combination of the approaches outlined in the previous two studies was recently undertaken for assessing alectinib distribution in murine brains. The quantitative MSI analysis, validated by LC-MS data, again showed correlation of compound to heme, but the researchers also analyzed the effect on PK profiles of the multidrug resistance protein-1 (Mdr1) using Mdr1a/b knockout mice (Aikawa et al., 2016).

13. BEYOND SMALL MOLECULES

Traditionally, therapeutics have been small molecules that fall within the Lipinski's rule of five (i.e., a molecule with a molecular mass less than 500 Da, no more than 5 hydrogen bond donors, no more than 10 hydrogen bond acceptors, and an octanol–water partition coefficient log P not greater than 5). Such molecules are often detected by the various MSI technologies as we have previously discussed. However, there are emerging new therapeutic modalities that are much more complex and have moved way beyond traditional low molecular weight compounds. Nevertheless, MSI can still have a role to play in their development. Macromolecule therapeutics range in size and include oligonucleotides, proteins, antibodies, and antibody–drug conjugates (ADCs). Most of these molecules are currently beyond the size of analytes detectable by MSI technologies. While MALDI time of flight mass spectrometers are able to detect antibodies >100 kDa, they are not currently able to detect them in tissue sections. The simplest role MSI currently has to play in large molecule therapeutics is to measure and monitor biomarkers for efficacy or safety in the same way as employed for traditional small molecule work. Attempts have been made to develop strategies for larger molecules. Proteins, for example, can be detected and mapped indirectly by performing on tissue tryptic digestion and subsequent detection of the resulting lower mass peptides. This unfortunately generates multiple peptide fragments for each protein, so the most abundant proteins will swamp the detectable mass range and risk the detection of any lower abundance target. Another complication is the delocalization of targets, as samples need to be stored in humid atmosphere for enzymatic reaction to take place. However, with carefully controlled protocols, researchers have reported high-quality data at 50 µm spatial resolution (Schober, Guenther, Spengler, & Römpp, 2012). To increase the sensitivity for lower abundance targets a labeling strategy is required. While this moves us away from the label-free analysis approach that is integral to the success of MSI, multiplexed label detection is possible, where traditional IHC or other probe methods are restricted to single-target detection. A recent development has been termed spectroimmunohistochemistry, which combines the use of specific antibodies against targets and mass spectrometric imaging detection (Longuespée et al., 2013). While further optimization would be required to assess the robustness of this technique and to allow quantitative measurements to be obtained, such a methodology has the potential to be applied to both clinical and pharmacological applications. The early data looks promising and is a novel solution to an obstacle for determining novel large molecule therapeutics. The same group developed a technique called Tag-Mass for mapping MSI-specific targets. The approach allows quantification and identification of biological macromolecules such as proteins and peptides by using photocleavable tag peptides of known sequences, attached to antibody probes on specific antibodies for a given biomarker (Stauber, Ayed, Wisztorski, Salzet, & Fournier, 2010). Once photocleaved, the peptide mass tags are detected by mass spectrometry. The researchers reported the distribution of 180-kDa carboxypeptidase D membrane protein at 50 μm resolution (Lemaire et al., 2007). The cleavage is induced by the MALDI laser, so this technique could have wide uptake as the most commonly employment mass spectrometer for MSI researchers has MALDI ionization. This method was reported to be highly efficient and reproducible, but the design of the linked antibodies has been considered expensive. However, for the specificity reported, such an approach could prove effective in large molecule drug development.

Finally, we should consider the first reported—and what may well be the most effective—tag-based MSI method for tissue-wide distribution of large molecule targets. The approach uses laser ablation-inductively coupled plasma mass spectrometry (ICP-MS). Researchers reported imaging of human gastric mucosa for a target cancer biomarker at 10 μm resolution (Fig. 2) (Seuma et al., 2008). This was achieved by coupling antibodies to metal tags. The ICP-MS has sensitivity for metals at subparts per billion when analyzing tissue samples. Therefore, this approach offers high spatial resolution with a large dynamic range for target quantitation. The use of this approach with a range of lanthanides offers real multiplexing and multitarget detection.

All the tag-based MSI methods are equally applicable to analysis of either large molecule therapeutics or endogenous large molecules that may be target receptors or biomarkers for efficacy or safety endpoints. There is one key point that should be considered before undertaking complex generation of the MS-labeled probe. Can adequate information be achieved by combining traditional IHC for target distribution with tissue homogenization for quantitation? This can allow higher spatial resolution IHC to be combined with

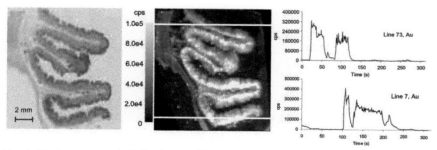

Fig. 2 Photomicrograph (*left*) of normal human gastric mucosa stained for MUC-1 antigen using the immunogold technique, and corresponding Au distribution by LA-ICP-MS (81 line rasters) (*center*). *Right*: Au single-line rasters. *Reproduced with permission from Seuma, J., Bunch, J., Cox, A., McLeod, C., Bell, J., & Murray, C. (2008). Combination of immunohistochemistry and laser ablation ICP mass spectrometry for imaging of cancer biomarkers. Proteomics, 8(18), 3775–3784.*

traditional proteomics or Western blots. However, MSI-labeled analysis offers the ability to multiplex significantly the number of targets mapped and quantified, simultaneously, while still allowing other nonlabeled molecular species to be detected in the same sample simultaneously. To date, such methods have not been widely used in pharmaceutical research, but it can be anticipated that with the increasing research into new larger molecule therapeutics, particularly within an oncology setting where tissue homogenization is a poor substitute for the complexity of a tumor microenvironment, the use of small molecule MSI analysis with large molecule labeling MSI will be increasingly performed.

Therapeutic ADCs are large molecule monoclonal antibodies coupled to potently cytotoxic drugs by chemical linkers with labile bonds. ADCs offer the possibility of tumor-targeted treatments. Targeting with the antibody allows lower dosing levels, reducing the risk of off-target effects of the cytotoxic payload. However, the bioanalysis, biodistribution, and monitoring of cytotoxic moiety required to mitigate toxicity remain challenging. The large molecule antibody and low copy number cytotoxic moiety push MS analysis to the limits. However, analysis of small molecule ADC catabolites in rat liver and tumor tissue by LESA coupled to microcapillary liquid chromatography/tandem mass spectrometry (LESA-µLC/MS/MS) has been reported. This alternative to tissue imaging using mass spectrometry is tissue profiling. Profiling MS can be described as direct and discrete position analysis used to compare different tissue regions. Imaging is like profiling but repeated at reparative distances to allow generation of molecular images. Therefore, the profiling of tissues, rather than high resolution imaging, offers researchers at least some direct tissue measurements of hard to detect targets (Lanshoeft et al., 2016). The analysis described generated both quantitative and qualitative information for the spatial distribution of ADCs and their related catabolites in tissue sections. The manuscript also highlighted how the LESA-µLC/MS/MS and QWBA were complementary. For complex bioanalysis, it is through a combination of traditional bioanalysis of plasma and tissue homogenates, related to traditional histopathological imaging, label-based assays, and a variety of tissue profiling and imaging by mass spectrometry that will allow researchers to discover and develop new safe and effective therapies. Finally, it is worth noting that LESA sampling but without LC separation has been used for direct tissue imaging but at low resolution (1 mm) for low molecular weight targets that are hard to detect using the more common higher resolution MALDI and DESI MSI methods (Swales et al., 2015, 2014).

14. CLINICAL TRANSLATION

What we have considered so far is how MSI can play an effective role in the discovery and development of new drugs. This work is predominantly performed using preclinical assays and models (in vitro, in vivo, and ex vivo samples) but to transform a compound from the lab bench to a medicine requires extensive clinical trials. This is where any compound attrition can result in loss of hundreds of millions of research dollars. Therefore, translation of preclinical to clinical data is vital and MSI has an important role to play in contributing to oncology translational science. One crucial area is in the analysis of both diseased tissue biopsies and tissue following surgical segmental resection. Application of MSI analysis to the latter is proving useful in understanding disease progression and metabolic phenotyping (Guenther et al., 2015) as well as providing molecular information for intraoperative MS analysis by techniques using rapid evaporative ionization mass spectrometry (REIMS and iKnife) (Balog et al., 2013). It is worth remember that any clinical analysis by MSI is best related to gold standard match IHC data for validation, as exemplified by the analysis for the discrimination of lymph node metastases using DESI MSI imaging (Abbassi-Ghadi et al., 2014). Clinical pathology using MSI endpoints is being explored, and this includes traditional tissue samples that have been formalin fixed and paraffin embedded. However, this work is beyond the remit of this discussion so will not be considered further now, but is well reviewed by Aichler and Walch (2015).

A natural extension from the tissue section MSI analysis of material collected during surgery is the collection and analysis of tissue biopsies. Such an approach needs higher resolution analysis for the smaller needle biopsies. However, with spatial resolutions now routinely sub-50 μm, no major technical obstacles remain. Work in this area is led by groups like that of Natalie Agar using MSI as for surgical decision making (Calligaris et al., 2013). Applying such methods to understanding drug penetration and localization within solid tumors in clinical trials will impact future drug development. Such ongoing work has been reported by researchers from National Cancer Center, Japan, where tumor biopsies analyzed for patients with solid tumors. In their analysis, biopsies were taken following treatment with olaparib (a PARP-1 inhibitor). They have reported that the distribution of drug was detected in the tumor region by MALDI MSI and the signal detected in areas of necrosis was higher than that observed in living cell areas (Shimoi et al., 2014). To date, this exciting example of drug distribution analysis by patient needle biopsy MSI analysis has yet to be fully published. A recent example of the effectiveness of needle biopsy analysis by MSI describes element bioimaging of liver needle biopsy specimens from patients with Wilson's disease by LA-ICP-MS (the ionization and mass analyzer method discussed earlier for detection of antibodies with MS detectable metal tags) (Hachmöller et al., 2016). In this study, there was no metal-tagged probe; rather patients had a rare genetic dysfunction of the copper metabolism, which causes the accumulation of the metal in different organs (including the liver and the CNS). This means that the LA-ICP-MS was able to detect the copper in liver needle biopsies at 10 μm spatial resolution as highlighted in Fig. 3. Due to the sensitive of the ionization and mass analyzer, this needle biopsy methodology would be highly effective at the analysis of metal-containing drugs such as cisplatin.

15. EMERGING APPLICATIONS: SPHEROIDS

With the ability to perform MSI at greater spatial resolutions and with increased sensitivities, the technologies are now being applied to the analysis of microphysiological systems. Microphysiological systems are 3D cell cultures increasingly used in early drug discovery for efficacy and safety screening. Early utility was demonstrated by the Hummon group who first demonstrated protein and peptide distributions in 3D culture systems (Li & Hummon, 2011) before the group reported the utility of the application to drug discovery (Liu & Hummon, 2015). Their proof-of-concept

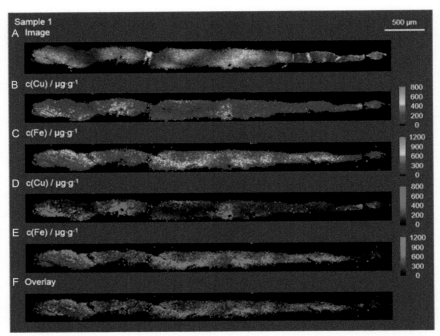

Fig. 3 Autofluorescence microscopic image of the Wilson's disease liver sample 1 (A) investigated by LA-ICP-MS, quantitative distribution maps of copper (B, D) and iron (C, E), and overlay of the copper and iron distribution (F). *Reproduced from Hachmöller, O., Aichler, M., Schwamborn, K., Lutz, L., Werner, M., Sperling, M., ... Karst, U. (2016). Element bioimaging of liver needle biopsy specimens from patients with Wilson's disease by laser ablation-inductively coupled plasma-mass spectrometry. Journal of Trace Elements in Medicine and Biology, 35, 97–102.*

experiment, using in HCT 116 colon carcinoma multicellular spheroids, was to assess the distribution of the anticancer drug, irinotecan (Liu, Weaver, & Hummon, 2013). They were able to demonstrate time-dependent penetration of drug, as well as simultaneously the abundance and distribution of three metabolites as seen in Fig. 4. For smaller dimension samples, analysts are able to combine and relate the MSI analysis with other spectroscopic technologies such as fluorescence or Raman microscopy, allowing 3D imaging studies to be performed (Ahlf, Masyuko, Hummon, & Bohn, 2014).

There has been a recent explosion in use and analysis of microphysiological system and MSI analysis endpoints, a flavor of which includes the use of MSI technologies like laser ablation ICP-MS imaging of multicellular tumor spheroids as a tool in the preclinical development of metal-based anticancer drugs (Theiner et al., 2016), or sample processing

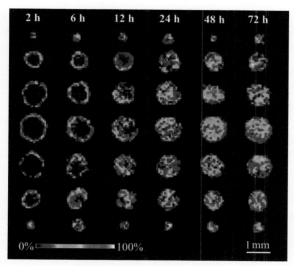

Fig. 4 Time-dependent penetration of irinotecan (*m/z* 587) in HCT 116 spheroids analyzed by MALDI-IMS. Spheroids were treated with 20.6 μM irinotecan for 2, 6, 12, 24, 48, and 72 h (from *left* to *right*). For every treatment duration, color gradient intensity maps were generated from seven consecutive 12 μm slices from a single spheroid in 120 μm vertical intervals. Reproduced from Liu, X., Weaver, E. M., & Hummon, A. B. (2013). Evaluation of therapeutics in three-dimensional cell culture systems by MALDI imaging mass spectrometry. Analytical Chemistry, 85(13), 6295–6302.

improvements using ultrathin matrix coating for tumor microenvironment model to study epithelial-to-mesenchymal transitions (Wang et al., 2015). To aid such analysis, there has also been development of new sample generation methods, such as the generation and functional assessment of 3D multicellular spheroids in droplet-based microfluidics platform (Sabhachandani et al., 2016) and the development of mixed hydrogel bead-based tumor spheroid formation for anticancer drug testing (Wang & Wang, 2014). Combined, such reported methods for spheroid generation and MSI analysis demonstrate that MSI and MPS can be efficiently scaled in production and analysis for effective screening of new therapeutics for efficacy and safety studies.

16. INCREASED SPATIAL RESOLUTION

One crucial aspect not yet considered in detail is that most MSI technologies do not allow intracellular analysis. Intracellular measurements are possible using secondary ion mass spectrometry (SIMS) and nanoSIMS but are not easily performed. Such methods reveal the relative abundance of drug or endogenous metabolites at a subcellular level (Dollery, 2013; Passarelli et al., 2015). To date, the various SIMS instruments have not widely been used in the bioanalysis of tissues or cells by pharma though this is likely to increase as awareness of the technologies is further disseminated and next-generation instruments arrive. However, they have found use in medicine formulation and production where higher abundance of target compensates for the smaller site of sample ionization (Bich, Touboul, & Brunelle, 2013; Pajander, Haugshøj, Bjørneboe, Wahlberg, & Rantanen, 2013; Qu et al., 2015).

17. METROLOGY FOR MS IMAGING

Oncology bioscience and drug discovery are a highly interdisciplinary field requiring measurements from many techniques. As described through this chapter, MSI is emerging as a powerful suite of methods for the fundamental study of cancer biology and efficacy of therapeutics. While we have discussed the advantages and opportunities these relatively new and highly complex methods afford, we have not considered the metrology and regulatory factors of applying new technologies in R&D. One limitation is the confidence to apply data generated by new techniques in a regulatory environment or to clinical samples. In the majority of methods discussed in this chapter, there is a strong interdependence between sample preparation, instrument-operating conditions, and resulting data and suitability of data mining approaches. The correctness, reproducibility, and repeatability of the data produced are of great consequence for effective use, interpretation, and decision making. There is an increasing need for metrology and standards to support the continued use and development of MSI in these research areas. Standards ensure that the measurements are fit for purpose between different systems, users, laboratories, and instrument models. The goal of many researchers is to elevate MSI to an analytical technique validated for use in a clinical setting for measuring drug and biomarkers and as a diagnostic. This aim undoubtedly requires further work. For example, in MALDI MSI, the wide variety of matrices, sample preparation methods, instrumentation, analytical parameters, and mass analyzers (Caprioli, Farmer, & Gile, 1997; Gusev, Vasseur, Proctor, Sharkey, & Hercules, 1995) means that there have been few attempts to standardize methodologies, create reference samples, or assess repeatability. The comparison of tissue preparation protocols by single groups has been carried out periodically (Gemperline, Rawson, & Li, 2014; Martin-Lorenzo et al., 2014; Seeley, Oppenheimer, Mi, Chaurand, & Caprioli, 2008) and are important in enabling methodological standardization. The validity of these comparisons must then be proven across different instrumentation and institutions. Multicenter studies are a vital step in this process and are beginning to be carried out by groups such as those of McDonnell, Walch, and Schmitt (Dekker et al., 2014). Key to a better understanding of a technique is a well-characterized reference sample, a particularly challenging requirement in many MSI application areas. Studies pursuing this aim include the recent development of a lateral resolution standard. Development and sharing of test and reference standards and materials will importantly allow the MSI community to benchmark and report their performance (Passarelli et al., 2014). Benchmarking and comparison to other analytical techniques also provide an important level of confirmation for MALDI MSI, examples of which include comparison with liquid chromatography tandem mass spectrometry (LC-MS/MS) (Hankin & Murphy, 2010; Takai, Tanaka, Inazawa, & Saji, 2012), desorption electrospray ionization (DESI) MSI (Eberlin et al., 2011), infrared matrix-assisted laser desorption electrospray ionization (IR-MALDESI) MSI (Barry, Groseclose, Robichaud, Castellino, & Muddiman, 2015), LESA MS (Swales et al., 2014), SIMS imaging, Raman imaging (Bocklitz et al., 2013), magnetic resonance imaging (Sinha et al., 2008), autoradiography (Hsieh et al., 2006; Solon, Schweitzer, Stoeckli, & Prideaux, 2010; Takai et al., 2012), histochemistry (Oppenheimer, Mi, Sanders, & Caprioli, 2010), and immunohistochemistry (Acquadro et al., 2014; Klein et al., 2014). The automated registration of MSI data sets to histology and on-line histological "atlases" has also recently been explored (Abdelmoula, Carreira, et al., 2014; Abdelmoula, Škrášková, et al., 2014; Verbeeck et al., 2014).

Within a field with such a variety of variables and sample types, the effective communication of information relating to the experimental process is particularly necessary. A recent publication entitled "Discussion point: reporting guidelines for MSI" (McDonnell et al., 2015) seeks to set out a framework to ensure the appropriate reporting of the "metadata" associated with MSI workflows. Furthermore, the common data format imzML (Race, Styles, & Bunch, 2012; Schramm et al., 2012) for MSI was recently introduced to better enable sharing and comparison of imaging data from multiple platforms within multiple software packages. New software tools have recently been developed, which allows researchers to evaluate and compare preprocessing and postprocessing (e.g., multivariate analysis) methods and to process large data from multiple mass spectrometer types. The need for measurements to be repeatable and reliable led to a VAMAS (Versailles Project on Advanced Materials and Standards) interlaboratory study. Participants from 20 different laboratories were provided with a DESI protocol and two samples: a thin film of Rhodamine B and a piece of double-sided tape (both deposited on glass). Repeatability of absolute intensity of Rhodamine B was assessed along with the consistency of size and shape of erosion craters. While some DESI experimental setups gave consistent and repeatable results (repeatability <20%), others were considerably worse and crater sizes varied due to inconsistency of spray and stage movement (Gurdak et al., 2014). Recent reports reconsider spray parameters for tissue imaging: Abbassi-Ghadi et al. reported that through the use of optimal DESI spray and geometry parameters, highly reproducible spectral profiles from a complex, biologically relevant sample could be obtained (Abbassi-Ghadi et al., 2015).

As MSI is used more widely in drug discovery, further research in metrology will be needed to underpin the technique development and to support academic and industrial research. Establishing this reliable framework with standardized approaches will be an important step in using the techniques more frequently in clinical research and in support of regulatory submissions in drug development.

18. CONCLUSION

For maximal effective impact in aiding oncology drug discovery, MSI will need to be used in combination with a range of bioanalysis and molecular technologies that allow researchers to understand the structure within the tumor microenvironment, tumor metabolism, efficacy of therapeutics, and measurements of the phenotypic features of tumors, all of which feeding into both diagnostic and prognostic data. However, MSI can be embedded within all areas of drug discovery. Some researchers are using MSI techniques to aid early lead identification, though most applications still apply MSI to in vivo efficacy and safety studies. However, increasingly MSI applications are being explored in the clinic and it can be assumed that in the near future MSI may well have a role in hospital pathology departments. The challenges of developing new, effective, and safe oncology medicines remain considerable. However, MSI encompasses a formidable array of technologies that are bringing insight and understanding to the molecular maelstrom in oncology, biology, drug discovery, and treatment.

REFERENCES

Abbassi-Ghadi, N., Jones, E. A., Veselkov, K. A., Huang, J., Kumar, S., Strittmatter, N., ... Takats, Z. (2015). Repeatability and reproducibility of desorption electrospray ionization-mass spectrometry (DESI-MS) for the imaging analysis of human cancer tissue: A gateway for clinical applications. *Analytical Methods*, 7(1), 71–80.

Abbassi-Ghadi, N., Veselkov, K., Kumar, S., Huang, J., Jones, E., Strittmatter, N., ... Hanna, G. B. (2014). Discrimination of lymph node metastases using desorption electrospray ionisation-mass spectrometry imaging. *Chemical Communications*, 50(28), 3661–3664.

Abdelmoula, W. M., Carreira, R. J., Shyti, R., Balluff, B., van Zeijl, R. J. M., Tolner, E. A., ... Dijkstra, J. (2014). Automatic registration of mass spectrometry imaging data sets to the allen brain atlas. *Analytical Chemistry*, 86(8), 3947–3954.

Abdelmoula, W. M., Škrášková, K., Balluff, B., Carreira, R. J., Tolner, E. A., Lelieveldt, B. P. F., ... Dijkstra, J. (2014). Automatic generic registration of mass spectrometry imaging data to histology using nonlinear stochastic embedding. *Analytical Chemistry*, 86(18), 9204–9211.

Acquadro, E., Caron, I., Tortarolo, M., Bucci, E. M., Bendotti, C., & Corpillo, D. (2014). Human SOD1-G93A specific distribution evidenced in murine brain of a transgenic model for amyotrophic lateral sclerosis by MALDI imaging mass spectrometry. *Journal of Proteome Research*, 13(4), 1800–1809.

Agarwal, S., Manchanda, P., Vogelbaum, M. A., Ohlfest, J. R., & Elmquist, W. F. (2013). Function of the blood-brain barrier and restriction of drug delivery to invasive glioma cells: Findings in an orthotopic rat xenograft model of glioma. *Drug Metabolism and Disposition*, 41(1), 33–39.

Ahlf, D. R., Masyuko, R. N., Hummon, A. B., & Bohn, P. W. (2014). Correlated mass spectrometry imaging and confocal Raman microscopy for studies of three-dimensional cell culture sections. *Analyst*, 139(18), 4578–4585.

Aichler, M., & Walch, A. (2015). MALDI imaging mass spectrometry: Current frontiers and perspectives in pathology research and practice. *Laboratory Investigation*, 95(4), 422–431.

Aikawa, H., Hayashi, M., Ryu, S., Yamashita, M., Ohtsuka, N., Nishide, M., ... Hamada, A. (2016). Visualizing spatial distribution of alectinib in murine brain using quantitative mass spectrometry imaging. *Scientific Reports*, 6, 23749.

Amstalden van Hove, E. R., Smith, D. F., & Heeren, R. M. A. (2010). A concise review of mass spectrometry imaging. *Journal of Chromatography. A*, 1217(25), 3946–3954.

Ashton, S., Song, Y. H., Nolan, J., Cadogan, E., Murray, J., Odedra, R., ... Barry, S. T. (2016). Aurora kinase inhibitor nanoparticles target tumors with favorable therapeutic index in vivo. *Science Translational Medicine*, 8(325), 325ra317.

Atkinson, S. J., Loadman, P. M., Sutton, C., Patterson, L. H., & Clench, M. R. (2007). Examination of the distribution of the bioreductive drug AQ4N and its active metabolite AQ4 in solid tumours by imaging matrix-assisted laser desorption/ionisation mass spectrometry. *Rapid Communications in Mass Spectrometry*, 21(7), 1271–1276.

Balluff, B., Rauser, S., Meding, S., Elsner, M., Schone, C., Feuchtinger, A., ... Walch, A. (2011). MALDI imaging identifies prognostic seven-protein signature of novel tissue markers in intestinal-type gastric cancer. *American Journal of Pathology*, 179(6), 2720–2729.

Balog, J., Sasi-Szabó, L., Kinross, J., Lewis, M. R., Muirhead, L. J., Veselkov, K., Takáts, Z. ... (2013). Intraoperative tissue identification using rapid evaporative ionization mass spectrometry. *Science Translational Medicine*, 5(194), 194ra193.

Bao, Y., Mukai, K., Hishiki, T., Kubo, A., Ohmura, M., Sugiura, Y., ... Minamishima, Y. A. (2013). Energy management by enhanced glycolysis in G1-phase in human colon cancer cells in vitro and in vivo. *Molecular Cancer Research*, 11(9), 973–985.

Barry, J. A., Groseclose, M. R., Robichaud, G., Castellino, S., & Muddiman, D. C. (2015). Assessing drug and metabolite detection in liver tissue by UV-MALDI and IR-MALDESI mass spectrometry imaging coupled to FT-ICR MS. *International Journal of Mass Spectrometry, 377*, 155–448.

Bauer, J. A., Chakravarthy, A. B., Rosenbluth, J. M., Mi, D., Seeley, E. H., De Matos Granja-Ingram, N., ... Pietenpol, J. A. (2010). Identification of markers of taxane sensitivity using proteomic and genomic analyses of breast tumors from patients receiving neoadjuvant paclitaxel and radiation. *Clinical Cancer Research, 16*(2), 681–690.

Berwick, D. C., Hers, I., Heesom, K. J., Moule, S. K., & Tavare, J. M. (2002). The identification of ATP-citrate lyase as a protein kinase B (Akt) substrate in primary adipocytes. *The Journal of Biological Chemistry, 277*(37), 33895–33900.

Bich, C., Touboul, D., & Brunelle, A. (2013). Cluster TOF-SIMS imaging as a tool for micrometric histology of lipids in tissue. *Mass Spectrometry Reviews, 33*(6), 442–451.

Bocklitz, T. W., Crecelius, A. C., Matthäus, C., Tarcea, N., von Eggeling, F., Schmitt, M., ... Popp, J. (2013). Deeper understanding of biological tissue: Quantitative correlation of MALDI-TOF and Raman imaging. *Analytical Chemistry, 85*(22), 10829–10834.

Cai, D., Mataraza, J. M., Qin, Z.-H., Huang, Z., Huang, J., Chiles, T. C., ... Ren, Z. (2005). Highly efficient molecular delivery into mammalian cells using carbon nanotube spearing. *Nature Methods, 2*(6), 449–454.

Calligaris, D., Caragacianu, D., Liu, X., Norton, I., Thompson, C. J., Richardson, A. L., ... Agar, N. Y. R. (2014). Application of desorption electrospray ionization mass spectrometry imaging in breast cancer margin analysis. *Proceedings of the National Academy of Sciences of the United States of America, 111*(42), 15184–15189.

Calligaris, D., Norton, I., Feldman, D. R., Ide, J. L., Dunn, I. F., Eberlin, L. S., ... Agar, N. Y. (2013). Mass Spectrometry Imaging as a tool for surgical decision-making. *Journal of Mass Spectrometry: JMS, 48*(11), 1178–1187.

Caprioli, R. M., Farmer, T. B., & Gile, J. (1997). Molecular imaging of biological samples: Localization of peptides and proteins using MALDI-TOF MS. *Analytical Chemistry, 69*(23), 4751–4760.

Castellino, S., Groseclose, M. R., Sigafoos, J., Wagner, D., de Serres, M., Polli, J. W., ... Hamilton, B. (2012). Central nervous system disposition and metabolism of fosdevirine (GSK2248761), a non-nucleoside reverse transcriptase inhibitor: An LC-MS and matrix-assisted laser desorption/ionization imaging MS investigation into central nervous system toxicity. *Chemical Research in Toxicology, 26*(2), 241–251.

Chen, S., Xiong, C., Liu, H., Wan, Q., Hou, J., He, Q., ... Nie, Z. (2015). Mass spectrometry imaging reveals the sub-organ distribution of carbon nanomaterials. *Nature Nanotechnology, 10*(2), 176–182.

Choo, E. F., Ly, J., Chan, J., Shahidi-Latham, S. K., Messick, K., Plise, E., ... Yang, L. (2014). Role of P-glycoprotein on the brain penetration and brain pharmacodynamic activity of the MEK inhibitor cobimetinib. *Molecular Pharmaceutics, 11*(11), 4199–4207.

Chughtai, K., Jiang, L., Greenwood, T. R., Glunde, K., & Heeren, R. M. A. (2013). Mass spectrometry images acylcarnitines, phosphatidylcholines, and sphingomyelin in MDA-MB-231 breast tumor models. *Journal of Lipid Research, 54*(2), 333–344.

Chumbley, C. W., Reyzer, M. L., Allen, J. L., Marriner, G. A., Via, L. E., Barry, C. E., & Caprioli, R. M. (2016). Absolute quantitative MALDI imaging mass spectrometry: A case of rifampicin in liver tissues. *Analytical Chemistry, 88*(4), 2392–2398.

Cobice, D. F., Goodwin, R. J. A., Andren, P. E., Nilsson, A., Mackay, C. L., & Andrew, R. (2015). Future technology insight: Mass spectrometry imaging as a tool in drug research and development. *British Journal of Pharmacology, 172*(13), 3266–3283.

Cobice, D. F., Mackay, C. L., Goodwin, R. J. A., McBride, A., Langridge-Smith, P. R., Webster, S. P., ... Andrew, R. (2013). Mass spectrometry imaging for dissecting steroid intracrinology within target tissues. *Analytical Chemistry, 85*(23), 11576–11584.

Cole, L. M., Djidja, M. C., Bluff, J., Claude, E., Carolan, V. A., Paley, M., ... Clench, M. R. (2011). Investigation of protein induction in tumour vascular targeted strategies by MALDI MSI. *Methods, 54*(4), 442–453.

Dekker, T. J. A., Balluff, B. D., Jones, E. A., Schöne, C. D., Schmitt, M., Aubele, M., ... McDonnell, L. A. (2014). Multicenter matrix-assisted laser desorption/ionization mass spectrometry imaging (MALDI MSI) identifies proteomic differences in breast-cancer-associated stroma. *Journal of Proteome Research, 13*(11), 4730–4738.

Deprez, J., Vertommen, D., Alessi, D. R., Hue, L., & Rider, M. H. (1997). Phosphorylation and activation of heart 6-phosphofructo-2-kinase by protein kinase B and other protein kinases of the insulin signaling cascades. *The Journal of Biological Chemistry, 272*(28), 17269–17275.

Dollery, C. T. (2013). Intracellular drug concentrations. *Clinical Pharmacology & Therapeutics, 93*(3), 263–266.

Drexler, D. M., Garrett, T. J., Cantone, J. L., Diters, R. W., Mitroka, J. G., Prieto Conaway, M. C., ... Sanders, M. (2007). Utility of imaging mass spectrometry (IMS) by matrix-assisted laser desorption ionization (MALDI) on an ion trap mass spectrometer in the analysis of drugs and metabolites in biological tissues. *Journal of Pharmacological and Toxicological Methods, 55*(3), 279–288.

Drexler, D. M., Tannehill-Gregg, S. H., Wang, L., & Brock, B. J. (2011). Utility of quantitative whole-body autoradiography (QWBA) and imaging mass spectrometry (IMS) by matrix-assisted laser desorption/ionization (MALDI) in the assessment of ocular distribution of drugs. *Journal of Pharmacological and Toxicological Methods, 63*(2), 205–208.

Eberlin, L. S., Liu, X., Ferreira, C. R., Santagata, S., Agar, N. Y. R., & Cooks, R. G. (2011). Desorption electrospray ionization then MALDI mass spectrometry imaging of lipid and protein distributions in single tissue sections. *Analytical Chemistry, 83*(22), 8366–8371.

Eberlin, L. S., Norton, I., Dill, A. L., Golby, A. J., Ligon, K. L., Santagata, S., ... Agar, N. Y. R. (2012). Classifying human brain tumors by lipid imaging with mass spectrometry. *Cancer Research, 72*(3), 645–654.

Eberlin, L. S., Tibshirani, R. J., Zhang, J., Longacre, T. A., Berry, G. J., Bingham, D. B., ... Poultsides, G. A. (2014). Molecular assessment of surgical-resection margins of gastric cancer by mass-spectrometric imaging. *Proceedings of the National Academy of Sciences of the United States of America, 111*(7), 2436–2441.

Elsner, M., Rauser, S., Maier, S., Schone, C., Balluff, B., Meding, S., ... Walch, A. (2012). MALDI imaging mass spectrometry reveals COX7A2, TAGLN2 and S100-A10 as novel prognostic markers in Barrett's adenocarcinoma. *Journal of Proteomics, 75*(15), 4693–4704.

Flavin, R., Peluso, S., Nguyen, P. L., & Loda, M. (2010). Fatty acid synthase as a potential therapeutic target in cancer. *Future Oncology (London, England), 6*(4), 551–562.

Fujimura, Y., & Miura, D. (2014). MALDI mass spectrometry imaging for visualizing in situ metabolism of endogenous metabolites and dietary phytochemicals. *Metabolites, 4*(2), 319–346.

Gabizon, A., & Papahadjopoulos, D. (1988). Liposome formulations with prolonged circulation time in blood and enhanced uptake by tumors. *Proceedings of the National Academy of Sciences of the United States of America, 85*(18), 6949–6953.

Gemperline, E., Rawson, S., & Li, L. (2014). Optimization and comparison of multiple MALDI matrix application methods for small molecule mass spectrometric imaging. *Analytical Chemistry, 86*(20), 10030–10035.

Golan, T., Grenader, T., Ohana, P., Amitay, Y., Shmeeda, H., La-Beck, N. M., ... Gabizon, A. A. (2015). Pegylated liposomal mitomycin C prodrug enhances tolerance of mitomycin C: A phase 1 study in advanced solid tumor patients. *Cancer Medicine, 4*(10), 1472–1483.

Goodwin, R. J. A. (2012). Sample preparation for mass spectrometry imaging: Small mistakes can lead to big consequences. *Journal of Proteomics, 75*(16), 4893–4911.

Goodwin, R. J. A., Dungworth, J. C., Cobb, S. R., & Pitt, A. R. (2008). Time-dependent evolution of tissue markers by MALDI-MS imaging. *Proteomics, 8*(18), 3801–3808.

Goodwin, R. J., Iverson, S. L., & Andren, P. E. (2012). The significance of ambient-temperature on pharmaceutical and endogenous compound abundance and distribution in tissues sections when analyzed by matrix-assisted laser desorption/ionization mass spectrometry imaging. *Rapid Communications in Mass Spectrometry, 26*(5), 494–498.

Goodwin, R. J. A., Pennington, S. R., & Pitt, A. R. (2008). Protein and peptides in pictures: Imaging with MALDI mass spectrometry. *Proteomics, 8*(18), 3785–3800.

Goodwin, R. J. A., & Webborn, P. J. H. (2015). Future directions of imaging MS in pharmaceutical R&D. *Bioanalysis, 7*(20), 2667–2673.

Gottlob, K., Majewski, N., Kennedy, S., Kandel, E., Robey, R. B., & Hay, N. (2001). Inhibition of early apoptotic events by Akt/PKB is dependent on the first committed step of glycolysis and mitochondrial hexokinase. *Genes and Development, 15*(11), 1406–1418.

Grime, K. H., Barton, P., & McGinnity, D. F. (2013). Application of in silico, in vitro and preclinical pharmacokinetic data for the effective and efficient prediction of human pharmacokinetics. *Molecular Pharmaceutics, 10*(4), 1191–1206.

Guenther, S., Muirhead, L. J., Speller, A. V. M., Golf, O., Strittmatter, N., Ramakrishnan, R., ... Takats, Z. (2015). Spatially resolved metabolic phenotyping of breast cancer by desorption electrospray ionization mass spectrometry. *Cancer Research, 75*(9), 1828–1837.

Gurdak, E., Green, F. M., Rakowska, P. D., Seah, M. P., Salter, T. L., & Gilmore, I. S. (2014). VAMAS interlaboratory study for desorption electrospray ionization mass spectrometry (DESI MS) intensity repeatability and constancy. *Analytical Chemistry, 86*(19), 9603–9611.

Gusev, A. I., Vasseur, O. J., Proctor, A., Sharkey, A. G., & Hercules, D. M. (1995). Imaging of thin-layer chromatograms using matrix-assisted laser desorption/ionization mass spectrometry. *Analytical Chemistry, 67*(24), 4565–4570.

Hachmöller, O., Aichler, M., Schwamborn, K., Lutz, L., Werner, M., Sperling, M., ... Karst, U. (2016). Element bioimaging of liver needle biopsy specimens from patients with Wilson's disease by laser ablation-inductively coupled plasma-mass spectrometry. *Journal of Trace Elements in Medicine and Biology, 35*, 97–102.

Hamm, G., Bonnel, D., Legouffe, R., Pamelard, F., Delbos, J.-M., Bouzom, F., & Stauber, J. (2012). Quantitative mass spectrometry imaging of propranolol and olanzapine using tissue extinction calculation as normalization factor. *Journal of Proteomics, 75*(16), 4952–4961.

Hankin, J. A., & Murphy, R. C. (2010). The relationship between MALDI IMS intensity and measured quantity of selected phospholipids in rat brain sections. *Analytical Chemistry, 82*(20), 8476–8484.

Hare, J. I., Lammers, T., Ashford, M. B., Puri, S., Storm, G., & Barry, S. T. (2016). Challenges and strategies in anti-cancer nanomedicine development: An industry perspective. *Advanced Drug Delivery Reviews*. pii: S0169-409X(16)30135-1, Epub ahead of print.

Hsieh, Y., Casale, R., Fukuda, E., Chen, J., Knemeyer, I., Wingate, J., ... Korfmacher, W. (2006). Matrix-assisted laser desorption/ionization imaging mass spectrometry for direct measurement of clozapine in rat brain tissue. *Rapid Communications in Mass Spectrometry, 20*(6), 965–972.

Jiunn, H. L. (2009). Pharmacokinetics of biotech drugs: Peptides, proteins and monoclonal antibodies. *Current Drug Metabolism, 10*(7), 661–691.

Källback, P., Shariatgorji, M., Nilsson, A., & Andren, P. E. (2012). Novel mass spectrometry imaging software assisting labeled normalization and quantitation of drugs and neuropeptides directly in tissue sections. *Journal of Proteomics, 75*(16), 4941–4951. http://dx.doi.org/10.1016/j.jprot.2012.07.034.

Kang, H. B., Fan, J., Lin, R., Elf, S., Ji, Q., Zhao, L., ... Chen, J. (2015). Metabolic rewiring by oncogenic BRAF V600E links ketogenesis pathway to BRAF-MEK1 signaling. *Molecular Cell, 59*(3), 345–358.

Kim, C. W., Yun, J. W., Bae, I. H., Lee, J. S., Kang, H. J., Joo, K. M., ... Lim, K. M. (2010). Determination of spatial distribution of melamine-cyanuric acid crystals in rat kidney tissue by histology and imaging matrix-assisted laser desorption/ionization quadrupole time-of-flight mass spectrometry. *Chemical Research in Toxicology, 23*(1), 220–227.

Klein, O., Strohschein, K., Nebrich, G., Oetjen, J., Trede, D., Thiele, H., ... Winkler, T. (2014). MALDI imaging mass spectrometry: Discrimination of pathophysiological regions in traumatized skeletal muscle by characteristic peptide signatures. *Proteomics, 14*(20), 2249–2260.

Kohn, A. D., Summers, S. A., Birnbaum, M. J., & Roth, R. A. (1996). Expression of a constitutively active Akt Ser/Thr kinase in 3 T3-L1 adipocytes stimulates glucose uptake and glucose transporter 4 translocation. *The Journal of Biological Chemistry, 271*(49), 31372–31378.

Kreye, F., Hamm, G., Karrout, Y., Legouffe, R., Bonnel, D., Siepmann, F., & Siepmann, J. (2012). MALDI-TOF MS imaging of controlled release implants. *Journal of Controlled Release, 161*(1), 98–108.

Kubinyi, H. (2003). Drug research: Myths, hype and reality. *Nature Reviews. Drug Discovery, 2*(8), 665–668.

Lanshoeft, C., Stutz, G., Elbast, W., Wolf, T., Walles, M., Stoeckli, M., ... Kretz, O. (2016). Analysis of small molecule antibody–drug conjugate catabolites in rat liver and tumor tissue by liquid extraction surface analysis micro-capillary liquid chromatography/tandem mass spectrometry. *Rapid Communications in Mass Spectrometry, 30*(7), 823–832.

Lemaire, R., Stauber, J., Wisztorski, M., Van Camp, C., Desmons, A., Deschamps, M., ... Fournier, I. (2007). Tag-mass: Specific molecular imaging of transcriptome and proteome by mass spectrometry based on photocleavable tag. *Journal of Proteome Research, 6*(6), 2057–2067.

Li, H., & Hummon, A. B. (2011). Imaging mass spectrometry of three-dimensional cell culture systems. *Analytical Chemistry, 83*(22), 8794–8801.

Liu, X., & Hummon, A. B. (2015). Mass spectrometry imaging of therapeutics from animal models to three-dimensional cell cultures. *Analytical Chemistry, 87*(19), 9508–9519.

Liu, X., Ide, J. L., Norton, I., Marchionni, M. A., Ebling, M. C., Wang, L. Y., ... Agar, N. Y. R. (2013). Molecular imaging of drug transit through the blood-brain barrier with MALDI mass spectrometry imaging. *Scientific Reports, 3*, 2859.

Liu, X., Weaver, E. M., & Hummon, A. B. (2013). Evaluation of therapeutics in three-dimensional cell culture systems by MALDI imaging mass spectrometry. *Analytical Chemistry, 85*(13), 6295–6302.

Longuespée, R., Boyon, C., Desmons, A., Kerdraon, O., Leblanc, E., Farré, I., ... Salzet, M. (2013). Spectroimmunohistochemistry: A novel form of MALDI mass spectrometry imaging coupled to immunohistochemistry for tracking antibodies. *OMICS: A Journal of Integrative Biology, 18*(2), 132–141.

Löscher, W., & Potschka, H. (2005). Role of drug efflux transporters in the brain for drug disposition and treatment of brain diseases. *Progress in Neurobiology, 76*(1), 22–76.

Maddocks, O. D., Berkers, C. R., Mason, S. M., Zheng, L., Blyth, K., Gottlieb, E., & Vousden, K. H. (2013). Serine starvation induces stress and p53-dependent metabolic remodelling in cancer cells. *Nature, 493*(7433), 542–546.

Maeda, H., Fang, J., Inutsuka, T., & Kitamoto, Y. (2003). Vascular permeability enhancement in solid tumor: Various factors, mechanisms involved and its implications. *International Immunopharmacology, 3*(3), 319–328.

Martin-Lorenzo, M., Balluff, B., Sanz-Maroto, A., van Zeijl, R. J. M., Vivanco, F., Alvarez-Llamas, G., & McDonnell, L. A. (2014). 30 μm spatial resolution protein MALDI MSI:

In-depth comparison of five sample preparation protocols applied to human healthy and atherosclerotic arteries. *Journal of Proteomics, 108,* 465–468.

Mascini, N. E., Cheng, M., Jiang, L., Rizwan, A., Podmore, H., Bhandari, D. R., ... Heeren, R. M. A. (2016). Mass spectrometry imaging of the hypoxia marker pimonidazole in a breast tumor model. *Analytical Chemistry, 88*(6), 3107–3114.

McDonnell, L. A., Römpp, A., Balluff, B., Heeren, R. M. A., Albar, J. P., Andrén, P. E., ... Stoeckli, M. (2015). Discussion point: Reporting guidelines for mass spectrometry imaging. *Analytical and Bioanalytical Chemistry, 407*(8), 2035–2045.

Meistermann, H., Norris, J. L., Aerni, H. R., Cornett, D. S., Friedlein, A., Erskine, A. R., ... Ducret, A. (2006). Biomarker discovery by imaging mass spectrometry: Transthyretin is a biomarker for gentamicin-induced nephrotoxicity in rat. *Molecular & Cellular Proteomics, 5*(10), 1876–1886.

Metallo, C. M., Gameiro, P. A., Bell, E. L., Mattaini, K. R., Yang, J., Hiller, K., ... Stephanopoulos, G. (2012). Reductive glutamine metabolism by IDH1 mediates lipogenesis under hypoxia. *Nature, 481*(7381), 380–384.

Mirnezami, R., Spagou, K., Vorkas, P. A., Lewis, M. R., Kinross, J., Want, E., ... Nicholson, J. K. (2014). Chemical mapping of the colorectal cancer microenvironment via MALDI imaging mass spectrometry (MALDI-MSI) reveals novel cancer-associated field effects. *Molecular Oncology, 8*(1), 39–49.

Miura, D., Fujimura, Y., & Wariishi, H. (2012). In situ metabolomic mass spectrometry imaging: Recent advances and difficulties. *Jouranl of Proteomics, 75*(16), 5052–5060.

Miura, D., Fujimura, Y., Yamato, M., Hyodo, F., Utsumi, H., Tachibana, H., & Wariishi, H. (2010). Ultrahighly sensitive in situ metabolomic imaging for visualizing spatiotemporal metabolic behaviors. *Analytical Chemistry, 82*(23), 9789–9796.

Miyako, E., Kono, K., Yuba, E., Hosokawa, C., Nagai, H., & Hagihara, Y. (2012). Carbon nanotube–liposome supramolecular nanotrains for intelligent molecular-transport systems. *Nature Communications, 3,* 1226.

Mokhtarzadeh, A., Alibakhshi, A., Yaghoobi, H., Hashemi, M., Hejazi, M., & Ramezani, M. (2016). Recent advances on biocompatible and biodegradable nanoparticles as gene carriers. *Expert Opinion on Biological Therapy, 16*(6), 771–785.

Nelson, K. A., Daniels, G. J., Fournie, J. W., & Hemmer, M. J. (2013). Optimization of whole-body zebrafish sectioning methods for mass spectrometry imaging. *Journal of Biomolecular Techniques: JBT, 24*(3), 119–127.

Nilsson, A., Fehniger, T. E., Gustavsson, L., Andersson, M., Kenne, K., Marko-Varga, G., & Andren, P. E. (2010). Fine mapping the spatial distribution and concentration of unlabeled drugs within tissue micro-compartments using imaging mass spectrometry. *PLoS One, 5*(7), e11411.

Nilsson, A., Forngren, B., Bjurström, S., Goodwin, R. J. A., Basmaci, E., Gustafsson, I., ... Lindberg, J. (2012). In situ mass spectrometry imaging and ex vivo characterization of renal crystalline deposits induced in multiple preclinical drug toxicology studies. *PLoS One, 7*(10)e47353.

Nilsson, A., Goodwin, R. J. A., Shariatgorji, M., Vallianatou, T., Webborn, P. J. H., & Andrén, P. E. (2015). Mass spectrometry imaging in drug development. *Analytical Chemistry, 87*(3), 1437–1455.

Oppenheimer, S. R., Mi, D., Sanders, M. E., & Caprioli, R. M. (2010). A molecular analysis of tumor margins by MALDI mass spectrometry in renal carcinoma. *Journal of Proteome Research, 9*(5), 2182–2190.

Pajander, J., Haugshøj, K. B., Bjørneboe, K., Wahlberg, P., & Rantanen, J. (2013). Foreign matter identification from solid dosage forms. *Journal of Pharmaceutical and Biomedical Analysis, 80,* 116–125.

Passarelli, M. K., Newman, C. F., Marshall, P. S., West, A., Gilmore, I. S., Bunch, J., ... Dollery, C. T. (2015). Single-cell analysis: Visualizing pharmaceutical and metabolite uptake in cells with label-free 3D mass spectrometry imaging. *Analytical Chemistry*, *87*(13), 6696–6702.

Passarelli, M. K., Wang, J., Mohammadi, A. S., Trouillon, R., Gilmore, I., & Ewing, A. G. (2014). Development of an organic lateral resolution test device for imaging mass spectrometry. *Analytical Chemistry*, *86*(19), 9473–9480.

Pellegatti, M., & Pagliarusco, S. (2011). Drug and metabolite concentrations in tissues in relationship to tissue adverse findings: A review. *Expert Opinion on Drug Metabolism & Toxicology*, *7*(2), 137–146.

Petersen, G. H., Alzghari, S. K., Chee, W., Sankari, S. S., & La-Beck, N. M. (2016). Meta-analysis of clinical and preclinical studies comparing the anticancer efficacy of liposomal versus conventional non-liposomal doxorubicin. *Journal of Controlled Release*, *232*, 255–264.

Porstmann, T., Santos, C. R., Griffiths, B., Cully, M., Wu, M., Leevers, S., ... Schulze, A. (2008). SREBP activity is regulated by mTORC1 and contributes to Akt-dependent cell growth. *Cell Metabolism*, *8*(3), 224–236.

Qu, L., Zhou, Q., Gengenbach, T., Denman, J. A., Stewart, P. J., Hapgood, K. P., ... Morton, D. A. V. (2015). Investigation of the potential for direct compaction of a fine ibuprofen powder dry-coated with magnesium stearate. *Drug Development and Industrial Pharmacy*, *41*(5), 825–837.

Race, A. M., Styles, I. B., & Bunch, J. (2012). Inclusive sharing of mass spectrometry imaging data requires a converter for all. *Journal of Proteomics*, *75*(16), 5111–5112.

Rathmell, J. C., Fox, C. J., Plas, D. R., Hammerman, P. S., Cinalli, R. M., & Thompson, C. B. (2003). Akt-directed glucose metabolism can prevent Bax conformation change and promote growth factor-independent survival. *Molecular and Cellular Biology*, *23*(20), 7315–7328.

Reyzer, M. L., Caldwell, R. L., Dugger, T. C., Forbes, J. T., Ritter, C. A., Guix, M., ... Caprioli, R. M. (2004). Early changes in protein expression detected by mass spectrometry predict tumor response to molecular therapeutics. *Cancer Research*, *64*(24), 9093–9100.

Sabhachandani, P., Motwani, V., Cohen, N., Sarkar, S., Torchilin, V., & Konry, T. (2016). Generation and functional assessment of 3D multicellular spheroids in droplet based microfluidics platform. *Lab on a Chip*, *16*(3), 497–505.

Schober, Y., Guenther, S., Spengler, B., & Römpp, A. (2012). High-resolution matrix-assisted laser desorption/ionization imaging of tryptic peptides from tissue. *Rapid Communications in Mass Spectrometry*, *26*(9), 1141–1146.

Schramm, T., Hester, A., Klinkert, I., Both, J.-P., Heeren, R. M. A., Brunelle, A., ... Römpp, A. (2012). imzML—A common data format for the flexible exchange and processing of mass spectrometry imaging data. *Journal of Proteomics*, *75*(16), 5106–5110.

Schuster, D., Laggner, C., & Langer, T. (2005). Why drugs fail—A study on side effects in new chemical entities. *Current Pharmaceutical Design*, *11*(27), 3545–3559.

Seeley, E. H., Oppenheimer, S. R., Mi, D., Chaurand, P., & Caprioli, R. M. (2008). Enhancement of protein sensitivity for MALDI imaging mass spectrometry after chemical treatment of tissue sections. *Journal of the American Society for Mass Spectrometry*, *19*(8), 1069–1077.

Seuma, J., Bunch, J., Cox, A., McLeod, C., Bell, J., & Murray, C. (2008). Combination of immunohistochemistry and laser ablation ICP mass spectrometry for imaging of cancer biomarkers. *Proteomics*, *8*(18), 3775–3784.

Shahidi-Latham, S. K., Dutta, S. M., Prieto Conaway, M. C., & Rudewicz, P. J. (2012). Evaluation of an accurate mass approach for the simultaneous detection of drug and metabolite distributions via whole-body mass spectrometric imaging. *Analytical Chemistry*, *84*(16), 7158–7165.

Shariatgorji, M., Källback, P., Gustavsson, L., Schintu, N., Svenningsson, P., Goodwin, R. J. A., & Andren, P. E. (2012). Controlled-pH tissue cleanup protocol for signal enhancement of small molecule drugs analyzed by MALDI-MS imaging. *Analytical Chemistry*, *84*(10), 4603–4607.

Shariatgorji, M., Nilsson, A., Goodwin, R. J., Kallback, P., Schintu, N., Zhang, X., ... Andren, P. E. (2014). Direct targeted quantitative molecular imaging of neurotransmitters in brain tissue sections. *Neuron*, *84*(4), 697–707.

Shariatgorji, M., Strittmatter, N., Nilsson, A., Källback, P., Alvarsson, A., Zhang, X., ... Andren, P. E. (2016). Simultaneous imaging of multiple neurotransmitters and neuroactive substances in the brain by desorption electrospray ionization mass spectrometry. *NeuroImage*, *136*, 129–138.

Shimoi, T., HAmada, A., Yonemori, K., Shimma, S., Osawa, S., Tanabe, Y.,Tamura, K. ... (2014). Imaging mass spectrometry of novel drug in human tumor specimens: Distribution of unlabeled drugs to support early phase clinical trial. *Annals of Oncology*, *25*(Suppl. 4), iv146.

Sinha, T. K., Khatib-Shahidi, S., Yankeelov, T. E., Mapara, K., Ehtesham, M., Cornett, D. S., ... Gore, J. C. (2008). Integrating spatially resolved three-dimensional MALDI IMS with in vivo magnetic resonance imaging. *Nature Methods*, *5*(1), 57–59.

Solon, E., Schweitzer, A., Stoeckli, M., & Prideaux, B. (2010). Autoradiography, MALDI-MS, and SIMS-MS imaging in pharmaceutical discovery and development. *The AAPS Journal*, *12*(1), 11–26.

Stauber, J., Ayed, M. E., Wisztorski, M., Salzet, M., & Fournier, I. (2010). Specific MALDI-MSI: TAG-MASS. In S. S. Rubakhin & V. J. Sweedler (Eds.), *Mass spectrometry imaging: Principles and protocols* (pp. 339–361). New York: Humana Press.

Strohalm, M., Strohalm, J., Kaftan, F., Krásný, L., Volný, M., Novák, P., ... Havlíček, V. (2011). Poly[N-(2-hydroxypropyl)methacrylamide]-based tissue-embedding medium compatible with MALDI mass spectrometry imaging experiments. *Analytical Chemistry*, *83*(13), 5458–5462.

Swales, J. G., Tucker, J. W., Spreadborough, M. J., Iverson, S. L., Clench, M. R., Webborn, P. J. H., & Goodwin, R. J. A. (2015). Mapping drug distribution in brain tissue using liquid extraction surface analysis mass spectrometry imaging. *Analytical Chemistry*, *87*(19), 10146–10152.

Swales, J. G., Tucker, J. W., Strittmatter, N., Nilsson, A., Cobice, D., Clench, M. R., ... Goodwin, R. J. A. (2014). Mass spectrometry imaging of cassette-dosed drugs for higher throughput pharmacokinetic and biodistribution analysis. *Analytical Chemistry*, *86*(16), 8473–8480.

Takai, N., Tanaka, Y., Inazawa, K., & Saji, H. (2012). Quantitative analysis of pharmaceutical drug distribution in multiple organs by imaging mass spectrometry. *Rapid Communications in Mass Spectrometry*, *26*(13), 1549–1556.

Tata, A., Zheng, J., Ginsberg, H. J., Jaffray, D. A., Ifa, D. R., & Zarrine-Afsar, A. (2015). Contrast agent mass spectrometry imaging reveals tumor heterogeneity. *Analytical Chemistry*, *87*(15), 7683–7689.

Theiner, S., Schreiber-Brynzak, E., Jakupec, M. A., Galanski, M., Koellensperger, G., & Keppler, B. K. (2016). LA-ICP-MS imaging in multicellular tumor spheroids—A novel tool in the preclinical development of metal-based anticancer drugs. *Metallomics*, *8*(4), 398–402.

Vander Heiden, M. G., Cantley, L. C., & Thompson, C. B. (2009). Understanding the Warburg effect: The metabolic requirements of cell proliferation. *Science*, *324*(5930), 1029–1033.

Verbeeck, N., Yang, J., De Moor, B., Caprioli, R. M., Waelkens, E., & Van de Plas, R. (2014). Automated anatomical interpretation of ion distributions in tissue: Linking imaging mass spectrometry to curated atlases. *Analytical Chemistry*, *86*(18), 8974–8982.

Wang, Y., & Wang, J. (2014). Mixed hydrogel bead-based tumor spheroid formation and anticancer drug testing. *Analyst, 139*(10), 2449–2458.

Wang, G., Zhao, T., Song, X., Zhong, W., Yu, L., Hua, W., … Qiu, X. (2015). A 3-D multicellular tumor spheroid on ultrathin matrix coated single cancer cells provides a tumor microenvironment model to study epithelial-to-mesenchymal transitions. *Polymer Chemistry, 6*(2), 283–293.

Warburg, O. (1956). On the origin of cancer cells. *Science, 123*(3191), 309–314.

Wise, D. R., Ward, P. S., Shay, J. E., Cross, J. R., Gruber, J. J., Sachdeva, U. M., … Thompson, C. B. (2011). Hypoxia promotes isocitrate dehydrogenase-dependent carboxylation of alpha-ketoglutarate to citrate to support cell growth and viability. *Proceedings of the National Academy of Sciences of the United States of America, 108*(49), 19611–19616.

Yasunaga, M., Furuta, M., Ogata, K., Koga, Y., Yamamoto, Y., Takigahira, M., & Matsumura, Y. (2013). The significance of microscopic mass spectrometry with high resolution in the visualisation of drug distribution. *Scientific Reports, 3*, 3050.

Ye, H., Mandal, R., Catherman, A., Thomas, P. M., Kelleher, N. L., Ikonomidou, C., & Li, L. (2014). Top-down proteomics with mass spectrometry imaging: A pilot study towards discovery of biomarkers for neurodevelopmental disorders. *PLoS One, 9*(4). e92831.

Ying, H., Kimmelman, A. C., Lyssiotis, C. A., Hua, S., Chu, G. C., Fletcher-Sananikone, E., … DePinho, R. A. (2012). Oncogenic Kras maintains pancreatic tumors through regulation of anabolic glucose metabolism. *Cell, 149*(3), 656–670.

Zhou, Z., & Lu, Z.-R. (2016). Molecular imaging of the tumor microenvironment. *Advanced Drug Delivery Reviews*. pii: S0169-409X(16)30232-0. http://dx.doi.org/10.1016/j.addr.2016.07.012. [Epub ahead of print].

CHAPTER SEVEN

MALDI IMS and Cancer Tissue Microarrays

R. Casadonte*, R. Longuespée*, J. Kriegsmann*,†,‡, M. Kriegsmann[§,1]
*Proteopath GmbH, Trier, Germany
†Institute of Molecular Pathology, Trier, Germany
‡Center for Histology, Cytology and Molecular Diagnostics, Trier, Germany
§Institute of Pathology, University of Heidelberg, Heidelberg, Germany
[1]Corresponding author: e-mail address: mark.kriegsmann@med.uni-heidelberg.de

Contents

1. Introduction 174
2. TMA Technology 175
    2.1 Preparation of the Donor Block 175
    2.2 FFPE TMA Construction 176
3. MALDI IMS Analysis of TMAs 178
    3.1 Sample Preparation 178
    3.2 MALDI IMS Analysis 181
    3.3 Data Analysis 183
4. Identification of Peptides 186
5. Application of MALDI IMS on FFPE TMAs 188
6. Perspectives and Concluding Remarks 192
Acknowledgments 193
References 193

Abstract

Matrix-assisted laser desorption/ionization imaging mass spectrometry (MALDI IMS) technology creates a link between the molecular assessment of numerous molecules and the morphological information about their special distribution. The application of MALDI IMS on formalin-fixed paraffin-embedded (FFPE) tissue microarrays (TMAs) is suitable for large-scale discovery analyses. Data acquired from FFPE TMA cancer samples in current research are very promising, and applications for routine diagnostics are under development. With the current rapid advances in both technology and applications, MALDI IMS technology is expected to enter into routine diagnostics soon. This chapter is intended to be comprehensive with respect to all aspects and considerations for the application of MALDI IMS on FFPE cancer TMAs with in-depth notes on technical aspects.

1. INTRODUCTION

Mass spectrometry (MS) and matrix-assisted laser desorption/ionization time-of-fight (MALDI TOF) technologies, in particular, are flexible methods used in clinical diagnostics and basic research. This method has been applied to blood (Deininger et al., 2016), serum (Roh et al., 2016), plasma (Yang, Rower, Koy, et al., 2015), cell colonies (Carannante, De Carolis, Vacca, et al., 2015), tissues (Kriegsmann et al., 2014), organs (Casadonte, Kriegsmann, Zweynert, et al., 2014), and organ systems (Swales, Tucker, Spreadborough, et al., 2015). Not only humans (Kriegsmann, Seeley, Schwarting, et al., 2012) but also animals (Khalil, Rompp, Pretzel, Becker, & Spengler, 2015) and plants (Ozawa, Osaka, Hamada, et al., 2016) have been studied. Microorganisms (van Belkum et al., 2015), proteins (Longuespee, Alberts, Pottier, et al., 2016), peptides (Kriegsmann, Kriegsmann, & Casadonte, 2015), lipids (Yalcin & de la Monte, 2015), carbohydrates (Harvey, 2015; Kailemia, Ruhaak, Lebrilla, & Amster, 2014), mutations in genes (Kriegsmann, Arens, Endris, Weichert, & Kriegsmann, 2015), drugs, and metabolites (Buck, Ly, Balluff, et al., 2015) have all been successfully detected by MS.

One of the earliest applications of MS was the evaluation and analysis of the composition of elements from rocks, soil, and crystals and it is noteworthy that the first space missions had a mass spectrometer on board (Fenselau et al., 2003). Later, MS was increasingly used in the field of medicine.

One of the first applications was the identification of unknown substances in forensic science (Bohn & Rucker, 1969) and the identification of drugs in drug abusers (Bonnichsen, Maehly, Marde, Ryhage, & Schubert, 1970). Recently, MS has been implemented in microbiology and has since gained official approval for the subtyping of bacterial species (Cheng, Chui, Domish, Hernandez, & Wang, 2016; van Belkum et al., 2015) and their resistances against antibiotics (Charretier & Schrenzel, 2016), fungi (Cassagne, Normand, L'Ollivier, Ranque, & Piarroux, 2016), and viruses (Cobo, 2013).

Today, the application of MS has been transferred to study human tissue samples, which is most challenging due to their higher complexity. However, MS methods are already routinely used in clinical diagnostics to subtype renal amyloidosis, a serious disease caused by misfolded proteins (Said, Sethi, Valeri, et al., 2013; Sethi, Theis, Vrana, et al., 2013).

First described by Caprioli et al., MALDI imaging mass spectrometry (IMS) has been developed to create a link between the molecular assessment of numerous molecules and morphological information about their special distribution (Caprioli, Farmer, & Gile, 1997). Since intact tissue samples contain the complete spatial proteomic and genomic information, something that cannot be replaced by the investigation of tissue lysates, blood, or its derivatives, they represent the best possible source to study cancer. Thus, the application of IMS on tissue samples is not only a logical step, but has the greatest potential to improve cancer diagnostics, prognosis, and prediction of response to therapy.

IMS can be performed on tissue material that has been snap frozen in −80°C liquid nitrogen shortly after excision. Although this technique allows an exact molecular assessment, it is not the diagnostic standard because the sample handling is expensive and requires special logistics and storage. Furthermore, sufficient fresh-frozen tissue material is not available for many human diseases.

Formalin-fixed paraffin-embedded (FFPE) tissue material is available in all institutes of pathology, and large clinically well-documented patient cohorts are collected and stored all over the world. The application of IMS is possible on FFPE tissue and therefore opens the path for the investigation of all known solid cancers (Hoos & Cordon-Cardo, 2001).

Moreover, small tissue cores from hundreds or thousands of tissue blocks can be transferred to only very few tissue blocks, called tissue microarrays (TMAs). These allow the analysis of multiple cancer patients under the same conditions during one single experiment. In 2011, Casadonte et al. have developed a protocol to investigate TMAs by IMS (Casadonte & Caprioli, 2011). This approach is particularly suitable for high-throughput biomarker research on FFPE tissue material. This chapter reviews the application of IMS on FFPE TMAs with in-depth notes on technical aspects.

2. TMA TECHNOLOGY
2.1 Preparation of the Donor Block

After surgical removal, tissue samples are usually fixed in neutral-buffered 10% formalin (according to 3.7% formaldehyde) to avoid degradation. Formalin can be prepared with 100 mL 40% formaldehyde, 900 mL distilled water, 4 g sodium dihydrogen phosphate, and 6.5 g anhydrous disodium hydrogen phosphate. For optimal fixation a ratio of 10:1 (formalin:tissue) is recommended. Since a low pH leads to degradation of tissue material, formalin should be stored in a dark place to avoid the oxidation of formic acid by the action of light. Fixation is a critical step as tissue degeneration starts as soon as cells are deprived of a blood supply. Thus, tissue material needs to be rapidly immerged in the fixative. Formalin induces inter- and intramolecular cross-links between proteins and other biomolecules and therefore preserves tissue integrity at the molecular level. Tissues are then paraffin-embedded by dehydration in a series of graded ethanol solutions, and a clearing procedure that uses a clearing agent, usually xylene, to displace the ethanol and allow paraffin embedding. Paraffin blocks are very stable and can be cut down to a thickness of 2 µm. The quality of the FFPE tissue block clearly affects the later analysis.

2.2 FFPE TMA Construction

The TMA is an idea that has been initially developed by Battifora et al. and was subsequently improved by Kononen et al. (Fedor & De Marzo, 2005). In principle, TMAs are created by removing multiple cylindrical cores from so-called donor blocks (Fig. 1B) and transferring them to a receiver block (Fig. 1C and D). Prior to TMA construction, a pathologist marks the area of interest on a H&E-stained tissue section (Fig. 1A). Furthermore, the TMA layout is created including the number of cores per patient, the

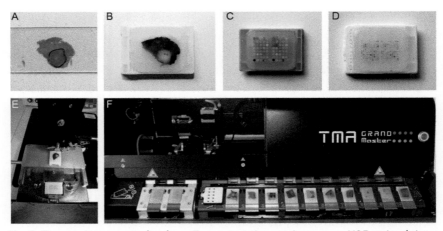

Fig. 1 Tissue microarray technology: To create a tissue microarray a H&E-stained tissue section from the donor block is evaluated by a pathologist and the region of interest is annotated (A). Cylindrical cores are subsequently transferred from a donor block (B) to a receiver block (C and D). Tissue microarrays may be constructed manually (E) or with an automated machine (F).

localization of each core, the diameter of the cores, and the spacing between the cores. This layout is safely stored since it is needed to link the results from individual cores on the TMA to patient data again. Up to 1000 tissue cores can be inserted in one TMA block. However, it is clear that with an increasing number of individual cores, the diameter and therewith the degree of tissue information decreases (Fig. 1B–D). One has to be aware of the fact that small cores from tumors with great morphological heterogeneity might not be representative of the original tumor. Thus, the questions intended to be answered guides the strategy of TMA creation and the size and number of cores used per tumor. For most analyses core diameters of 0.5–1 mm are used and two to four cores are extracted per patient, resulting in a TMA with around 50–200 patients per block. Thus, large cohorts with more than 1000 patients can be assembled in only a handful of TMAs (Kriegsmann, Muley, Harms, et al., 2015).

For the TMA assembly two methods are available: (i) the use of a manual TMA machine and (ii) the use of an automated TMA machine. The advantage of a manual TMA machine is that it is inexpensive. However, manual TMA building is time consuming and imprecise compared to automated TMA machines, which are able to create high-quality multitissue blocks with high throughput (Fig. 1E and F; Zlobec, Koelzer, Dawson, Perren, & Lugli, 2013).

TMA blocks can then be cut with a microtome and mounted onto either standard or special glass slides for subsequent analyses. Depending on the core depth, 100–500 sections can be generated. In biomarker studies, TMAs provide a valuable tool for fast screening (Warth, Muley, Herpel, et al., 2012). A wide range of possible analyses may be performed (Jacquemier, Ginestier, Charafe-Jauffret, et al., 2003) including immunohistochemistry (IHC), fluorescence in situ hybridization, RNA in situ hybridization, or IMS (Casadonte & Caprioli, 2011). Mostly, IHC is performed on TMAs.

TMA offers a high degree of standardization as the study conditions for the different tumors remain constant when they are assembled onto a single slide. Furthermore, useful biomarkers stratify patients into groups with high-/low- and often absent biomarker expression. Thus, internal positive and negative controls should also be present in the TMA. A major advantage is that approximately the same amount of reagents is needed to analyze one whole tissue section as would be needed to analyze one TMA slide. Therefore, the analysis of TMAs can be regarded as cost effective. Additionally, the handling of hundreds of samples becomes very convenient as compared to the organization and storage of the same amount of whole slides.

In summary, TMAs are easy to handle and cost-effective tools and perfectly suitable for multiplex analysis of biomarkers.

3. MALDI IMS ANALYSIS OF TMAs

A typical IMS procedure for the analysis of an FFPE TMA consists of three main steps: sample preparation (including cutting, deparaffinization, antigen retrieval (AR), in situ enzymatic digestion, and matrix application), IMS analysis, and data analysis (Fig. 2).

3.1 Sample Preparation
3.1.1 Cutting, Deparaffinization, and AR

FFPE TMA blocks are usually cut into ~5 μm thick sections using a microtome. Sections are then placed onto warm water to be flattened, collected onto conductive slides (indium thin oxide-coated glass slides), and completely dehydrated to avoid distortion and unlikely lift during further steps. To recover proteins, the paraffin must be removed from the tissue and the tissue rehydrated for enzymatic digestion. This is typically carried out by a previous heating step that allows drainage of melting paraffin before tissue immersion in xylene and graded ethanol baths. Dewaxed tissue sections are then immerged in a buffer solution (10 mM Tris–HCl, pH 9 (Casadonte & Caprioli, 2011) or 10 mM citrate pH 6.0 (Gustafsson, Oehler, McColl, & Hoffmann, 2010)) for further analysis.

Cross-link reversal of formalin-fixed proteins is performed by an AR step that typically uses high temperature to recover, at least partially, the

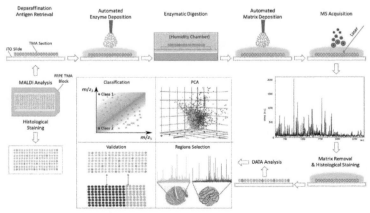

Fig. 2 Schematic workflow for the analysis of FFPE TMA samples by MALDI IMS.

antigenicity of the tissues. Several heating equipments have been described for AR, including microwave oven, vegetable steamer, or pressure cooker. We propose a decloacker device (e.g., Biocare Medical), that is a commercial pressure cooker with electronic controls for temperature and time, which has the advantage to maintain a constant temperature independent of the external pressure, for a more efficient cross-linkage reversal condition. AR procedures can cause sections to come off slides especially when are cut from lipid-rich tissues like breast and brain. Loss of tissue is a relevant issue for valuable TMA sections. Preheating of the slides at 60°C prior to deparaffinization will prevent tissue detachment. Alternatively, a polycation compound, like the poly-L-lysine, can be used as a coating substrate for slides to enhance electrostatic interaction between negatively charged ions of the cell membranes and the slide surface. Although such coating helps to avoid loss of cores, possible components related to the polyl-L-lysine solution can interfere with MALDI MS analysis with suppression effects. From the analysis of serial sections cut from the same FFPE specimen and mounted onto noncoated and coated poly-L-lysine slides, we have observed different unique signals present only in the spectra acquired from the coated slide (data not shown). Further investigations to minimize or eliminate coating contaminants are needed. However, in order to ensure the true biological contribution to the peaks detected, it is highly recommended to avoid any polymeric interferences.

Combining heat-induced AR with proteolytic enzymes has proved efficient for protein recovery in MS-based proteomic analyses on FFPE tissues (Nirmalan, Harnden, Selby, & Banks, 2008). Additionally, enzymatic digestion offers possibilities to analyze larger molecular weight proteins that are difficult to detect without proteolysis due to their low detection efficiency in the high mass range (>30 kDa). Trypsin is commonly used for protein digestion. Moreover, other enzymes, such as the recombinant Peptide NGlycosidase F (PNGaseF) for N-glycan release (Powers, Jones, Betesh, et al., 2013) or combination of enzymes (Heijs, Holst, Briaire-de Bruijn, et al., 2016), can be used to get more information from FFPE tissues.

3.1.2 In Situ Enzymatic Digestion
The deposition of the enzyme on the surface of the tissue section is particularly important since the area of enzyme droplets directly affects the subsequent digestion reaction. This zone has to be as little as the intended spatial resolution of the imaging analysis. Commercial automatic spraying apparatus including ImagePrep (Bruker Daltonik) (Gustafsson et al.,

2010), TM Sprayer (HTX Technologies) (Cornett, 2015), SunCollect (SunChrom) (Beine, Diehl, Meyer, & Henkel, 2016), or robotic spotters such as the ChIP-1000 (Shimadzu Scientific Instruments) (Groseclose, Andersson, Hardesty, & Caprioli, 2007) or the Portrait® 630 Spotter (Labcyte Inc.) can be used for the enzyme deposition. Spotting relies on the ejection of droplets (~160 pL per drop) in a well-defined array onto the tissue section. Thus, the resolution of images is limited by the size of the droplets, which vary from 150 to 200 µm in diameter. Spotting is the most appropriated method of choice to perform histology-directed IMS, where proteins are extracted from specific morphological features within the section. Briefly, a histopathological examination of a tissue section is performed by a pathologist, and specific areas of interest are digitally marked. One can then correlate those areas with a photomicrograph of a serial section that will be analyzed by IMS. Mass spectra from the marked areas are then extracted as profiles specific to that histopathological entity. Each acquisition spot on discrete areas of the tissue takes only a few seconds to profile, thus large cohort of samples, such as TMAs, can be analyzed and compared in a high-throughput manner.

Higher spatial resolutions can be achieved using spray-coating devices (30–50 µm in FFPE tissues), depending on the matrix and the preparation conditions used a very fine reagent aerosol can be homogeneously deposited on the sample surface. To retain enzyme activity and improve digestion efficiency, the enzyme-coated tissue is incubated in a humidity and temperature-controlled environment for a set time (e.g., 2–12 h) depending on the tissue type, thickness, and the protocol used. Several trypsin proteases were tested at different temperatures and incubation times. The recommended trypsin enzyme for the analysis of bladder, prostate, and lung cancer tissues was found to be the "Trypsin Gold, Mass Spectrometry Grade," manufactured by Promega, at 37°C with an incubation time of 2 h (Diehl, Beine, Elm, et al., 2015).

3.1.3 Matrix Deposition

It is well known that the detection of molecules by MALDI MS highly depends on the matrix selection and its preparation. Applying a certain amount of matrix in several iterations and maintaining the wetness at which the matrix is applied affects the quality of the results obtained. The most common matrix employed for the analysis of peptides is α-cyano-4-hydroxycinnamic acid (CHCA), as it generally produces small homogeneous crystals leading to a good resolution in MALDI analysis (Chaurand, Schriver, &

Caprioli, 2007). Similar to the enzyme deposition, the matrix can be applied using automatic spotting or spraying, depending on the results desired. Spatial resolution is limited by the size of the crystal, which should be smaller than the laser beam diameter to avoid oversampling (Zavalin, Yang, Hayden, Vestal, & Caprioli, 2015). High resolution is required when molecular information at the cellular level (~10 μm) is needed and when regions of interest are small as in TMAs. Automatic spraying is considered the best approach to analyze large sample sets like TMAs, as the coating process is faster than the robotic spotting (minutes compared to hours); however, care must be taken to avoid delocalization or migration of analytes during the application process, which affect both the reproducibility and the image quality. Thus, parameters such as matrix composition spray rate, drying time, and amount of matrix need to be optimized. To our knowledge, a well-adapted and reproducible methodology that uses spraying devices for matrix coating of FFPE tissues is needed to further improve the sensitivity of IMS. With robust equipment and protocols, there will be a rapid growth of IMS applications in the clinical field.

3.2 MALDI IMS Analysis

IMS technology analyzes and localizes biomolecules directly in tissue specimens without any prior knowledge of target specific reagents such as antibodies. MALDI uses a laser to desorb and ionize molecules in a sample that have been cocrystallized with a suitable matrix. The resulting ions are separated according to their mass-to-charge ratio (m/z values) and displayed on a mass spectrum. A typical average spectrum acquired from an FFPE TMA is represented in Fig. 3.

Several MS instruments can be used to acquire mass spectra at peptide m/z values (600–5000) using time-of-flight (TOF) or other mass analyzers, e.g., TOF-TOF, hybrid trap-TOF and quadrupole-TOF (Qq-TOF), ion trap. Depending on the biological question, specific MS parameters are required that can vary between the different systems; therefore care must be taken to select a suitable instrument. For example, Fourier transform ion cyclotron resonance (FT-ICR) mass spectrometers are particularly useful because of the high mass accuracy (~3 ppm) and high resolution (mass resolving power > 100.000) they can provide. Such instruments are very well suited for the identification of compounds directly from tissue. In a study of nonsmall-cell lung carcinoma (NSCLC), in situ protein identification was performed directly from a TMA section using a MALDI

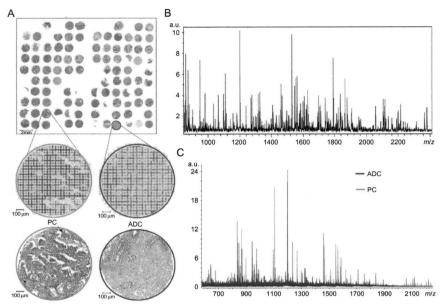

Fig. 3 (A) Representative TMA section with the laser-irradiated spots arrayed across the entire section. Zoomed view of two tissue cores with different tumor phenotype, pleomorphic carcinoma (PC), and adenocarcinoma (ADC), respectively. (B) Typical average spectrum acquired from an FFPE TMA section. (C) Overlay of average spectra from the adenocarcinoma (ADC) tissue core (*blue*) and the pleomorphic carcinoma (PC) tissue core (*orange*) showed in (A).

TOF-TOF (Groseclose, Massion, Chaurand, & Caprioli, 2008) and lately an FT-ICR MS instrument.

There has been a significant improvement in image resolution. Modern MALDI MS systems offer better laser focus (~5 μm in diameter; Spraggins et al., 2016) for imaging resolutions of less than 10 μm. It is tempting to speculate that further instrumental improvements will allow unraveling subcellular information soon. Single-cell imaging can support the use of proteomic-based approaches to potentially aid personalized medicine for cancer patients, where each individual's disease is treated according to their underlying unique molecular pattern (Chaurand, Sanders, Jensen, & Caprioli, 2004). Improvements with regard to desorption and ionization techniques as well as accelerated speed of acquisition have enhanced our ability to analyze proteins. The analysis of TMAs and large tissue sections (~4 cm^2) at high spatial resolution was previously impaired by the slow acquisition speed of existing platforms. However, new imaging systems allow quasisquare, discrete pixels, and continuous stage movement with acquisition speed of >30 pixel/s, contributing to consistent data quality even at small pixel size. We applied this technology for the analysis of FFPE TMAs in a study of lung cancer. Two serial sections from the same FFPE TMA block including 116 NSCLC tissue core biopsies were treated with the same exact sample preparation conditions. MALDI data were generated from two different platforms, AutoflexSpeed and rapifleX TissueTyper, evaluated and compared based on the acquisition speed, time, and maintenance of the tissue integrity. The TMA section analyzed with the novel high-speed instrument allowed a measurement of 51,932 spectra at 50 μm pixel resolution with a total acquisition time of <1 h (Fig. 4).

3.3 Data Analysis

The large amount of collected data in IMS represents a high degree of complexity that can be partially overcome by adequate dataset processing. The heterogeneous nature of tissues and/or matrix deposition issues can be an important source of artifacts. A step of data preprocessing, which includes smoothing, baseline correction, alignment, normalization, and peak

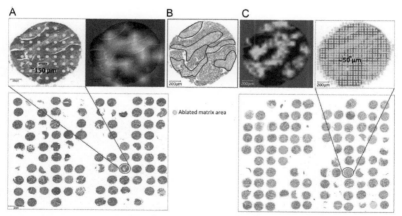

Fig. 4 (A) IMS analysis on a TMA section performed with an AutofleX speed with 2 kHz repetition rate laser at 150 μm spatial distribution. (B) 2–3 pixel/s was acquired generating 51,932 spectra within 5 h. (B) H&E after MALDI of the adenocarcinoma tissue core. (C) IMS analysis on a serial section of the same TMA used in (A) performed with a rapifleX TissueTyper with 10 kHz repetition rate laser at 50 μm spatial resolution. 40–50 pixel/s was acquired generating 160,694 spectra within ~1.5 h. In the upper panels (A) and (B), zoomed view of a lung squamous cell carcinoma tissue core with the ablated matrix areas and the image distribution of Cytokeratin 5 protein (peptide at m/z 1410.7) are shown.

detection, is therefore highly recommended after image spectra collection to reduce experimental variance within the dataset. Several software packages exist for IMS data processing. Several published papers and reviews describe extensively processing applications, and the reader is referred to these for further information (Chaurand, 2012; Chughtai & Heeren, 2010; Deininger, Cornett, Paape, et al., 2011; Jones, Deininger, Hogendoorn, Deelder, & McDonnell, 2012; McCombie, Staab, Stoeckli, & Knochenmuss, 2005).

Many hundreds of molecular images can be obtained from a single tissue section. The major advantage of this approach is the capability to directly correlate mass spectra with anatomical or pathological features since tissue integrity is preserved under the MALDI measurement. The same tissue section can be stained after MS analysis for histomorphological interpretation. Signal intensities of the tryptic peptides can vary across the dataset because of inhomogeneity of the tissue section. Nonetheless, comparable peak profiles from similar histological regions can be obtained demonstrating reproducible methodology. An example is illustrated in Fig. 5. Comparison of the

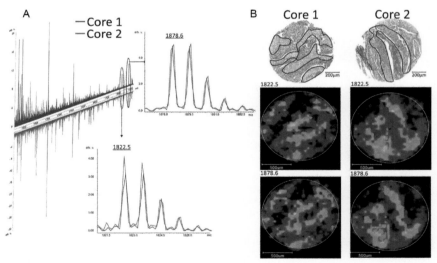

Fig. 5 Evaluation of tryptic profile of two duplicate tissue cores in the same TMA section. (A) Tryptic profile comparison between two duplicate cores collected from a patient with squamous cell carcinoma lung tumor. (B) H&E of two duplicate cores collected from a patient with squamous cell carcinoma lung tumor. Ion density map of two peptides peaks (*m/z* 1822.5, 1878.6) identified as Cytokeratin15 protein shows distribution in the tumor area (*black line*).

relative intensities at each m/z value between regions of interest can be used to determine the relative amounts of a given compound across a single section or a series of samples. For the diagnosis of a disease, candidate peaks should be reproducible and have a good signal quality, especially when a larger cohort of samples is analyzed. The development of a diagnostic algorithm depends on reliable classification of specimens that serve as training and evaluation sets. A common method for a rapid comparison of spectra from selected regions of interest is principal component analysis (PCA). This multivariate analysis computes linearly uncorrelated variables from the multidimensional dataset, called "components," each having the highest possible variance while still being orthogonal to the previous components. This calculation subsequently allows the representation of spectra as dots in a 3D space using the first three components (McCombie et al., 2005). In a study of a pancreatic tumor TMA, including 90 needle core biopsies, distinct peptide profile pattern was obtained discriminating tumor from nontumor tissues using a principal component analysis-discriminant analysis (PCA-DA) approach (Djidja, Claude, Snel, et al., 2010). Classification algorithms such as the support vector machine (SVM; Groseclose et al., 2008), random forest (Balluff, Elsner, Kowarsch, et al., 2010; Hanselmann, Kothe, Kirchner, et al., 2009), neural networks (Rauser, Marquardt, Balluff, et al., 2010), and genetic algorithm (Schwamborn et al., 2007) are used to discriminate samples of different phenotypes. A novel software used for classification purpose has recently been proposed, which uses a linear discriminant analysis (LDA) algorithm (SCiLS Lab). This algorithm has been applied in a study to classify different pancreatic cancer phenotypes according to their molecular profiles (Steurer, Borkowski, Odinga, et al., 2013). Fig. 6 shows an example of a LDA-based classification between colon and lung adenocarcinoma TMAs.

IMS can be considered as any other classical proteomic method for biomarker studies, and statistical processes may be applied in order to overcome any experimental bias. As in clinical trials, samples can be compared and the sensitivity and specificity of the biomarkers assessed. These assessments are typically performed by receiver operating characteristic (ROC) curves (Pallua, Schaefer, Seifarth, et al., 2013) for categorical comparisons (e.g., diagnosis) and Kaplan–Meier analyses for patient survival (Elsner, Rauser, Maier, et al., 2012). These studies prove evidence that MSI may be incorporated into clinical trials.

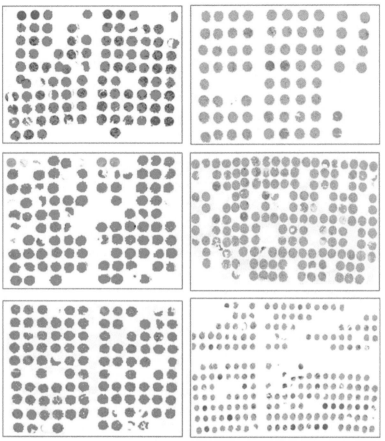

Fig. 6 Representative image classification of lung (*red*) and colon (*yellow*) adenocarcinoma TMAs. Classification was performed using linear discriminant analysis (LDA) algorithm provided by SCiLS Lab software (SCiLS GmbH, Bremen, Germany).

4. IDENTIFICATION OF PEPTIDES

Many techniques are used to validate the presence of proteins in tissue, such as western blot analysis, reverse phase protein arrays (Becker, Schott, Hipp, et al., 2007; Ikeda, Monden, Kanoh, et al., 1998), and IHC (Kriegsmann, Muley, et al., 2015). However, these methods do not allow the identification of peptides or proteins. MS-based methods are most appropriate for the identification of peptides, in parallel to their mapping by IMS. The choice of the method applied highly depends on the need to retain spatial information and on the tissue source (fresh-frozen- or FFPE tissue).

The most immediate method for identification of compounds from tissues is in situ MS/MS. In MS/MS using TOF/TOF devices, peptide fragments are usually generated by postsource decay (Kaufmann, Chaurand, Kirsch, & Spengler, 1996). The series of peptide fragment ions in the MS/MS spectrum are then compared against public repositories of protein sequences for identification (Groseclose et al., 2008; Casadonte et al., 2015). The quality and the exploitability of the generated MS/MS spectra greatly depend on the initial in situ abundance of the proteolytic peptide to fragment and on the sensitivity of the instrumentation used.

Alternative methods to identify peptides and proteins from tissues consist in the previous separation of compounds by electrophoresis (Elsner et al., 2012; Longuespee, Tastet, Desmons, et al., 2014; Meding, Balluff, Elsner, et al., 2012) or liquid chromatography (LC) (Balluff, Rauser, Meding, et al., 2011) before their analysis by MS/MS. For FFPE tissues, LC-MS/MS is generally used since it is suitable for the analysis of proteolytic peptides. After LC separation, the analysis can be made either online (LC device coupled to the mass spectrometer) or offline (separated chromatograph and mass spectrometer; Camerini & Mauri, 2015; Edwards & Thomas-Oates, 2005).

Only few studies addressed the identification of peptides from FFPE tissues in parallel to IMS. One possible workflow for peptide identification in parallel to IMS consists in the preparation of two serial sections: one for IMS and one for parallel identification. Both sections are sprayed with trypsin before incubation for digestion. One section is sprayed with matrix solution and is analyzed by IMS, the other section is used for identification, where the digested peptides are extracted with a solvent solution from the surface of the tissue. Trypsin can be deposited either using spraying devices or manually in the form of a droplet. The resulting trypsin/digest droplets can subsequently be collected for further analysis. Using manual deposition, additional solvent extraction is no longer useful for the collection of the digest (Longuespee, Fleron, Pottier, et al., 2014; Longuespee, Gagnon, Boyon, et al., 2013). A recent improvement of the method consists in the use of hydrogels for both digestion and extraction of proteolytic peptides from tissue sections (Harris, Nicklay, & Caprioli, 2013; Taverna, Pollins, Nanney, Sindona, & Caprioli, 2016). After desalting, the peptide solution can be analyzed by online or offline LC-MS. The separation of peptides by LC followed by the collection of fractions that are deposited onto the target for MALDI MS/MS analysis has been shown earlier (Casadonte et al., 2014; Gravius et al., 2015). In this study, LC-MALDI allowed the identification of only a limited number of peptides. This might be due to the low fragmentation yields obtained by PSD, which are also encountered during in situ identification. Online approaches usually allow better peptide identification yields. New methods were developed for the creation of repositories of peptides identifications to match with IMS datasets. In the work of Gustafsson et al., FFPE TMA cores were collected by microdissection and processed to collect molecular information from multiple samples (Gustafsson et al., 2013; Meding, Martin, Gustafsson, et al., 2013). However, the method for sample preparation presented some drawbacks, especially concerning extraction and identification yields. It was recently proved that limiting the steps for the processing of laser microdissected FFPE tissues and performing the digestion of tissue pieces instead of proteins extracts greatly improve peptide and protein identification yields from small samples (Longuespee, Alberts, et al., 2016). This updated method may represent an adequate tool for the identification of peptides from small regions of interest of tissues, in parallel to IMS analyses. The workflow has already been proven to be robust enough for biomarker discovery studies and can be used to explore molecular heterogeneity in tissues (Longuespee, Casadonte, Kriegsmann, et al., 2016). Laser microdissection-based microproteomics has great potential to build up molecular classifications for various diseases. In this context, TMAs will have a major role for the collection of a large panel of tissues that represent the molecular variability between patients.

Information of protein and peptide identification can be correlated with m/z values from IMS datasets and assigned to the correct protein through public databases such as MSiMass list (McDonnell, Walch, Stoeckli, & Corthals, 2014) and MaTisse (Maier, Hahne, Gholami, et al., 2013).

After protein identification, validation is generally performed by independent methods such as IHC or western blot in order to confirm the presence of the proteins of interest.

5. APPLICATION OF MALDI IMS ON FFPE TMAs

The direct analysis of FFPE TMA using IMS allows the analysis of multiple tissue samples in a single experiment (Casadonte & Caprioli, 2011; Groseclose et al., 2008). This high-throughput method was used to detect peptide patterns in gastric cancer (Morita, Ikegami, Goto-Inoue, et al., 2010), nonsmall-cell lung cancer (NSCLC; Groseclose et al.,

2008), ovarian cancer (Meding et al., 2013), endometrial cancer (Mittal, Klingler-Hoffmann, Arentz, et al., 2016), renal cancer (Morgan, Seeley, Fadare, Caprioli, & Clark, 2013; Steurer, Seddiqi, Singer, et al., 2014), pancreatic and breast cancer (Casadonte et al., 2014), oesophageal cancer (Quaas, Bahar, von Loga, et al., 2013), bladder cancer (Steurer, Singer, Rink, et al., 2014), and prostate cancer (Steurer et al., 2013; Table 1). In the largest study, more than 1000 patient samples have been analyzed. Besides peptides, also glycans have been studied on TMAs. The investigation of these molecules is described in chapter "MALDI Mass Spectrometry Imaging of N-Linked Glycans in Cancer Tissues" by Drake et al.

The analysis of TMAs allows the investigation of diagnostic, prognostic, and predictive problems.

Diagnostic studies aim to differentiate subtypes of diseases. For example, in the context of a liver metastasis of an adenocarcinoma with unknown primary, the origin of the neoplasm determines treatment stratification and prognosis. In most cases diagnosis can be made based on morphological appearance alone; however, in a substantial subset of cases IHC stains are

Table 1 Overview of MALDI IMS-Based Studies and Findings Using FFPE TMA in the Field of Cancer Research

PMID	Tumor Type	Question Addressed	n=	Remarks
19961487	Gastric cancer	Diagnosis	3	
18712763	Nonsmall-cell lung cancer	Diagnosis	26	
20204332	Pancreatic adenocarcinoma	Diagnosis	30	
23214983	Ovarian cancer	Reference dataset	33	LC-MS/MS
27061135	Endometrial carcinoma	Prognosis	43	
23009866	Clear cell renal cell carcinoma	Diagnosis	70	
24482424	Breast and pancreatic cancer	Diagnosis	188	
23855813	Oesophageal cancer	Diagnosis, prognosis	477	
25131659	Bladder cancer	Prognosis	697	
24778028	Kidney cancer	Diagnosis, prognosis	789	
23381989	Prostate cancer	Diagnosis	1044	

necessary and in some cases correct diagnosis cannot be achieved until autopsy (Pavlidis & Pentheroudakis, 2012). While little treatment options are available for patients with metastasized pancreatic cancer (Koh, Chok, Zheng, Tan, & Goh, 2014), there are several for patients with breast cancer, depending on the breast cancer subtype (Foulkes, Smith, & Reis-Filho, 2010; Verma, Miles, Gianni, et al., 2012). Therefore, correct diagnosis is mandatory.

To address this problem, Casadonte et al. have analyzed 96 breast and 92 pancreatic primary cancer samples and 6 blinded liver metastasis from breast and pancreatic cancer (Casadonte et al., 2014). 29 breast and 32 pancreatic cancer samples were used to train an SVM algorithm model. This model was tested on two independent validation sets consisting of the remaining samples. Overall, an accuracy of 85% was reached for the discrimination of breast and pancreatic tumor samples. Relevant peptides for discrimination have been identified by MS/MS: namely heat-shock protein 27, heterogeneous nuclear ribonucleoprotein A1 and A2/B1, filamin-A, and SH3 domain-binding glutamic acid-rich-like protein.

Another study performed for diagnostic purposes was performed on NSCLC by Groseclose et al. The authors investigated peptide patterns from 14 patients with squamous cell carcinoma and 12 patients with adenocarcinoma of the lung, the two most common subtypes of NSCLC. They found statistically significant differences comparing both cancer subtypes but a true diagnostic value could not be reached since the number of patient samples used for the classification study was small. Nevertheless, this study showed that peptides from proteins can be detected that are currently used in clinical diagnostics, such as cytokeratins (5, 6, 7, and 19; Warth et al., 2012). Moreover, other interesting peptides were identified that anticipate validation on large patient cohorts.

Differences in the peptide pattern of adenocarcinomas and squamous cell carcinomas have also been described in oesophageal cancer. Quaas et al. have analyzed 300 and 177 patient samples from both tumor entities, respectively. From the 72 peaks detected in tumor cells, 48 were found to be associated with squamous cell and 12 were associated with adenocarcinomas. Additionally, in squamous cell carcinomas, a total of 24 peptide signals were linked to clinical features such as tumor stage, histological grade, and the presence of lymph node metastasis. In the adenocarcinomas, six peaks were identified in early-stage tumors and related to the presence or absence of lymph node metastasis. Among the peptides identified there were collagen subunits, filamin-A, keratin, desmin, hemoglobin subunit alpha, and heat-shock protein beta-1 proteins. Actin isoforms were associated with survival in squamous cell carcinomas.

A study conducted by Morgan et al. (2013) analyzed two TMAs with 70 patients with clear cell renal cell carcinoma and matched normal kidney tissue by IMS. They identified a peptide pattern that could discriminate both classes with 85.5–92.4% accuracy (depending on the TMA examined). Identification by MALDI MS/MS revealed Histone H2A, hemoglobin subunits, vimentin, alpha enolase, and Ig alpha-1 chain C region to be relevant classifiers.

Studies have been performed to address not only diagnostic but also prognostic problems. Predictive investigations have not been performed to date. Whereas a prognostic biomarker provides information about the likely course of the disease in an individual, a predictive biomarker allows to identify a subpopulation of patients who are most likely to respond to a given therapy.

Mittal et al. (2016) have analyzed tissue samples from endometrial cancer patients. They investigated if patients with ($n=16$) and without ($n=27$) lymph node metastases show discriminatory peptide patterns. Using MALDI IMS data, they found a number of m/z values that could predict lymph node metastases with an overall accuracy of 88.4%. Downregulation of alpha-actin-2 identified by LC-MS/MS has been recognized as one of the most potent classifiers toward lymph node metastases in these patients.

The largest investigations of peptide patterns on TMAs have been performed by Steurer et al. on bladder cancer ($n=697$), renal cell cancer ($n=789$), and prostate cancer ($n=1044$). In their study on bladder cancer, they found 40 peptides associated to epithelial structures. 30-peptide peaks were found to correlate with prognostic markers as stage, grade, or nodal status, 2 peptides were linked to tumor recurrence in noninvasive tumors, and the absence of one peak was related to decreased survival in a subset of muscle-invasive bladder cancer. In total, eight peptide peaks could be identified by MS/MS analysis. These were histone H2A-type 1A, hemoglobin subunits, cytokeratin 7 and 19, collagen alpha, and heat-shock protein beta-1. In their study on renal cell cancer, the same authors found peaks related to papillary and clear cell histology. Within the latter group correlations to tumor stage, grade, and the presence of lymph node metastasis were detected. The prostate cancer study was performed on more than 1044 patients and represents the largest study regarding peptide profiles on TMAs by IMS to date. 15 signals showed a correlation to decreased grade, stage, and low proliferation and 4 signals were associated to high proliferation. 1-peptide peak was demonstrated to indicate prolonged time to recurrence. Three isoforms of actin and hemoglobin subunits were identified.

6. PERSPECTIVES AND CONCLUDING REMARKS

Although, clinical imaging techniques such as computed tomography or magnetic resonance tomography allow to localize a lesion in the human body, these clinical imaging methods cannot make a sufficiently specific and secure diagnosis. Currently, a tissue biopsy with subsequent evaluation by an experienced pathologist is the gold standard to diagnose cancer, many infections, and autoimmune diseases. Often a standard hematoxylin and eosin (H&E)-stained tissue slide is sufficient to render a diagnosis. Sometimes, subsequent additional histochemical and IHC stains are required to (i) secure the diagnosis, (ii) to determine prognosis, or (iii) to predict the response to certain therapies. An example for the use of histochemical stains is the Berlin-Blue Reaction to visualize iron in the liver in patients with hemochromatosis, a disease characterized by massive iron overload leading to severe end-organ damage. An example for the use of IHC stains is the determination of the estrogen receptor, the progesterone receptor, the epidermal growth factor receptor-2 (HER-2/neu), and the proliferative activity (ki-67 index) in breast cancer, all of which are markers to stratify patients into prognostic and predictive subgroups. However, for each single histochemical and IHC staining, a new tissue section is required. Thus, relatively large quantities of tissue are necessary that are not available for subsequent molecular analyses such as genetic testing. Especially in the biopsy situation where often only tiny amounts of tissue are resected, it would be highly beneficial for clinicians and patients to save as much tissue material as possible. By IMS, only one single section is required for the analysis of hundreds of peptides or proteins and this single section can even be stained by H&E after the MS analysis. This makes IMS a promising candidate to supplement or even replace the currently used approaches. Interestingly, ferritin heavy and light chains as a surrogate for iron have been identified also by IMS (Kriegsmann et al., 2014) and it is tempting to speculate that some histochemical stains as the Berlin-Blue reaction could be replaced by MS analysis. Moreover, IMS does not require any staining or labeling by antibodies and not even prior knowledge of the target structures. That would make stepwise immunohistochemical approaches as they are used today, for example, to identify the primary tumor in a patient with metastatic disease obsolete since all relevant proteins/peptides to identify the origin of a neoplasia may also be detected by MS. It has been shown that, indeed, relevant diagnostic proteins such as cytokeratins can be analyzed by IMS (Groseclose et al., 2008).

IMS analysis is highly specific and provides detailed information about the protein sequence when used in the MS/MS mode for identification purposes. As previously mentioned, with the advent of new instruments on the market, resolutions of <10 μm can be achieved that have the capacity to analyze subcellular structures (Ogrinc Potocnik, Porta, Becker, Heeren, & Ellis, 2015).

Data acquired on cancer samples in current research are very promising and applications for routine diagnostics are under development. However, up to now, the application in routine diagnostics is hampered by (i) high instrumental costs, (ii) limitations of the mass range depending on the matrix, the tissue type, and other preanalytic conditions, (iii) lack of standardization and quality control, (iv) difficulties in data interpretation, and (v) storage of large datasets.

However, most of the methodological limitations, for example, regarding sample preparation, resolution, sensitivity and data analysis have greatly improved during the past several years.

Thus, broader applications of IMS in routine diagnostics are to be expected in the next years.

ACKNOWLEDGMENTS

M.K. has been supported by the Post-Doc Program of the Medical Faculty of the University of Heidelberg. The authors would like to acknowledge the following funding: BMBF Grant FKZ 131A029F as part of the Leading—Edge Cluster Ci3 (Cluster for Individualized Immune Intervention), Zim Grant KF3342501SB4 (Development of a digital staining method as pathological–histological diagnostic tool based on the MALDI Imaging Technology; Development of a formalin-based sample preparation for fixation of tissue as sample preparation for MALDI imaging method), and BMBF Grant 13GW0081B sponsored by the Federal Ministry of Education and Research. KMU-innovativ: Medizintechnik.

REFERENCES

Balluff, B., Elsner, M., Kowarsch, A., et al. (2010). Classification of HER2/neu status in gastric cancer using a breast-cancer derived proteome classifier. *Journal of Proteome Research*, 9(12), 6317–6322. http://dx.doi.org/10.1021/pr100573s [published Online First: Epub Date].

Balluff, B., Rauser, S., Meding, S., et al. (2011). MALDI imaging identifies prognostic seven-protein signature of novel tissue markers in intestinal-type gastric cancer. *The American Journal of Pathology*, 179(6), 2720–2729. http://dx.doi.org/10.1016/j.ajpath.2011.08.032 [published Online First: Epub Date].

Becker, K. F., Schott, C., Hipp, S., et al. (2007). Quantitative protein analysis from formalin-fixed tissues: Implications for translational clinical research and nanoscale molecular diagnosis. *The Journal of Pathology, 211*(3), 370–378. http://dx.doi.org/10.1002/path.2107 [published Online First: Epub Date].

Beine, B., Diehl, H. C., Meyer, H. E., & Henkel, C. (2016). Tissue MALDI mass spectrometry imaging (MALDI MSI) of peptides. *Methods in Molecular Biology (Clifton, N.J.), 1394*, 129–150. http://dx.doi.org/10.1007/978-1-4939-3341-9_10 [published Online First: Epub Date].

Bohn, G., & Rucker, G. (1969). On mass spectrometry detection of barbituric acid derivatives in autopsy material after separation by thin layer chromatography. *Archiv für Toxikologie, 25*(1), 95–101.

Bonnichsen, R., Maehly, A. C., Marde, Y., Ryhage, R., & Schubert, B. (1970). Determination and identification of sympathomimetic amines in blood samples from drivers by a combination of gas chromatography and mass spectrometry. *Zeitschrift für Rechtsmedizin. Journal of Legal Medicine, 67*(1), 19–26.

Buck, A., Ly, A., Balluff, B., et al. (2015). High-resolution MALDI-FT-ICR MS imaging for the analysis of metabolites from formalin-fixed, paraffin-embedded clinical tissue samples. *The Journal of Pathology, 237*(1), 123–132. http://dx.doi.org/10.1002/path.4560 [published Online First: Epub Date].

Camerini, S., & Mauri, P. (2015). The role of protein and peptide separation before mass spectrometry analysis in clinical proteomics. *Journal of Chromatography. A, 1381*, 1–12. http://dx.doi.org/10.1016/j.chroma.2014.12.035 [published Online First: Epub Date].

Caprioli, R. M., Farmer, T. B., & Gile, J. (1997). Molecular imaging of biological samples: Localization of peptides and proteins using MALDI-TOF MS. *Analytical Chemistry, 69*(23), 4751–4760.

Carannante, A., De Carolis, E., Vacca, P., et al. (2015). Evaluation of matrix-assisted laser desorption ionization-time of flight mass spectrometry (MALDI-TOF MS) for identification and clustering of Neisseria gonorrhoeae. *BMC Microbiology, 15*, 142. http://dx.doi.org/10.1186/s12866-015-0480-y [published Online First: Epub Date].

Casadonte, R., & Caprioli, R. M. (2011). Proteomic analysis of formalin-fixed paraffin-embedded tissue by MALDI imaging mass spectrometry. *Nature Protocols, 6*(11), 1695–1709. http://dx.doi.org/10.1038/nprot.2011.388 [published Online First: Epub Date].

Casadonte, R., Kriegsmann, M., Deininger, S.-O., Amann, K., Paape, R., Belau, E., et al. (2015). Imaging mass spectrometry analysis of renal amyloidosis biopsies reveals protein co-localization with amyloid deposits. *Analytical and Bioanalytical Chemistry, 407*(18), 5323–5331.

Casadonte, R., Kriegsmann, M., Zweynert, F., et al. (2014). Imaging mass spectrometry to discriminate breast from pancreatic cancer metastasis in formalin-fixed paraffin-embedded tissues. *Proteomics, 14*(7–8), 956–964. http://dx.doi.org/10.1002/pmic.201300430 [published Online First: Epub Date].

Cassagne, C., Normand, A. C., L'Ollivier, C., Ranque, S., & Piarroux, R. (2016). Performance of MALDI-TOF MS platforms for fungal identification. *Mycoses, 59*, 678–690. http://dx.doi.org/10.1111/myc.12506[published Online First: Epub Date].

Charretier, Y., & Schrenzel, J. (2016). Mass spectrometry methods for predicting antibiotic resistance. *Proteomics. Clinical Applications, 10*, 964–981. http://dx.doi.org/10.1002/prca.201600041 [published Online First: Epub Date].

Chaurand, P. (2012). Imaging mass spectrometry of thin tissue sections: A decade of collective efforts. *Journal of Proteomics, 75*(16), 4883–4892. http://dx.doi.org/10.1016/j.jprot.2012.04.005 [published Online First: Epub Date].

Chaurand, P., Sanders, M. E., Jensen, R. A., & Caprioli, R. M. (2004). Proteomics in diagnostic pathology: Profiling and imaging proteins directly in tissue sections. *The American Journal of Pathology, 165*(4), 1057–1068. http://dx.doi.org/10.1016/s0002-9440(10)63367-6 [published Online First: Epub Date].

Chaurand, P., Schriver, K. E., & Caprioli, R. M. (2007). Instrument design and characterization for high resolution MALDI-MS imaging of tissue sections. *Journal of Mass Spectrometry: JMS, 42*(4), 476–489. http://dx.doi.org/10.1002/jms.1180 [published Online First: Epub Date].

Cheng, K., Chui, H., Domish, L., Hernandez, D., & Wang, G. (2016). Recent development of mass spectrometry and proteomics applications in identification and typing of bacteria. *Proteomics. Clinical Applications, 10*(4), 346–357. http://dx.doi.org/10.1002/prca.201500086 [published Online First: Epub Date].

Chughtai, K., & Heeren, R. M. (2010). Mass spectrometric imaging for biomedical tissue analysis. *Chemical Reviews, 110*(5), 3237–3277. http://dx.doi.org/10.1021/cr100012c [published Online First: Epub Date].

Cobo, F. (2013). Application of MALDI-TOF mass spectrometry in clinical virology: A review. *The Open Virology Journal, 7*, 84–90. http://dx.doi.org/10.2174/1874357920130927003 [published Online First: Epub Date].

Cornett, S. (2015). M-JMioo-tdahsrPspattACoMSaA. MALDI imaging of on-tissue digests at high spatial resolution. In *Poster session presented at the conference on mass spectrometry and allied topics; St. Louis, MO.*

Deininger, S. O., Cornett, D. S., Paape, R., et al. (2011). Normalization in MALDI-TOF imaging datasets of proteins: Practical considerations. *Analytical and Bioanalytical Chemistry, 401*(1), 167–181. http://dx.doi.org/10.1007/s00216-011-4929-z [published Online First: Epub Date].

Deininger, L., Patel, E., Clench, M. R., Sears, V., Sammon, C., & Francese, S. (2016). Proteomics goes forensic: Detection and mapping of blood signatures in fingermarks. *Proteomics, 16*(11–12), 1707–1717. http://dx.doi.org/10.1002/pmic.201500544 [published Online First: Epub Date].

Diehl, H. C., Beine, B., Elm, J., et al. (2015). The challenge of on-tissue digestion for MALDI MSI—A comparison of different protocols to improve imaging experiments. *Analytical and Bioanalytical Chemistry, 407*(8), 2223–2243. http://dx.doi.org/10.1007/s00216-014-8345-z [published Online First: Epub Date].

Djidja, M. C., Claude, E., Snel, M. F., et al. (2010). Novel molecular tumour classification using MALDI-mass spectrometry imaging of tissue micro-array. *Analytical and Bioanalytical Chemistry, 397*(2), 587–601. http://dx.doi.org/10.1007/s00216-010-3554-6 [published Online First: Epub Date].

Edwards, E., & Thomas-Oates, J. (2005). Hyphenating liquid phase separation techniques with mass spectrometry: On-line or off-line. *The Analyst, 130*(1), 13–17.

Elsner, M., Rauser, S., Maier, S., et al. (2012). MALDI imaging mass spectrometry reveals COX7A2, TAGLN2 and S100-A10 as novel prognostic markers in Barrett's adenocarcinoma. *Journal of Proteomics, 75*(15), 4693–4704. http://dx.doi.org/10.1016/j.jprot.2012.02.012 [published Online First: Epub Date].

Fedor, H. L., & De Marzo, A. M. (2005). Practical methods for tissue microarray construction. *Methods in Molecular Medicine, 103*, 89–101.

Fenselau, C., Caprioli, R., Nier, A. O., Hanson, W. B., Seiff, A., Mcelroy, M. B., et al. (2003). Mass spectrometry in the exploration of Mars. *Journal of Mass Spectrometry, 38*(1), 1–10.

Foulkes, W. D., Smith, I. E., & Reis-Filho, J. S. (2010). Triple-negative breast cancer. *The New England Journal of Medicine, 363*(20), 1938–1948. http://dx.doi.org/10.1056/NEJMra1001389 [published Online First: Epub Date].

Gravius, S., Randau, T. M., Casadonte, R., Kriegsmann, M., Friedrich, M. J., & Kriegsmann, J. (2015). Investigation of neutrophilic peptides in periprosthetic tissue by matrix-assisted laser desorption ionisation time-of-flight imaging mass spectrometry. *International Orthopaedics, 39*(3), 559–567. http://dx.doi.org/10.1007/s00264-014-2544-2 [published Online First: Epub Date].

Groseclose, M. R., Andersson, M., Hardesty, W. M., & Caprioli, R. M. (2007). Identification of proteins directly from tissue: In situ tryptic digestions coupled with imaging mass spectrometry. *Journal of Mass Spectrometry: JMS, 42*(2), 254–262. http://dx.doi.org/10.1002/jms.1177 [published Online First: Epub Date].

Groseclose, M. R., Massion, P. P., Chaurand, P., & Caprioli, R. M. (2008). High-throughput proteomic analysis of formalin-fixed paraffin-embedded tissue microarrays using MALDI imaging mass spectrometry. *Proteomics, 8*(18), 3715–3724. http://dx.doi.org/10.1002/pmic.200800495 [published Online First: Epub Date].

Gustafsson, O. J., Eddes, J. S., Meding, S., McColl, S. R., Oehler, M. K., & Hoffmann, P. (2013). Matrix-assisted laser desorption/ionization imaging protocol for in situ characterization of tryptic peptide identity and distribution in formalin-fixed tissue. *Rapid Communications in Mass Spectrometry: RCM, 27*(6), 655–670. http://dx.doi.org/10.1002/rcm.6488 [published Online First: Epub Date].

Gustafsson, J. O., Oehler, M. K., McColl, S. R., & Hoffmann, P. (2010). Citric acid antigen retrieval (CAAR) for tryptic peptide imaging directly on archived formalin-fixed paraffin-embedded tissue. *Journal of Proteome Research, 9*(9), 4315–4328. http://dx.doi.org/10.1021/pr9011766 [published Online First: Epub Date].

Hanselmann, M., Kothe, U., Kirchner, M., et al. (2009). Toward digital staining using imaging mass spectrometry and random forests. *Journal of Proteome Research, 8*(7), 3558–3567. http://dx.doi.org/10.1021/pr900253y[published Online First: Epub Date].

Harris, G. A., Nicklay, J. J., & Caprioli, R. M. (2013). Localized in situ hydrogel-mediated protein digestion and extraction technique for on-tissue analysis. *Analytical Chemistry, 85*(5), 2717–2723. http://dx.doi.org/10.1021/ac3031493 [published Online First: Epub Date].

Harvey, D. J. (2015). Analysis of carbohydrates and glycoconjugates by matrix-assisted laser desorption/ionization mass spectrometry: An update for 2011–2012. *Mass Spectrometry Reviews, 34*, 268–422. http://dx.doi.org/10.1002/mas.21471 [published Online First: Epub Date].

Heijs, B., Holst, S., Briaire-de Bruijn, I. H., et al. (2016). Multimodal mass spectrometry imaging of N-glycans and proteins from the same tissue section. *Analytical Chemistry, 88*, 7745–7753. http://dx.doi.org/10.1021/acs.analchem.6b01739 [published Online First: Epub Date].

Hoos, A., & Cordon-Cardo, C. (2001). Tissue microarray profiling of cancer specimens and cell lines: Opportunities and limitations. *Laboratory Investigation: A Journal of Technical Methods and Pathology, 81*(10), 1331–1338.

Ikeda, K., Monden, T., Kanoh, T., et al. (1998). Extraction and analysis of diagnostically useful proteins from formalin-fixed, paraffin-embedded tissue sections. *The Journal of Histochemistry and Cytochemistry: Official Journal of the Histochemistry Society, 46*(3), 397–403.

Jacquemier, J., Ginestier, C., Charafe-Jauffret, E., et al. (2003). Small but high throughput: How "tissue-microarrays" became a favorite tool for pathologists and scientists. *Annales de Pathologie, 23*(6), 623–632.

Jones, E. A., Deininger, S. O., Hogendoorn, P. C., Deelder, A. M., & McDonnell, L. A. (2012). Imaging mass spectrometry statistical analysis. *Journal of Proteomics, 75*(16), 4962–4989. http://dx.doi.org/10.1016/j.jprot.2012.06.014 [published Online First: Epub Date].

Kailemia, M. J., Ruhaak, L. R., Lebrilla, C. B., & Amster, I. J. (2014). Oligosaccharide analysis by mass spectrometry: A review of recent developments. *Analytical Chemistry, 86*(1), 196–212. http://dx.doi.org/10.1021/ac403969n [published Online First: Epub Date].

Kaufmann, R., Chaurand, P., Kirsch, D., & Spengler, B. (1996). Post-source decay and delayed extraction in matrix-assisted laser desorption/ionization-reflectron time-of-flight mass spectrometry. Are there trade-offs? *Rapid Communications in Mass Spectrometry: RCM*, *10*(10), 1199–1208. http://dx.doi.org/10.1002/(SICI)1097-0231(19960731)10:10<1199::AID-RCM643>3.0.CO;2-F [published Online First: Epub Date].

Khalil, S. M., Rompp, A., Pretzel, J., Becker, K., & Spengler, B. (2015). Phospholipid topography of whole-body sections of the anopheles stephensi mosquito, characterized by high-resolution atmospheric-pressure scanning microprobe matrix-assisted laser desorption/ionization mass spectrometry imaging. *Analytical Chemistry*, *87*(22), 11309–11316. http://dx.doi.org/10.1021/acs.analchem.5b02781 [published Online First: Epub Date].

Koh, Y. X., Chok, A. Y., Zheng, H. L., Tan, C. S., & Goh, B. K. (2014). Systematic review and meta-analysis comparing the surgical outcomes of invasive intraductal papillary mucinous neoplasms and conventional pancreatic ductal adenocarcinoma. *Annals of Surgical Oncology*, *21*(8), 2782–2800. http://dx.doi.org/10.1245/s10434-014-3639-0 [published Online First: Epub Date].

Kriegsmann, M., Arens, N., Endris, V., Weichert, W., & Kriegsmann, J. (2015). Detection of KRAS, NRAS and BRAF by mass spectrometry—A sensitive, reliable, fast and cost-effective technique. *Diagnostic Pathology*, *10*, 132. http://dx.doi.org/10.1186/s13000-015-0364-3 [published Online First: Epub Date].

Kriegsmann, M., Casadonte, R., Randau, T., Gravius, S., Pennekamp, P., Strauss, A., et al. (2014). MALDI imaging of predictive ferritin, fibrinogen and proteases in haemophilic arthropathy. *Haemophilia*, *20*(3), 446–453.

Kriegsmann, J., Kriegsmann, M., & Casadonte, R. (2015). MALDI TOF imaging mass spectrometry in clinical pathology: A valuable tool for cancer diagnostics (review). *International Journal of Oncology*, *46*(3), 893–906. http://dx.doi.org/10.3892/ijo.2014.2788 [published Online First: Epub Date].

Kriegsmann, M., Muley, T., Harms, A., et al. (2015). Differential diagnostic value of CD5 and CD117 expression in thoracic tumors: A large scale study of 1465 non-small cell lung cancer cases. *Diagnostic Pathology*, *10*, 210. http://dx.doi.org/10.1186/s13000-015-0441-7 [published Online First: Epub Date].

Kriegsmann, M., Seeley, E. H., Schwarting, A., et al. (2012). MALDI MS imaging as a powerful tool for investigating synovial tissue. *Scandinavian Journal of Rheumatology*, *41*(4), 305–309. http://dx.doi.org/10.3109/03009742.2011.647925 [published Online First: Epub Date].

Longuespee, R., Alberts, D., Pottier, C., et al. (2016). A laser microdissection-based workflow for FFPE tissue microproteomics: Important considerations for small sample processing. *Methods*, *104*, 154–162. http://dx.doi.org/10.1016/j.ymeth.2015.12.008 [published Online First: Epub Date].

Longuespee, R., Casadonte, R., Kriegsmann, M., et al. (2016). MALDI mass spectrometry imaging: A cutting-edge tool for fundamental and clinical histopathology. *Proteomics Clinical Applications*, *10*, 701–719. http://dx.doi.org/10.1002/prca.201500140 [published Online First: Epub Date].

Longuespee, R., Fleron, M., Pottier, C., et al. (2014). Tissue proteomics for the next decade? Towards a molecular dimension in histology. *Omics: A Journal of Integrative Biology*, *18*(9), 539–552. http://dx.doi.org/10.1089/omi.2014.0033 [published Online First: Epub Date].

Longuespee, R., Gagnon, H., Boyon, C., et al. (2013). Proteomic analyses of serous and endometrioid epithelial ovarian cancers—Cases studies—Molecular insights of a possible histological etiology of serous ovarian cancer. *Proteomics. Clinical Applications*, *7*(5–6), 337–354. http://dx.doi.org/10.1002/prca.201200079. [published Online First: Epub Date].

Longuespee, R., Tastet, C., Desmons, A., et al. (2014). HFIP extraction followed by 2D CTAB/SDS-PAGE separation: A new methodology for protein identification from tissue sections after MALDI mass spectrometry profiling for personalized medicine research. *Omics: A Journal of Integrative Biology*, *18*(6), 374–384. http://dx.doi.org/10.1089/omi.2013.0176 [published Online First: Epub Date].

Maier, S. K., Hahne, H., Gholami, A. M., et al. (2013). Comprehensive identification of proteins from MALDI imaging. *Molecular & Cellular Proteomics: MCP*, *12*(10), 2901–2910. http://dx.doi.org/10.1074/mcp.M113.027599 [published Online First: Epub Date].

McCombie, G., Staab, D., Stoeckli, M., & Knochenmuss, R. (2005). Spatial and spectral correlations in MALDI mass spectrometry images by clustering and multivariate analysis. *Analytical Chemistry*, *77*(19), 6118–6124. http://dx.doi.org/10.1021/ac051081q [published Online First: Epub Date].

McDonnell, L. A., Walch, A., Stoeckli, M., & Corthals, G. L. (2014). MSiMass list: A public database of identifications for protein MALDI MS imaging. *Journal of Proteome Research*, *13*(2), 1138–1142. http://dx.doi.org/10.1021/pr400620y [published Online First: Epub Date].

Meding, S., Balluff, B., Elsner, M., et al. (2012). Tissue-based proteomics reveals FXYD3, S100A11 and GSTM3 as novel markers for regional lymph node metastasis in colon cancer. *The Journal of Pathology*, *228*(4), 459–470. http://dx.doi.org/10.1002/path.4021 [published Online First: Epub Date].

Meding, S., Martin, K., Gustafsson, O. J., et al. (2013). Tryptic peptide reference data sets for MALDI imaging mass spectrometry on formalin-fixed ovarian cancer tissues. *Journal of Proteome Research*, *12*(1), 308–315. http://dx.doi.org/10.1021/pr300996x [published Online First: Epub Date].

Mittal, P., Klingler-Hoffmann, M., Arentz, G., et al. (2016). Lymph node metastasis of primary endometrial cancers: Associated proteins revealed by MALDI imaging. *Proteomics*, *16*(11–12), 1793–1801. http://dx.doi.org/10.1002/pmic.201500455 [published Online First: Epub Date].

Morgan, T. M., Seeley, E. H., Fadare, O., Caprioli, R. M., & Clark, P. E. (2013). Imaging the clear cell renal cell carcinoma proteome. *The Journal of Urology*, *189*(3), 1097–1103. http://dx.doi.org/10.1016/j.juro.2012.09.074 [published Online First: Epub Date].

Morita, Y., Ikegami, K., Goto-Inoue, N., et al. (2010). Imaging mass spectrometry of gastric carcinoma in formalin-fixed paraffin-embedded tissue microarray. *Cancer Science*, *101*(1), 267–273. http://dx.doi.org/10.1111/j.1349-7006.2009.01384.x [published Online First: Epub Date].

Nirmalan, N. J., Harnden, P., Selby, P. J., & Banks, R. E. (2008). Mining the archival formalin-fixed paraffin-embedded tissue proteome: Opportunities and challenges. *Molecular BioSystems*, *4*(7), 712–720. http://dx.doi.org/10.1039/b800098k [published Online First: Epub Date].

Ogrinc Potocnik, N., Porta, T., Becker, M., Heeren, R. M., & Ellis, S. R. (2015). Use of advantageous, volatile matrices enabled by next-generation high-speed matrix-assisted laser desorption/ionization time-of-flight imaging employing a scanning laser beam. *Rapid Communications in Mass Spectrometry: RCM*, *29*(23), 2195–2203. http://dx.doi.org/10.1002/rcm.7379 [published Online First: Epub Date].

Ozawa, T., Osaka, I., Hamada, S., et al. (2016). Direct imaging mass spectrometry of plant leaves using surface-assisted laser desorption/ionization with sputter-deposited platinum film. *Analytical Sciences: The International Journal of the Japan Society for Analytical Chemistry*, *32*(5), 587–591. http://dx.doi.org/10.2116/analsci.32.587 [published Online First: Epub Date].

Pallua, J. D., Schaefer, G., Seifarth, C., et al. (2013). MALDI-MS tissue imaging identification of biliverdin reductase B overexpression in prostate cancer. *Journal of Proteomics*, *91*, 500–514. http://dx.doi.org/10.1016/j.jprot.2013.08.003 [published Online First: Epub Date].

Pavlidis, N., & Pentheroudakis, G. (2012). Cancer of unknown primary site. *Lancet*, *379*(9824), 1428–1435. http://dx.doi.org/10.1016/S0140-6736(11)61178-1 [published Online First: Epub Date].

Powers, T. W., Jones, E. E., Betesh, L. R., et al. (2013). Matrix assisted laser desorption ionization imaging mass spectrometry workflow for spatial profiling analysis of N-linked glycan expression in tissues. *Analytical Chemistry*, *85*(20), 9799–9806. http://dx.doi.org/10.1021/ac402108x [published Online First: Epub Date].

Quaas, A., Bahar, A. S., von Loga, K., et al. (2013). MALDI imaging on large-scale tissue microarrays identifies molecular features associated with tumour phenotype in oesophageal cancer. *Histopathology*, *63*(4), 455–462. http://dx.doi.org/10.1111/his.12193 [published Online First: Epub Date].

Rauser, S., Marquardt, C., Balluff, B., et al. (2010). Classification of HER2 receptor status in breast cancer tissues by MALDI imaging mass spectrometry. *Journal of Proteome Research*, *9*(4), 1854–1863. http://dx.doi.org/10.1021/pr901008d [published Online First: Epub Date].

Roh, K., Yeo, S. G., Yoo, B. C., Kim, K. H., Kim, S. Y., & Kim, M. J. (2016). Seven low-mass ions in pretreatment serum as potential predictive markers of the chemoradiotherapy response of rectal cancer. *Anti-Cancer Drugs*, *27*, 787–793. http://dx.doi.org/10.1097/CAD.0000000000000391 [published Online First: Epub Date].

Said, S. M., Sethi, S., Valeri, A. M., et al. (2013). Renal amyloidosis: Origin and clinicopathologic correlations of 474 recent cases. *Clinical Journal of the American Society of Nephrology: CJASN*, *8*(9), 1515–1523. http://dx.doi.org/10.2215/CJN.10491012 [published Online First: Epub Date].

Schwamborn, K., Krieg, R. C., Reska, M., Jakse, G., Knuechel, R., & Wellmann, A. (2007). Identifying prostate carcinoma by MALDI-imaging. *International Journal of Molecular Medicine*, *20*(2), 155–159.

Sethi, S., Theis, J. D., Vrana, J. A., et al. (2013). Laser microdissection and proteomic analysis of amyloidosis, cryoglobulinemic GN, fibrillary GN, and immunotactoid glomerulopathy. *Clinical Journal of the American Society of Nephrology: CJASN*, *8*(6), 915–921. http://dx.doi.org/10.2215/CJN.07030712 [published Online First: Epub Date].

Spraggins, J. M., Rizzo, D. G., Moore, J. L., Noto, M. J., Skaar, E. P., & Caprioli, R. M. (2016). Next-generation technologies for spatial proteomics: Integrating ultra-high speed MALDI-TOF and high mass resolution MALDI FTICR imaging mass spectrometry for protein analysis. *Proteomics*, *16*(11–12), 1678–1689. http://dx.doi.org/10.1002/pmic.201600003 [published Online First: Epub Date].

Steurer, S., Borkowski, C., Odinga, S., et al. (2013). MALDI mass spectrometric imaging based identification of clinically relevant signals in prostate cancer using large-scale tissue microarrays. *International Journal of Cancer*, *133*(4), 920–928. http://dx.doi.org/10.1002/ijc.28080 [published Online First: Epub Date].

Steurer, S., Seddiqi, A. S., Singer, J. M., et al. (2014). MALDI imaging on tissue microarrays identifies molecular features associated with renal cell cancer phenotype. *Anticancer Research*, *34*(5), 2255–2261.

Steurer, S., Singer, J. M., Rink, M., et al. (2014). MALDI imaging-based identification of prognostically relevant signals in bladder cancer using large-scale tissue microarrays. *Urologic Oncology*, *32*(8), 1225–1233. http://dx.doi.org/10.1016/j.urolonc.2014.06.007 [published Online First: Epub Date].

Swales, J. G., Tucker, J. W., Spreadborough, M. J., et al. (2015). Mapping drug distribution in brain tissue using liquid extraction surface analysis mass spectrometry imaging. *Analytical Chemistry*, *87*(19), 10146–10152. http://dx.doi.org/10.1021/acs.analchem.5b02998 [published Online First: Epub Date].

Taverna, D., Pollins, A. C., Nanney, L. B., Sindona, G., & Caprioli, R. M. (2016). Histology-guided protein digestion/extraction from formalin-fixed and paraffin-embedded pressure ulcer biopsies. *Experimental Dermatology*, *25*(2), 143–146. http://dx.doi.org/10.1111/exd.12870 [published Online First: Epub Date].

van Belkum, A., Chatellier, S., Girard, V., Pincus, D., Deol, P., & Dunne, W. M., Jr. (2015). Progress in proteomics for clinical microbiology: MALDI-TOF MS for microbial species identification and more. *Expert Review of Proteomics*, *12*(6), 595–605. http://dx.doi.org/10.1586/14789450.2015.1091731 [published Online First: Epub Date].

Verma, S., Miles, D., Gianni, L., et al. (2012). Trastuzumab emtansine for HER2-positive advanced breast cancer. *The New England Journal of Medicine*, *367*(19), 1783–1791. http://dx.doi.org/10.1056/NEJMoa1209124 [published Online First: Epub Date].

Warth, A., Muley, T., Herpel, E., et al. (2012). Large-scale comparative analyses of immunomarkers for diagnostic subtyping of non-small-cell lung cancer biopsies. *Histopathology*, *61*(6), 1017–1025. http://dx.doi.org/10.1111/j.1365-2559.2012.04308.x [published Online First: Epub Date].

Yalcin, E. B., & de la Monte, S. M. (2015). Review of matrix-assisted laser desorption ionization-imaging mass spectrometry for lipid biochemical histopathology. *Journal of Histochemistry & Cytochemistry*, *63*(10), 762–771.

Yang, J., Rower, C., Koy, C., et al. (2015). Mass spectrometric characterization of limited proteolysis activity in human plasma samples under mild acidic conditions. *Methods*, *89*, 30–37. http://dx.doi.org/10.1016/j.ymeth.2015.02.013 [published Online First: Epub Date].

Zavalin, A., Yang, J., Hayden, K., Vestal, M., & Caprioli, R. M. (2015). Tissue protein imaging at 1 mum laser spot diameter for high spatial resolution and high imaging speed using transmission geometry MALDI TOF MS. *Analytical and Bioanalytical Chemistry*, *407*(8), 2337–2342. http://dx.doi.org/10.1007/s00216-015-8532-6 [published Online First: Epub Date].

Zlobec, I., Koelzer, V. H., Dawson, H., Perren, A., & Lugli, A. (2013). Next-generation tissue microarray (ngTMA) increases the quality of biomarker studies: An example using CD3, CD8, and CD45RO in the tumor microenvironment of six different solid tumor types. *Journal of Translational Medicine*, *11*, 104. http://dx.doi.org/10.1186/1479-5876-11-104 [published Online First: Epub Date].

CHAPTER EIGHT

Mass Spectrometry Imaging for the Investigation of Intratumor Heterogeneity

B. Balluff*,[1],[2], M. Hanselmann[†],[3], R.M.A. Heeren*

*Maastricht University, Maastricht MultiModal Molecular Imaging institute (M4I), Maastricht, The Netherlands
†Heidelberg Collaboratory for Image Processing (HCI), Interdisciplinary Center for Scientific Computing (IWR), University of Heidelberg, Heidelberg, Germany
[2]Corresponding author: e-mail address: b.balluff@maastrichtuniversity.nl

Contents

1. Tumor Heterogeneity	202
1.1 Intratumor Heterogeneity	203
1.2 The Study of ITH	204
1.3 Clinical Relevance of ITH	206
1.4 Techniques to Study Spatial Organization of ITH	206
2. MSI to Study Tumor Heterogeneity	207
3. Multivariate Data Analysis Strategies in MSI	209
3.1 Unsupervised Analysis	211
3.2 Supervised Classification	212
3.3 Projection Methods	212
4. MSI Applications for the Investigation of ITH	213
4.1 Revealing ITH by Clustering	213
4.2 Supervised Classification of ITH	218
4.3 Investigating the Degree of ITH	219
4.4 Investigation of ITH on Different Molecular Levels	222
5. Future Applications of MSI in ITH Research	223
6. Perspective	224
References	224

Abstract

One of the big clinical challenges in the treatment of cancer is the different behavior of cancer patients under guideline therapy. An important determinant for this phenomenon has been identified as inter- and intratumor heterogeneity. While intertumor heterogeneity refers to the differences in cancer characteristics between patients,

---

[1] Senior author.
[3] Now at Robert Bosch GmbH, Robert-Bosch-Campus 1, Renningen, Germany.

intratumor heterogeneity refers to the clonal and nongenetic molecular diversity within a patient. The deciphering of intratumor heterogeneity is recognized as key to the development of novel therapeutics or treatment regimens.

The investigation of intratumor heterogeneity is challenging since it requires an untargeted molecular analysis technique that accounts for the spatial and temporal dynamics of the tumor. So far, next-generation sequencing has contributed most to the understanding of clonal evolution within a cancer patient. However, it falls short in accounting for the spatial dimension.

Mass spectrometry imaging (MSI) is a powerful tool for the untargeted but spatially resolved molecular analysis of biological tissues such as solid tumors. As it provides multidimensional datasets by the parallel acquisition of hundreds of mass channels, multivariate data analysis methods can be applied for the automated annotation of tissues. Moreover, it integrates the histology of the sample, which enables studying the molecular information in a histopathological context.

This chapter will illustrate how MSI in combination with statistical methods and histology has been used for the description and discovery of intratumor heterogeneity in different cancers. This will give evidence that MSI constitutes a unique tool for the investigation of intratumor heterogeneity, and could hence become a key technology in cancer research.

1. TUMOR HETEROGENEITY

In a meeting in 2015, which was organized by Nature Medicine, Nature Biotechnology and the Volkswagen Foundation, 20 leading cancer research scientists came together and to the conclusion that "… tumor heterogeneity is likely to influence—for some time to come—all aspects of cancer research, including how tumor biology is perceived, how techniques to study tumors are developed and how patients are treated." (Alizadeh et al., 2015).

Cancer is not a homogeneous disease. Classification systems distinguish dozens of different types and hundreds of subtypes (Fletcher, World Health Organization, & International Agency for Research on Cancer, 2013; World Health Organization, 1992). Patients can further be stratified on the level of their tumor's progression, termed staging—most solid tumors are staged according to the TNM system (Edge & American Joint Committee on Cancer, 2010)—and the status of certain molecular markers. But even patients that share the same cancer type, subtype, stage, and molecular marker status will differ in their clinical course and response to therapy. The disease-related reason for the diversity in clinical outcome among patients is known as tumor heterogeneity. Consequently, there is the need for understanding tumor heterogeneity to enable highly personalized therapeutic strategies.

1.1 Intratumor Heterogeneity

Heterogeneity between individuals with the same tumor type is termed intertumor heterogeneity, whereas intratumor heterogeneity (ITH) refers to the mix of different tumor cells within a patient (Fig. 1A) (Burrell, McGranahan, Bartek, & Swanton, 2013). Although interconnected, it is important to distinguish between histological and molecular ITH. Molecular ITH is based on genetic and nongenetic diversity in the cells, and histological ITH on morphological and cytological difference between tumor cells.

Differences in cell morphology between tumor cells are known since the times of Virchow. This was followed by observations of molecular variance

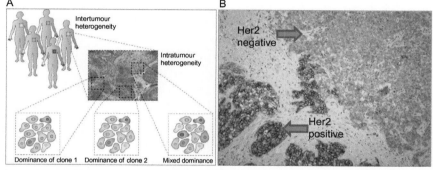

Fig. 1 (A) Tumor heterogeneity can be of two kinds: inter- and intratumor heterogeneity. The first refers to differences in tumor type, subtype, stage, molecular status, etc. between patients, and the latter to the mix of histologically or molecularly different tumor cells within a patient. These distinct subpopulations are unequally distributed over space and time and are known to influence the tumor's progression. (B) Heterogeneous HER2 expression is commonly observed as shown in this breast cancer tissue by immunohistochemistry. Cancer cells overexpressing HER2 are believed to be the tumor drivers. However, patients with sufficiently HER2-positive cells can be treated by targeted therapy which leads to a better prognosis compared to patients with HER2-negative breast cancer (Dawood, Broglio, Buzdar, Hortobagyi, & Giordano, 2010). *Panel (A) reprinted by permission from Macmillan Publishers Ltd.: Nature Reviews Cancer. Marusyk, A., Almendro, V., & Polyak, K. (2012). Intra-tumour heterogeneity: A looking glass for cancer? Nature Reviews. Cancer, 12(5), 323–334. http://dx.doi.org/10.1038/nrc3261, Copyright (2012). Panel (B) reprinted from Wu, J. M., Halushka, M. K., & Argani, P. (2010). Intratumoral heterogeneity of HER-2 gene amplification and protein overexpression in breast cancer. Human Pathology, 41(6), 914–917. http://dx.doi.org/10.1016/j.humpath.2009.10.022, Copyright 2010, with permission from Elsevier.*

within tumors such as genetic abnormalities and differential expression of cell surface markers, indicating the existence of subclonal structures within tumors (Caiado, Silva-Santos, & Norell, 2016). An example for molecular ITH in a breast cancer tissue section is depicted in Fig. 1B. Then in 1976 Peter Nowell was first to publish in Science the theory of clonal evolution in cancer to explain the existence and development of different tumor cell populations based on classical Darwinian evolutionary mechanisms (Nowell, 1976). He postulated that genetic diversity is created through the continuous acquisition of mutations during replication of the cells. Then, selective pressures lead to the natural selection of the clones with the highest fitness as determined by their molecular setup (Greaves & Maley, 2012) (Fig. 2A). The selective pressures can be multiple: clonal competition for nutrients, hypoxia and evasion of apoptosis, cytotoxic agents during chemotherapy, attacks of the immune system, and new microenvironmental conditions as the tumor outgrows its natural ecosystem, which may ultimately lead to metastasis. Advantageous mutations are called "driver" mutations and selectively neutral mutations, "passenger" mutations. The impact of a mutation is defined by the spatio-temporal dynamics surrounding the process of clonal evolution (de Bruin et al., 2014; Swanton, 2012).

1.2 The Study of ITH

Several decades later with the introduction of high-throughput next-generation sequencing (NGS) technologies, a myriad of studies have demonstrated the validity of the clonal evolution hypothesis (Gerlinger et al., 2014, 2012), see Fig. 2B, and proposed alternative processes for the creation of clonal diversity based on multilineage differentiation of cancer stem cells (Shibata, 2006). Moreover, it has been confirmed that ITH not only plays an important role in the tumor's progression, but also determines the disease outcome of the patient including prognosis (Maley et al., 2006), response to chemotherapy (Ding et al., 2012; Turner & Reis-Filho, 2012), or relapse (Ding et al., 2012).

Most of the studies investigating ITH were genetic studies, as phenotypic ITH is generally considered a reflection of the underlying genetic diversity (Caiado et al., 2016). But it has been shown that tumor cells with equal genetic setup can display functional variability, which contributes to both cancer growth and therapy tolerance (Kreso et al., 2013). This has been

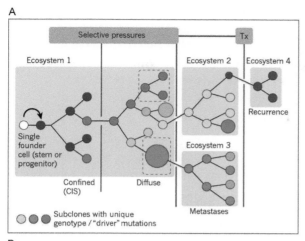
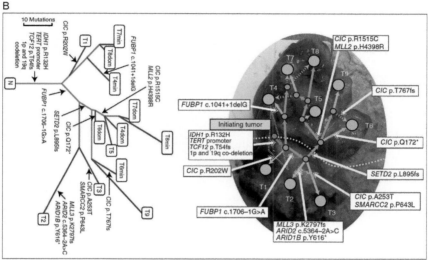

Fig. 2 (A) The theory of clonal evolution of cancer suggests a sequential acquisition of mutations which results in a diversity of molecularly distinct tumor populations, termed subclones. Selective pressures such as competition for nutrients, exposure to chemotherapy, or new microenvironmental conditions will select the subclones with the highest fitness. These so-called tumor driver clones will promote the tumor's progression, ultimately leading to growth, recurrence, or metastasis of the tumor. (B) For instance, a study on gliomas investigated clonal evolution and expansion through multiregion sampling and next-generation sequencing. This allowed mapping the phylogenetic evolution of driver and parallel mutations (*left*, phylogenic tree) into a spatial context (*right*, surgical specimen with sampling sites in *blue*). *Panel (A) adapted by permission from Macmillan Publishers Ltd.: Nature. Greaves, M., & Maley, C. C. (2012). Clonal evolution in cancer. Nature, 481(7381), 306–313. http://dx.doi.org/10.1038/nature10762, Copyright (2012). Panel (B) adapted by permission from Macmillan Publishers Ltd.: Nature Genetics. Suzuki, H., Aoki, K., Chiba, K., Sato, Y., Shiozawa, Y., Shiraishi, Y., et al. (2015). Mutational landscape and clonal architecture in grade II and III gliomas. Nature Genetics, 47(5), 458–468. http://dx.doi.org/10.1038/ng.3273, Copyright (2015).*

shown by studies on single-cell transcriptomal data from glioblastomas and colorectal carcinomas which indicated ITH of equal degree on a protein expression level (Dalerba et al., 2011; Patel et al., 2014).

1.3 Clinical Relevance of ITH

In fact, in classical histopathological investigations and clinical routine, which use immunohistochemistry (IHC) to visualize a specific protein's expression in tissue sections, ITH is encountered recurrently. An example is the heterogeneous expression of HER2 in several cancer types, among them Barrett's esophagus, gastric, and breast cancer (Fig. 1B) (Kanayama, Imai, Yoneda, Hirokawa, & Shiraishi, 2016; Walch et al., 2000; Wu, Halushka, & Argani, 2010). The correct determination of the HER2 status of a patient is of importance for therapy decision-making because patients with HER2 overexpression are eligible for the treatment with the monoclonal antibody trastuzumab. Clinical guidelines account for ITH in HER2 expression by considering a patient HER2-positive if at least 10% of the tumor cells are affected by HER2 overexpression (Ruschoff et al., 2012). The benefit of targeting this tumor driver clone is backed by clinical trials which show that HER2-positive patients that are treated with trastuzumab exhibit a statistically significant higher overall survival probability compared to HER2-negative and HER2-positive but untreated patients (Dawood et al., 2010). HER2 is an excellent example that illustrates how promising it can be for personalized medicine to discover new phenotypic clones of clinical relevance (potential tumor drivers), and to characterize them on a molecular level in order to develop new therapeutic strategies to target those clones.

In that light, it will become essential for cancer research to get a full picture on the landscape of ITH, in both its spatial and temporal extent, and on all molecular levels. In addition, the clinical relevance of the clonal composition has to be determined. This will expand the predictability horizon for precision cancer medicine (Lipinski et al., 2016).

1.4 Techniques to Study Spatial Organization of ITH

As presented in Fig. 1B, the spatial organization of molecular heterogeneity can be visualized by targeted imaging assays such as IHC or fluorescence in situ hybridization. However, targeted assays require a priori knowledge of the target to be studied and are therefore unsuited to discovery-based research of novel tumor subpopulations. On the other hand, untargeted discovery techniques such as NGS usually lack the capacity to account for the spatial information. This is currently circumvented by exploring the spatial ITH through multiregion sampling from the primary or metastatic tumor, or through multiple histology-directed cutouts from tissue sections (Turajlic, McGranahan, & Swanton, 2015) (Fig. 2B). This approach, called macroheterogeneity profiling, however, is unsuitable for ITH that is confined to small subclones of up to a few thousand cancer cells (Lipinski et al., 2016).

In consequence, the investigation of the microheterogeneity requires an unlabeled yet spatially resolved read-out of the molecular information of the tumor at a micrometer scale. An emerging technology that fulfills these requirements and could hence help in creating a map of molecular ITH in solid cancers is mass spectrometry imaging (MSI).

2. MSI TO STUDY TUMOR HETEROGENEITY

MSI is a molecular imaging modality, which extends the analytical capabilities of mass spectrometry to the spatially resolved analysis of tissue sections. The basic principle is the discrete or continuous acquisition of mass spectra from the surface of the tissue section (Fig. 3, upper part). The coordinates of the individual spectra are recorded, which allows the reconstruction of images for the single mass channels and hence the imaging of different molecules. There are different MSI systems mainly defined by the type of ion source and mass analyzer, which are explained and compared in detail elsewhere (Addie, Balluff, Bovee, Morreau, & McDonnell, 2015; Rubakhin, Jurchen, Monroe, & Sweedler, 2005).

The three most commonly used MSI ion sources are secondary ion mass spectrometry (SIMS), matrix-assisted laser desorption/ionization (MALDI) MSI, and desorption electrospray ionization (DESI) MSI. SIMS offers spatial resolution at the nanometer scale, but the detection is limited to lipids, other metabolites, and elements. MALDI extends the range of detectable molecular classes to biomolecules of higher mass such as proteins. The spatial resolution of MALDI-MSI is now at the single-cell level, although for the price of loss of sensitivity. DESI is an interesting technique as it works under ambient conditions and hence does not require sample preparation as compared to MALDI, which requires the coating of the tissue section with a matrix.

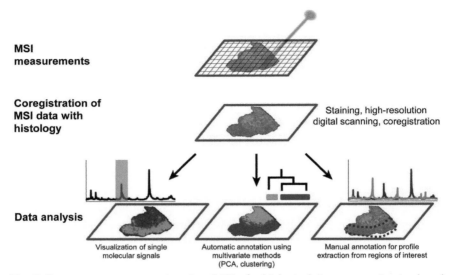

Fig. 3 For mass spectrometry imaging (MSI) of a biological tissue, a section is placed onto a glass slide. Then individual mass spectra are obtained across the tissue section while recording their position. For the in situ acquisition of the mass spectra different ionization methods can be used. Here, the application of a laser is shown in *red*, which is used in matrix-assisted laser desorption/ionization MSI (*upper part*). The use of a glass slide allows staining the very same tissue section after the experiment for a subsequent digital coregistration of the optical image to the spatially resolved MSI data (*center*). This facilitates investigating the distributions of single molecules within the section's histology (*lower left*), or conversely, the extraction of histology-specific molecular profiles (*lower right*). At last, multivariate methods can be applied for the automated annotation of the tissue (*lower center*). Reprinted and adapted from Balluff, B., Rauser, S., Ebert, M. P., Siveke, J. T., Hofler, H., & Walch, A. (2012). Direct molecular tissue analysis by MALDI imaging mass spectrometry in the field of gastrointestinal disease. Gastroenterology, 143(3), 544–549, e541–e542. http://dx.doi.org/10.1053/j.gastro.2012.07.022, Copyright 2012, with permission from Elsevier.

DESI offers good detection of lipids and metabolites at spatial resolutions comparable to MALDI-MSI. The strengths of MSI are:
- Unlabeled imaging of compounds
- Simultaneous imaging of hundreds to thousands of molecules
- Unmatched range of molecular classes that can be imaged (metabolites, lipids, drugs, peptides, proteins, protein modifications)
- Integration with other imaging modalities

With respect to the last advantage, the biggest strength of MSI is the possibility of combining the spatially resolved mass spectrometric data with the microscopic information of the analyzed tissues. This is possible, as the MSI experiment does not harm the tissue section's integrity, which allows the use of the very same section for coregistration. Therefore, the tissue section is stained after the MSI experiment and scanned with a high-resolution optical slide scanner (Fig. 3, vertical center).

The integration of the histological information enables the interpretation of the molecular MSI images within their histological context (Fig. 3, lower left). Furthermore, histological annotations can be used to extract molecular profiles from regions of interest and hence allocate these profiles to tumor, preneoplastic, or inflammatory cells (Fig. 3, lower right). This is called virtual microdissection. In cancer research virtual microdissection has been extensively used to obtain tumor cell-specific molecular profiles in order to decrease the contribution of confounding factors from surrounding tissues. This differentiated analysis of diseased tissues enabled a more specific biomarker discovery resulting in markers for diagnostic purposes (Guenther et al., 2015; Lazova, Seeley, Keenan, Gueorguieva, & Caprioli, 2012), for prognostic purposes (Elsner et al., 2012), or to predict response to therapy (Aichler et al., 2013). Further reviews have summarized the achievements in biomarker discovery by MSI (Addie et al., 2015; McDonnell et al., 2010; Schone, Hofler, & Walch, 2013). All these studies give evidence of the power of MSI to study intertumor heterogeneity by focusing on the tumor profiles of the patients only.

But MSI also constitutes a unique technique to discover ITH. Several studies observed nonhomogeneous distributions of single molecules within tumor regions (Abramowski et al., 2015; Buck et al., 2015; Le Faouder et al., 2014) (Fig. 4A). However, a big step forward for describing ITH in MSI data of tumors came with the application of advanced data analysis methods.

3. MULTIVARIATE DATA ANALYSIS STRATEGIES IN MSI

A central strategy for the analysis of within-sample MSI data is the application of multivariate machine learning methods for automatic annotation of the tissue (Fig. 3, lower center). Multivariate machine learning or analysis (MVA) refers to the process of automatically detecting patterns in data (machine learning) by the simultaneous and combined consideration of multiple features (multivariate). In mass spectrometry these features are

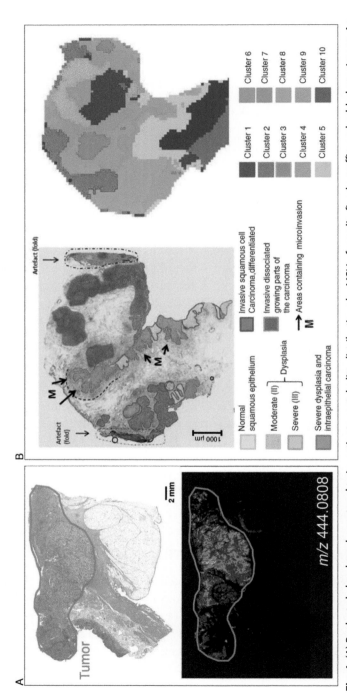

Fig. 4 (A) Buck et al. developed a protocol to investigate metabolite distributions by MSI in formalin-fixed paraffin-embedded specimens. An example for a metabolite (m/z 444.0808) is shown in a gastric cancer sample that has a heterogeneous visualization pattern within the tumor (red delineated area). (B) A complex larynx carcinoma specimen containing many histological entities was submitted to protein MSI analysis. A microscopic image of a larynx carcinoma section was obtained after MSI experiments (left). Regions of interest were manually annotated by an expert pathologist. A segmentation map of 10 clusters was generated (right) by multivariate clustering of the MSI data, which shows correspondence to the annotation (left). Colors are arbitrarily assigned and are not interrelated between manual and automated annotation. Panel (A) reprinted and adapted from Buck, A., Ly, A., Balluff, B., Sun, N., Gorzolka, K., Feuchtinger, A., et al. (2015). High-resolution MALDI-FT-ICR MS imaging for the analysis of metabolites from formalin-fixed, paraffin-embedded clinical tissue samples. The Journal of Pathology, 237(1), 123–132. http://dx.doi.org/10.1002/path.4560, Copyright 2015, John Wiley and Sons. Panel (B) reprinted with permission of Springer from Alexandrov, T., Becker, M., Guntinas-Lichius, O., Ernst, G., & von Eggeling, F. (2013). MALDI-imaging segmentation is a powerful tool for spatial functional proteomic analysis of human larynx carcinoma. Journal of Cancer Research and Clinical Oncology, 139(1), 85–95. http://dx.doi.org/10.1007/s00432-012-1303-2, Copyright 2012, Springer-Verlag.

the different mass channels of the molecules observed. The rational of employing MVA methods in biological questions is the fact that many biological processes involve many molecules, which are related and the isolated univariate analysis would not reveal any effect. It is beyond the scope of this chapter to explain all multivariate machine learning techniques, except for those that are commonly employed in MSI; for a comprehensive overview and explanation on that topic please refer the book by Murphy (2012). The methods used by the MSI community so far can be categorized into three groups:
- Unsupervised analysis (clustering)
- Supervised classification
- Projection methods

3.1 Unsupervised Analysis

Unsupervised analysis methods are used to reveal structures between objects by looking at their relationship in a multidimensional feature space. The major advantage of this approach is that nothing needs to be known beforehand about the data. A classical application of unsupervised analysis is clustering which assigns objects to clusters based on their similarity in the feature space.

In within-sample MSI, clustering is used to group the spatially resolved spectra according to their spectral similarity. The result can be visualized as segmentation maps of the tissue where each pixel is assigned to a cluster (Fig. 4B). This is used to show expected or reveal unexpected similarities or dissimilarities between histological entities within tissues (Fig. 4B).

Many clustering/segmentation methods are available. Conceptually, there are methods that need the predefinition of the expected number of clusters while others do not; k-means clustering is a well-known example for the first and hierarchical clustering for the latter. Clustering algorithms have not only been applied to MSI data (Bemis et al., 2016), but have also been adapted according to the nature of MSI data by, for example, considering the spatial neighborhood of a spectrum (Alexandrov et al., 2010; Alexandrov & Kobarg, 2011). Comparative studies on those techniques have revealed differences in performances, which depend on the type of MSI data (e.g., on its level and structure of intervariable correlation and noise) and the biological problem (Jones et al., 2011; Sarkari, Kaddi, Bennett, Fernandez, & Wang, 2014).

3.2 Supervised Classification

Supervised classification is a two-step procedure, which is divided into classifier training and application. The aim of this process is to train a mapping from inputs to outputs (Murphy, 2012). Unlike unsupervised analysis, the learning process of the supervised classification requires a priori knowledge on the output value assigned to the objects; for instance, their group membership, termed class or label. Once the classifier has been trained it can take the inputs from an unknown object and determine its class. Supervised classification is very powerful but its prediction precision depends highly on the availability of a sufficient number of representative training samples and the use of a classifier that generalizes well to previously unseen data. Known algorithms are support vector machines, neural networks, and (ensembles of) decision trees.

In MSI supervised classification has mostly been employed to determine the disease status of a patient (Cazares et al., 2009; Mao et al., 2016; Rauser et al., 2010) and, to a lesser extent, for the annotation of samples on a pixel-level (Eberlin et al., 2012; Willems et al., 2010).

3.3 Projection Methods

Another group of techniques routinely used in MSI data analysis are projection methods. These are called projection methods as they reduce the dimensionality of the data by projecting the data to a lower dimensional subspace, while maintaining the essence of the original data (Murphy, 2012). The variables of the new lower dimensional space can be thought of as combinations of the original features and are called components. Projection methods can work with and without labeled data. While principal component analysis, probabilistic latent semantic analysis (pLSA), and nonnegative matrix factorization allow a label-free reduction to a defined number of components, partial-least-squares discriminant analysis will project the data to a lower dimensional space by maximizing the differences between the labeled objects. All of these methods have been applied in MSI on either a patient-level to find new cancer patient subgroups (Djidja et al., 2010), or on a pixel-level to annotate the tissue by visualization of the components (Dill et al., 2011; Veselkov et al., 2014). In the latter it therefore serves a similar purpose as clustering.

More details on statistical data analysis in MSI can be found in two excellent reviews by Jones, Deininger, Hogendoorn, Deelder, and McDonnell (2012) and Alexandrov (2012).

4. MSI APPLICATIONS FOR THE INVESTIGATION OF ITH

In this section it will be demonstrated how these MVA techniques, when applied to MSI data, can be used to disclose and investigate ITH.

4.1 Revealing ITH by Clustering

The first publication to show the potential of MSI to uncover ITH was by Deininger, Ebert, Futterer, Gerhard, and Rocken (2008). They introduced hierarchical clustering of MSI data as a new tool for the interpretation of complex human cancers. Hierarchical clustering is a procedure which iteratively merges single objects and aggregations of these objects (clusters) based on their similarity into a tree-like structure, termed a dendrogram. When applied to MSI, each branch of the tree can be considered a class and all spectra (pixels) that belong to this class can be visualized with a single color (Deininger et al., 2008). For instance, Fig. 5A.C shows that the visualization of the first three branches (Fig. 5A.D) of a gastric cancer sample is consistent with its histology (Fig. 5A.B) by accurately delineating solid tumor, tumor stroma, and healthy connective tissue. The dendrogram also shows that the tumor stroma and the connective tissue exhibit a higher spectral similarity to each other than to the tumor (Fig. 5A.D).

In that same example further expansion and visualization of the sub-branches belonging to the solid tumor revealed a nodal-like structure of clusters within the tumor. As there was no histological correlate to that observation—moreover, the solid tumor appeared histologically homogeneous—the researchers surmised them being molecularly distinct but histologically undistinguishable tumor subclones (Deininger et al., 2008). This example shows how hierarchical clustering can be used to interactively investigate MSI datasets of tumors and their ITH.

But hierarchical clustering also has its drawbacks. The results strongly depend on the linkage method and the distance metric for quantifying mass spectral similarity (the data processing software MATLAB includes 77 possible combinations of these two parameters). The second disadvantage is that it does not optimize any well-defined objective function and will hence always create a full tree (Murphy, 2012). The question then arises at which depths of the tree are differences between spectra based on noise rather than on real biological differences.

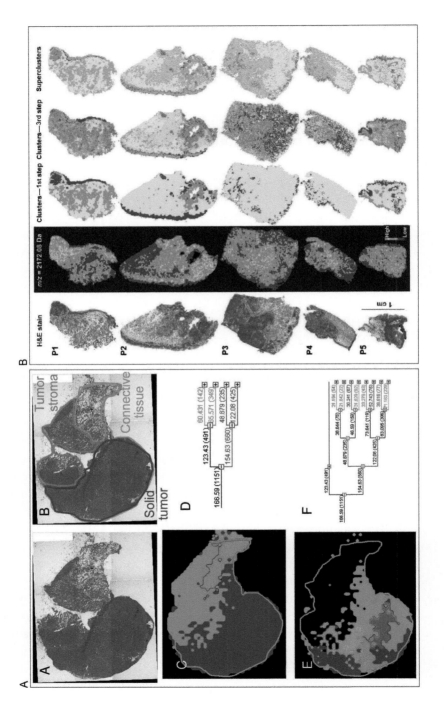

Fig. 5 See legend on opposite page.

Many of these issues have been addressed in the last years by the creation of novel or adaptations of multivariate segmentation techniques. Alexandrov and coworkers developed clustering algorithms that account for the spatial nature of MSI data by considering the spatial neighborhood of a spectrum for clustering (Alexandrov et al., 2010; Alexandrov & Kobarg, 2011), whereas others proposed that the clustering should maximize the discriminating power of the molecular signatures of the resulting segments (Bruand et al., 2011). Alternative approaches were based on dimensionality reduction prior to clustering (Palmer, Bunch, & Styles, 2015), the characterization of the certainty of the segmentation (Bemis et al., 2016), or the assumption and modeling of nonlinear relationships by t-distributed stochastic neighbor embedding (Fonville et al., 2013). Finally, the usage of pLSA was proposed as nonnegative decomposition of the MSI data which provides a probability distribution over the spectral dimension for each tissue type (Hanselmann et al., 2008). The regions with equal mixing coefficient are then interpreted as clusters. All of this led to the availability of a myriad of multivariate clustering and spectral decomposition algorithms.

The fact that these algorithms can produce different results on the same MSI data has been addressed by Jones et al. (2011). The reported approach, termed agreement analysis, combined the individual results of five different

Fig. 5 Multivariate clustering of MSI data is a powerful tool to investigate intratumor molecular heterogeneity without previous knowledge. (A) The MSI spectra of a gastric cancer sample (A.A), consisting of solid tumor (*blue*), tumor stroma (*green*), and connective tissue (*red*) (A.B), were grouped by hierarchical clustering (A.D). The top-level branches represented these three visible histological features (A.C). Expanding the tumor branch (A.D., *blue*) resulted in a nodal-like segmentation of the tumor region (A.E. and A.F), which was speculated to represent the different tumor populations (Deininger et al., 2008), Copyright 2008, American Chemical Society. (B) The next step was to extend the segmentation to multiple samples at the same time. Here, a MSI dataset is shown consisting of five oral squamous cell cancer samples (P1–P5). Besides proposing a novel clustering algorithm, Widlak et al. observed a higher proportion of the *red* supercluster (*rightmost column*) in patients with an advanced disease state (P1–P2), suggesting that segments found by MSI clustering could represent phenotypic tumor populations of clinical relevance. *Reprinted from Widlak, P., Mrukwa, G., Kalinowska, M., Pietrowska, M., Chekan, M., Wierzgon, J., et al. (2016). Detection of molecular signatures of oral squamous cell carcinoma and normal epithelium—Application of a novel methodology for unsupervised segmentation of imaging mass spectrometry data. Proteomics, 16(11–12), 1613–1621. http://dx.doi.org/10.1002/pmic.201500458, Copyright 2016, John Wiley and Sons.*

clustering methods to a consensus clustering. The strength of the agreement analysis is not only to deliver results that are corroborated, but also to consider clusters that are found only by a subset of MVA methods, as long as they are sufficiently congruent. This increases the likeliness of not missing a significant cluster. The same researchers were also the first to perform a simultaneous clustering of several soft tissue sarcoma samples at the same time at a pixel level (Willems et al., 2010). Also Widlak et al. segmented a MSI dataset consisting of five oral squamous cell carcinomas with the aim of studying the presence of the same clusters in different samples (Widlak et al., 2016) (Fig. 5B). The clustering of more than 45,000 spectra revealed two superclusters heterogeneously distributed across the tumor regions. Interestingly, a higher proportion of one of those superclusters was detected in two patients with an advanced disease status. This raises the general question if there is a connection between the observation of a particular cluster in a subset of patients and a certain clinical characteristic.

This question can definitively only be answered by functional assays in order to study the biological behavior of the cells belonging to that cluster by, for example, using patient-derived tumor xenografts. Another way of gaining insight is by investigating the statistical association between the presence of a cluster and the disease outcome of the patients using a sufficiently large patient cohort.

We have pursued the latter idea for several different cancer types and clinical endpoints by first using MALDI-MSI to obtain spatially resolved proteomic data from primary tumor specimens and then segmenting the histologically homogeneous tumor areas within each sample after virtual microdissection of these. Compared to genetic studies, MSI takes advantage of the fact that cellular selection operates on phenotypes by detecting phenotypic information in the form of mass spectral proteomic or metabolic profiles (Aparicio & Caldas, 2013). We hypothesized that statistical correlation of the patients' clinical data with the detected segments may enable the identification of these phenotypic subpopulations.

An inherent problem for all clustering methods is that the number of clusters has to be specified. To solve this, a very common solution is the use of statistical measures to determine the optimal number of clusters (Hanselmann et al., 2008; Widlak et al., 2016). Our approach instead was centered on clinical outcome, where the clinical effect determined the number of clusters; i.e., the number of clusters of a segmentation is regarded as the correct one when an association to the clinical data is observed.

The presence of lymph nodes metastasis is one of the strongest predictors for the disease outcome of a cancer patient. The capacity of a cancer cell to migrate and proliferate in the lymphatic system requires special molecular equipment (Das et al., 2013). In order to identify the cells with metastatic potential, the above-described procedure was applied to a MSI dataset comprising 32 patients with invasive ductal breast carcinoma of which 21 patients had lymph node metastasis and 11 were metastasis-free (Balluff et al., 2015) (Fig. 6A). Segmentation was performed with a varying number of expected clusters, ranging from 2 to 10. A statistically significant association of the presence of a cluster with the metastatic status of the patients was detected in four molecularly distinct tumor subpopulations, where one of the clusters was exclusively present in patients with affected lymph nodes (Fig. 6D, bar plot). This tumor subpopulation exhibited ITH (Fig. 6C) and was mainly characterized by epigenetic alterations on a histone level (Fig. 6D, phylo-plot).

The general applicability of this data analysis strategy was confirmed by the application to other cancer types and other clinical endpoints. For instance, in a cohort of 63 patients with intestinal-type gastric cancer the segmentation revealed substantial ITH and a statistical significant association of the overall survival to the presence of one particular cluster (Balluff et al., 2015).

That the presence of a molecularly distinct tumor subpopulation is indicative of a poor prognosis for the patients was also observed in a study on sarcomas (Lou et al., 2016). The particularity of that study was that sarcomas are characterized by a high intratumor histological heterogeneity with respect to the grade of differentiation of the tumor cells. This required a more differentiated virtual microdissection of the tumor regions by accounting for the different grades. Only in this manner could a prognostic value be assigned to a tumor region within moderately differentiated osteosarcoma cells.

This demonstrates that clustering combined with histological annotation is able to pinpoint microscopically indistinct tumor subpopulations that can have an adverse impact on clinical outcome. The potential is huge, since this will enable their further molecular characterization for deeper insights into the biological processes of cancer. But MSI can not only be used for the de novo discovery of ITH, but also for the description of ITH with respect to known clinically relevant features. Therefore, supervised classification algorithms are employed.

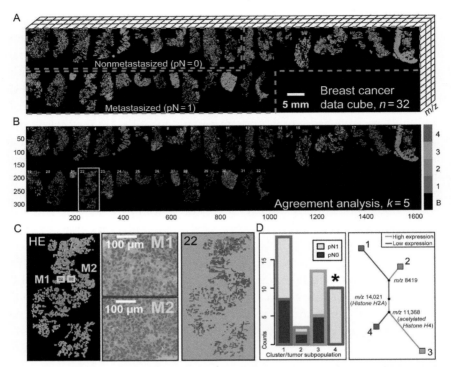

Fig. 6 (A) To find tumor populations associated with metastasis in breast cancer, protein profiles from histologically uniform regions of 21 metastasized and 11 nonmetastasized breast cancer tissues were obtained. (B) Clustering was performed using the agreement analysis with different values for the number of expected clusters ($k=2–10$) and the results for $k=5$ are shown; the agreement analysis might return a slightly different number of clusters based on the consensus quality, here four. (C) Magnifications within patient 22 (M1 & M2) prove the histological homogeneity within the measured tumor area, despite the detected molecular heterogeneity represented by tumor subpopulations 1 (*blue*) and 4 (*red*). (D) Moreover, relating the presence of the clusters in the samples to the metastatic status of the patients, revealed tumor subpopulation 4 to be significantly correlated with a positive metastatic status ($p=0.036$) (*bar plot*). It was characterized by changes in m/z 11,368 (acetylated histone H4), 8419 and 14,021 (histone H2A) (*phylogenetic plot*). *Reprinted from Balluff, B., Frese, C. K., Maier, S. K., Schone, C., Kuster, B., Schmitt, M., et al. (2015). De novo discovery of phenotypic intratumour heterogeneity using imaging mass spectrometry. The Journal of Pathology, 235(1), 3–13. http://dx.doi.org/10.1002/path.4436, Copyright 2015, John Wiley and Sons.*

4.2 Supervised Classification of ITH

As mentioned earlier, supervised classification is a two-step procedure, which is divided into classifier learning and application. The learning step associates the input data with the labels of the input. In clinical MSI, the input is the mass spectrometric profiles and the labels can be anything that is associated to the sample/patient, such as clinical data, or associated to the specific region from where the profiles were obtained, such as histological or molecular properties. The trained classifier can then be used to predict the label of an unknown sample. If applied on a pixel level of tumor samples, supervised classification can be used to describe the within-sample molecular diversity with respect to that label.

Eberlin and coworkers have used supervised classification of lipid MSI data from human brain tumors with the aim of speeding up histological diagnose. For the rapid detection of spatially resolved lipid profiles, DESI-MSI was carried out on 36 human glioma samples including oligodendrogliomas, astrocytomas, and oligoastrocytomas of different grades and varying tumor cell concentrations (Eberlin et al., 2012). Classifiers were generated for glioma subtype, grade, and concentration using support vector machines on selected samples for the training set and tested on the remaining samples as the validation set. As the prediction was carried out on a pixel level, significant molecular heterogeneity with regard to the trained labels could be observed in some of the samples. Biochemical ITH associated to different tumor grades was also observed by a similar study using MALDI-MSI proteins in myxofibrosarcomas (Willems et al., 2010).

Both studies focused on classifying histopathological properties of the tumor such as grade and subtype, rather than on molecular properties of the tumor. But as shown at the beginning of this chapter with the HER2 example (Fig. 1B), knowing the expression status of certain molecules can have important consequences for disease outcome and the chosen therapy scheme.

Using MSI on a patient-level, we have found a protein signature that is able to accurately discriminate HER2-positive from HER2-negative breast cancer patients using supervised classification (Rauser et al., 2010). In subsequent unpublished work, this classifier was then extended to the classification of single pixel spectra by adding connective tissue as a third label. When applied to breast cancer tissue sections, the classifier uncovered heterogeneous HER2 expression in some of the samples, which was confirmed by IHC (Fig. 7). And as reported in Section 1.3, the degree of ITH, in this case the fraction of HER2-positive cells, is important for the clinical decision-making (Ruschoff et al., 2012).

4.3 Investigating the Degree of ITH

The degree of clonal diversity itself has been the focus of several studies that have shown that it is related to the clinical outcome for patients. Maley et al.

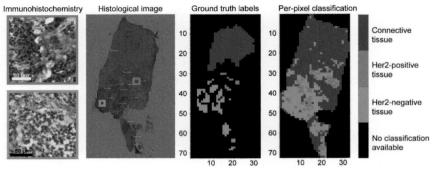

Fig. 7 Supervised classification allows the annotation of MSI datasets of unknown status after training of a classifier on MSI datasets with known status such as tumor type, stage, or some histological or molecular feature. Rauser et al. used MSI to obtain protein profiles from 15 HER2-positive and 15 HER2-negative breast cancer samples (Rauser et al., 2010). We used those profiles to build a classifier for the per-pixel annotation of the HER2 status within breast cancer samples. Moreover, the classifier was also trained to detect connective tissue. In the example shown, this three-class classifier was applied to a breast cancer sample with known HER2 expression heterogeneity (*center images*) as determined by immunohistochemistry on a consecutive section. The leftmost images show magnifications of the HER2-positive (*green*) and HER2-negative (*red*) regions. The annotation by the classifier (*rightmost*) was able to uncover the heterogeneous expression of HER2 and delineate tumor and connective tissue accurately.

investigated genetic heterogeneity in multisite tumor biopsies from patients suffering from Barrett's esophagus, a preneoplastic condition (Maley et al., 2006). They observed that a high clonal diversity was predictive for the progression of the disease to esophageal adenocarcinoma. They used several diversity measures, among it Shannon's index, a common diversity index in ecological literature.

Other studies followed, with the aim of defining simple, quantitative, generally applicable measures of genetic ITH that could be useful for clinical trial design and decision-making (Mroz, Tward, Hammon, Ren, & Rocco, 2015). For instance, Mroz et al. introduced the mutant-allele tumor heterogeneity (MATH) value to describe intratumor genetic heterogeneity based on whole-exome sequencing. When applied to samples from head and neck cancer, high MATH values were found to correlate with a decreased overall survival probability (Mroz et al., 2015).

Some of the proposed diversity measures can also be applied to MSI data. An example is shown in Fig. 8B, where Shannon's index is used to describe the molecular ITH in two different gastric cancer samples after multivariate

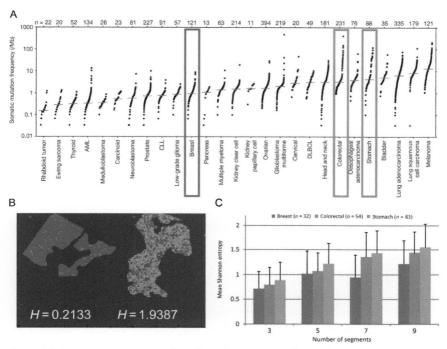

Fig. 8 (A) Tumor types vary significantly in their genetic diversity. This graph shows the differences in somatic mutation frequencies between 27 different tumor types obtained by exome sequencing of more than 3000 tumor-normal samples. (B) Diversity can also be assessed in the multidimensional MSI data. Here, two MSI gastric cancer datasets were clustered ($k=4$) and Shannon's index (H) as a diversity measure was calculated. High H values represent a higher diversity (*right sample*) and lower H values homogeneity (*left sample*). (C) Protein-level MSI datasets were available for breast, colorectal, and stomach cancers, with a total of 149 patients. Shannon's index was calculated for each sample individually after clustering. Irrespective of the number of expected segments chosen, stomach cancer exhibited on average the highest diversity, followed by colorectal and breast cancer, suggesting an overall correlation between genetic and proteomic diversity. *Panel (A) reprinted by permission from Macmillan Publishers Ltd.: Nature. Lawrence, M. S., Stojanov, P., Polak, P., Kryukov, G. V., Cibulskis, K., Sivachenko, A., et al. (2013). Mutational heterogeneity in cancer and the search for new cancer-associated genes. Nature, 499(7457), 214–218. http://dx.doi.org/10.1038/nature12213, Copyright (2013).*

clustering. We have found indications that the levels of nongenetic intratumor diversity also differs between tumor types (Fig. 8C). This comparison embraced MSI protein data of tumor regions from 32 breast, 54 colorectal, and 63 gastric cancer tissues, which were clustered and the Shannon's index applied. The results show that irrespective of the number of segments chosen, stomach cancer exhibited the highest MSI ITH, followed by colorectal cancer and breast cancer (Fig. 8C). This is in line with the order of genetic diversity as determined by several somatic mutations studies (Lawrence et al., 2013; Vogelstein et al., 2013) (Fig. 8A).

It would be interesting to see if the application of these heterogeneity measures in MSI studies also translates into clinical implications for a certain cancer type.

4.4 Investigation of ITH on Different Molecular Levels

As described earlier, MSI allows the detection of a broad range of molecular classes, depending on the technology and the sample preparation used. Consequently, ITH can be assessed on different molecular levels.

So far most of the work for investigating ITH has been done on a protein or peptide level using MALDI-MSI instrumentation, including gastric cancer (Balluff et al., 2015; Deininger et al., 2008), sarcomas (Jones et al., 2011; Lou et al., 2016; Willems et al., 2010), oral squamous cell carcinomas (Widlak et al., 2016), and breast cancer (Balluff et al., 2015).

Other studies have highlighted the importance of characterizing the heterogeneous distribution of lipids in tumor. Cimino et al. used MALDI-Fourier transform ion cyclotron resonance-MSI to study the ITH of low-abundant lipid species in human breast cancer xenografts (Cimino et al., 2013) and Eberlin et al. observed clinically relevant lipid ITH with DESI-MSI in gliomas, as described earlier (Eberlin et al., 2012).

Some MSI studies also indicate the presence of ITH on a metabolic level. Dekker et al., for example, was able to identify regions within tumor tissue samples with distinct metabolic signatures that were consistent with known tumor biology, such as the Warburg effect (Dekker et al., 2015). That metabolic studies are not restricted to frozen tissues has been proven by Buck et al. who also observed ITH when visualizing the distributions of single mass channels belonging to endogenous small molecules in formalin-fixed paraffin-embedded samples (Buck et al., 2015) (Fig. 4A).

But MSI can also be used to display exogenous, i.e., administered, compounds in tissues such as drugs or tracers. For the latter, MSI can complement existing in vivo techniques such as PET or MRI with cellular specificity (Buck & Walch, 2014). Tata et al. have studied the localization of Gadoteridol, a MRI contrast agent, by DESI-MSI in human breast cancer xenografts. They observed a heterogeneous spatial distribution pattern of Gadoteridol inside the solid tumor with a preference for the periphery of the tumor (Tata et al., 2015).

5. FUTURE APPLICATIONS OF MSI IN ITH RESEARCH

Another important application area of MSI is drug imaging as it combines the advantages of analytical methods such as high chemical specificity with the capability of imaging (Balluff, Schone, Hofler, & Walch, 2011). MSI has therefore already been extensively used to investigate the distribution and concentration levels of drugs and their metabolites in tissues, including pharmaceuticals in oncology (Prideaux & Stoeckli, 2012).

Although heterogeneous concentrations of drugs within cancers have been observed, a link to the molecular ITH has not been done yet. That it is important to question heterogeneous drug distributions in cancers has been shown by Gruner et al. (2016). They investigated the drug levels of the anticancer drug erlotinib in mouse models of pancreatic ductal adenocarcinoma and their effect on the survival time of the mice. Surprisingly, survival did not correlate with overall levels of erlotinib in the tumor, but with its location in atypical glands within the tumor. In this example, drug levels could be related to histologically visible entities. But histologically undistinguishable tumor cells with dissimilar molecular setup might metabolize drugs differently. Hence, we foresee not only the need to correlate drug distributions with histological information but also with information about the (invisible) molecular ITH.

But the investigation of the ITH should not be restricted to tumor cells. The tumor microenvironment and its heterogeneity are known to exhibit selection pressures that influence both the tumor's evolution as well as a drug's efficacy such as vascularization and immune cell densities (Junttila & de Sauvage, 2013). Few studies in MSI have incorporated the microenvironmental information into their analysis. For instance, another drug imaging study on tamoxifen-incubated human breast cancer tissues found differences in tumor/stroma drug concentration ratios between estrogen-positive and estrogen-negative patients (Vegvari et al., 2016). And Dekker et al. have investigated proteomic heterogeneity in tumor stroma by comparing intra- and extratumoral stroma (Dekker et al., 2014).

Despite all the strengths of MSI, one should not ignore the limitations of MSI for all future applications: the lack of unequivocal identification of high mass molecules (Balluff et al., 2011), an upper mass range limit of 30,000 Da (exceptionally 150,000 Da) (Mainini et al., 2013), and the presence of ionization bias which favors the detection of more easily ionizable molecules (Lanekoff, Stevens, Stenzel-Poore, & Laskin, 2014).

An alternative MSI technology is CyTOF mass cytometry. It combines targeted IHC with mass spectrometry where rare earth metals are used as reporters on antibodies (Giesen et al., 2014). Bodenmiller and workers have applied this approach for the simultaneous imaging of up to 32 proteins and protein modifications at subcellular resolution in human breast cancer samples, allowing among others the highlighting of tumor heterogeneity (Giesen et al., 2014).

6. PERSPECTIVE

Although CyTOF mass cytometry overcomes some of the limitations of untargeted MSI, both share the fact that they are invasive techniques, and hence not applicable for in vivo imaging such as PET, CT, or MRI. The need of sampling will, analogous to the genomics research performed in ITH, most likely underestimate the true phenotypic composition of the whole tumor. For instance, the Swanton lab observed that a single biopsy was able to maximally cover one-third of all nonsynonymous somatic mutations (Swanton, 2012). A solution would be to guide sampling and MSI by in vivo techniques that can operate more or less on the same scale, as shown by Kurczy et al. who combined by X-ray micro-CT with nanostructure imaging mass spectrometry (Kurczy et al., 2015). Suzuki et al. have done that already for genetic profiling guided by MRI (Suzuki et al., 2015).

Ultimately, it will be the integration of data from different imaging and nonimaging sources that will satisfy the needs for future studies on ITH. The challenge to an understanding of tumor evolution is huge, as it has to be studied over time, space, and at different molecular levels and in the context of its micro- and macroenvironment.

But also the potential is huge since recent studies have shown that with existing knowledge and data fascinating insights of the spatiotemporal dynamics of ITH and its mechanics can be obtained (Lipinski et al., 2016; Waclaw et al., 2015). It is therefore expected that more integrated and comprehensive data on cancer will help toward an understanding and exploitation of tumor heterogeneity with respect to an improved precision medicine.

REFERENCES

Abramowski, P., Kraus, O., Rohn, S., Riecken, K., Fehse, B., & Schluter, H. (2015). Combined application of RGB marking and mass spectrometric imaging facilitates detection of tumor heterogeneity. *Cancer Genomics & Proteomics*, *12*(4), 179–187.

Addie, R. D., Balluff, B., Bovee, J. V., Morreau, H., & McDonnell, L. A. (2015). Current state and future challenges of mass spectrometry imaging for clinical research. *Analytical Chemistry*, *87*(13), 6426–6433. http://dx.doi.org/10.1021/acs.analchem.5b00416.

Aichler, M., Elsner, M., Ludyga, N., Feuchtinger, A., Zangen, V., Maier, S. K., et al. (2013). Clinical response to chemotherapy in oesophageal adenocarcinoma patients is linked to defects in mitochondria. *The Journal of Pathology, 230*(4), 410–419. http://dx.doi.org/10.1002/path.4199.

Alexandrov, T. (2012). MALDI imaging mass spectrometry: Statistical data analysis and current computational challenges. *BMC Bioinformatics, 13*(Suppl. 16), S11. http://dx.doi.org/10.1186/1471-2105-13-s16-s11.

Alexandrov, T., Becker, M., Deininger, S. O., Ernst, G., Wehder, L., Grasmair, M., et al. (2010). Spatial segmentation of imaging mass spectrometry data with edge-preserving image denoising and clustering. *Journal of Proteome Research, 9*(12), 6535–6546. http://dx.doi.org/10.1021/pr100734z.

Alexandrov, T., & Kobarg, J. H. (2011). Efficient spatial segmentation of large imaging mass spectrometry datasets with spatially aware clustering. *Bioinformatics (Oxford, England), 27*(13), i230–i238. http://dx.doi.org/10.1093/bioinformatics/btr246.

Alizadeh, A. A., Aranda, V., Bardelli, A., Blanpain, C., Bock, C., Borowski, C., et al. (2015). Toward understanding and exploiting tumor heterogeneity. *Nature Medicine, 21*(8), 846–853. http://dx.doi.org/10.1038/nm.3915.

Aparicio, S., & Caldas, C. (2013). The implications of clonal genome evolution for cancer medicine. *The New England Journal of Medicine, 368*(9), 842–851. http://dx.doi.org/10.1056/NEJMra1204892.

Balluff, B., Frese, C. K., Maier, S. K., Schone, C., Kuster, B., Schmitt, M., et al. (2015). De novo discovery of phenotypic intratumour heterogeneity using imaging mass spectrometry. *The Journal of Pathology, 235*(1), 3–13. http://dx.doi.org/10.1002/path.4436.

Balluff, B., Schone, C., Hofler, H., & Walch, A. (2011). MALDI imaging mass spectrometry for direct tissue analysis: Technological advancements and recent applications. *Histochemistry and Cell Biology, 136*(3), 227–244. http://dx.doi.org/10.1007/s00418-011-0843-x.

Bemis, K. D., Harry, A., Eberlin, L. S., Ferreira, C. R., van de Ven, S. M., Mallick, P., et al. (2016). Probabilistic segmentation of mass spectrometry (MS) images helps select important ions and characterize confidence in the resulting segments. *Molecular & Cellular Proteomics: MCP, 15*(5), 1761–1772. http://dx.doi.org/10.1074/mcp.O115.053918.

Bruand, J., Alexandrov, T., Sistla, S., Wisztorski, M., Meriaux, C., Becker, M., et al. (2011). AMASS: Algorithm for MSI analysis by semi-supervised segmentation. *Journal of Proteome Research, 10*(10), 4734–4743. http://dx.doi.org/10.1021/pr2005378.

Buck, A., Ly, A., Balluff, B., Sun, N., Gorzolka, K., Feuchtinger, A., et al. (2015). High-resolution MALDI-FT-ICR MS imaging for the analysis of metabolites from formalin-fixed, paraffin-embedded clinical tissue samples. *The Journal of Pathology, 237*(1), 123–132. http://dx.doi.org/10.1002/path.4560.

Buck, A., & Walch, A. (2014). In situ drug and metabolite analysis [corrected] in biological and clinical research by MALDI MS imaging. *Bioanalysis, 6*(9), 1241–1253. http://dx.doi.org/10.4155/bio.14.88.

Burrell, R. A., McGranahan, N., Bartek, J., & Swanton, C. (2013). The causes and consequences of genetic heterogeneity in cancer evolution. *Nature, 501*(7467), 338–345. http://dx.doi.org/10.1038/nature12625.

Caiado, F., Silva-Santos, B., & Norell, H. (2016). Intra-tumour heterogeneity—Going beyond genetics. *The FEBS Journal, 283*(12), 2245–2258. http://dx.doi.org/10.1111/febs.13705.

Cazares, L. H., Troyer, D., Mendrinos, S., Lance, R. A., Nyalwidhe, J. O., Beydoun, H. A., et al. (2009). Imaging mass spectrometry of a specific fragment of mitogen-activated protein kinase/extracellular signal-regulated kinase kinase kinase 2 discriminates cancer from uninvolved prostate tissue. *Clinical Cancer Research: An Official Journal of the American Association for Cancer Research, 15*(17), 5541–5551. http://dx.doi.org/10.1158/1078-0432.ccr-08-2892.

Cimino, J., Calligaris, D., Far, J., Debois, D., Blacher, S., Sounni, N. E., et al. (2013). Towards lipidomics of low-abundant species for exploring tumor heterogeneity guided by high-resolution mass spectrometry imaging. *International Journal of Molecular Sciences*, *14*(12), 24560–24580. http://dx.doi.org/10.3390/ijms141224560.

Dalerba, P., Kalisky, T., Sahoo, D., Rajendran, P. S., Rothenberg, M. E., Leyrat, A. A., et al. (2011). Single-cell dissection of transcriptional heterogeneity in human colon tumors. *Nature Biotechnology*, *29*(12), 1120–1127. http://dx.doi.org/10.1038/nbt.2038.

Das, S., Sarrou, E., Podgrabinska, S., Cassella, M., Mungamuri, S. K., Feirt, N., et al. (2013). Tumor cell entry into the lymph node is controlled by CCL1 chemokine expressed by lymph node lymphatic sinuses. *The Journal of Experimental Medicine*, *210*(8), 1509–1528. http://dx.doi.org/10.1084/jem.20111627.

Dawood, S., Broglio, K., Buzdar, A. U., Hortobagyi, G. N., & Giordano, S. H. (2010). Prognosis of women with metastatic breast cancer by HER2 status and trastuzumab treatment: An institutional-based review. *Journal of Clinical Oncology: Official Journal of the American Society of Clinical Oncology*, *28*(1), 92–98. http://dx.doi.org/10.1200/jco.2008.19.9844.

de Bruin, E. C., McGranahan, N., Mitter, R., Salm, M., Wedge, D. C., Yates, L., et al. (2014). Spatial and temporal diversity in genomic instability processes defines lung cancer evolution. *Science (New York, N.Y.)*, *346*(6206), 251–256. http://dx.doi.org/10.1126/science.1253462.

Deininger, S. O., Ebert, M. P., Futterer, A., Gerhard, M., & Rocken, C. (2008). MALDI imaging combined with hierarchical clustering as a new tool for the interpretation of complex human cancers. *Journal of Proteome Research*, *7*(12), 5230–5236. http://dx.doi.org/10.1021/pr8005777.

Dekker, T. J., Balluff, B. D., Jones, E. A., Schone, C. D., Schmitt, M., Aubele, M., et al. (2014). Multicenter matrix-assisted laser desorption/ionization mass spectrometry imaging (MALDI MSI) identifies proteomic differences in breast-cancer-associated stroma. *Journal of Proteome Research*, *13*(11), 4730–4738. http://dx.doi.org/10.1021/pr500253j.

Dekker, T. J., Jones, E. A., Corver, W. E., van Zeijl, R. J., Deelder, A. M., Tollenaar, R. A., et al. (2015). Towards imaging metabolic pathways in tissues. *Analytical and Bioanalytical Chemistry*, *407*(8), 2167–2176. http://dx.doi.org/10.1007/s00216-014-8305-7.

Dill, A. L., Eberlin, L. S., Costa, A. B., Zheng, C., Ifa, D. R., Cheng, L., et al. (2011). Multivariate statistical identification of human bladder carcinomas using ambient ionization imaging mass spectrometry. *Chemistry (Weinheim an der Bergstrasse, Germany)*, *17*(10), 2897–2902. http://dx.doi.org/10.1002/chem.201001692.

Ding, L., Ley, T. J., Larson, D. E., Miller, C. A., Koboldt, D. C., Welch, J. S., et al. (2012). Clonal evolution in relapsed acute myeloid leukaemia revealed by whole-genome sequencing. *Nature*, *481*(7382), 506–510. http://dx.doi.org/10.1038/nature10738.

Djidja, M. C., Claude, E., Snel, M. F., Francese, S., Scriven, P., Carolan, V., et al. (2010). Novel molecular tumour classification using MALDI-mass spectrometry imaging of tissue micro-array. *Analytical and Bioanalytical Chemistry*, *397*(2), 587–601. http://dx.doi.org/10.1007/s00216-010-3554-6.

Eberlin, L. S., Norton, I., Dill, A. L., Golby, A. J., Ligon, K. L., Santagata, S., et al. (2012). Classifying human brain tumors by lipid imaging with mass spectrometry. *Cancer Research*, *72*(3), 645–654. http://dx.doi.org/10.1158/0008-5472.can-11-2465.

Edge, S. B., & American Joint Committee on Cancer. (2010). *AJCC cancer staging manual* (7th ed.). New York: Springer.

Elsner, M., Rauser, S., Maier, S., Schone, C., Balluff, B., Meding, S., et al. (2012). MALDI imaging mass spectrometry reveals COX7A2, TAGLN2 and S100-A10 as novel prognostic markers in Barrett's adenocarcinoma. *Journal of Proteomics*, *75*(15), 4693–4704. http://dx.doi.org/10.1016/j.jprot.2012.02.012.

Fletcher, C. D. M., World Health Organization, & International Agency for Research on Cancer (2013). *WHO classification of tumours of soft tissue and bone* (4th ed.). Lyon: IARC Press.

Fonville, J. M., Carter, C. L., Pizarro, L., Steven, R. T., Palmer, A. D., Griffiths, R. L., et al. (2013). Hyperspectral visualization of mass spectrometry imaging data. *Analytical Chemistry*, *85*(3), 1415–1423. http://dx.doi.org/10.1021/ac302330a.

Gerlinger, M., Horswell, S., Larkin, J., Rowan, A. J., Salm, M. P., Varela, I., et al. (2014). Genomic architecture and evolution of clear cell renal cell carcinomas defined by multiregion sequencing. *Nature Genetics*, *46*(3), 225–233. http://dx.doi.org/10.1038/ng.2891.

Gerlinger, M., Rowan, A. J., Horswell, S., Larkin, J., Endesfelder, D., Gronroos, E., et al. (2012). Intratumor heterogeneity and branched evolution revealed by multiregion sequencing. *The New England Journal of Medicine*, *366*(10), 883–892. http://dx.doi.org/10.1056/NEJMoa1113205.

Giesen, C., Wang, H. A., Schapiro, D., Zivanovic, N., Jacobs, A., Hattendorf, B., et al. (2014). Highly multiplexed imaging of tumor tissues with subcellular resolution by mass cytometry. *Nature Methods*, *11*(4), 417–422. http://dx.doi.org/10.1038/nmeth.2869.

Greaves, M., & Maley, C. C. (2012). Clonal evolution in cancer. *Nature*, *481*(7381), 306–313. http://dx.doi.org/10.1038/nature10762.

Gruner, B. M., Winkelmann, I., Feuchtinger, A., Sun, N., Balluff, B., Teichmann, N., et al. (2016). Modeling therapy response and spatial tissue distribution of erlotinib in pancreatic cancer. *Molecular Cancer Therapeutics*, *15*(5), 1145–1152. http://dx.doi.org/10.1158/1535-7163.mct-15-0165.

Guenther, S., Muirhead, L. J., Speller, A. V., Golf, O., Strittmatter, N., Ramakrishnan, R., et al. (2015). Spatially resolved metabolic phenotyping of breast cancer by desorption electrospray ionization mass spectrometry. *Cancer Research*, *75*(9), 1828–1837. http://dx.doi.org/10.1158/0008-5472.can-14-2258.

Hanselmann, M., Kirchner, M., Renard, B. Y., Amstalden, E. R., Glunde, K., Heeren, R. M., et al. (2008). Concise representation of mass spectrometry images by probabilistic latent semantic analysis. *Analytical Chemistry*, *80*(24), 9649–9658. http://dx.doi.org/10.1021/ac801303x.

Jones, E. A., Deininger, S. O., Hogendoorn, P. C., Deelder, A. M., & McDonnell, L. A. (2012). Imaging mass spectrometry statistical analysis. *Journal of Proteomics*, *75*(16), 4962–4989. http://dx.doi.org/10.1016/j.jprot.2012.06.014.

Jones, E. A., van Remoortere, A., van Zeijl, R. J., Hogendoorn, P. C., Bovee, J. V., Deelder, A. M., et al. (2011). Multiple statistical analysis techniques corroborate intratumor heterogeneity in imaging mass spectrometry datasets of myxofibrosarcoma. *PLoS One*, *6*(9). e24913. http://dx.doi.org/10.1371/journal.pone.0024913.

Junttila, M. R., & de Sauvage, F. J. (2013). Influence of tumour micro-environment heterogeneity on therapeutic response. *Nature*, *501*(7467), 346–354. http://dx.doi.org/10.1038/nature12626.

Kanayama, K., Imai, H., Yoneda, M., Hirokawa, Y. S., & Shiraishi, T. (2016). Significant intratumoral heterogeneity of human epidermal growth factor receptor 2 status in gastric cancer: A comparative study of immunohistochemistry, FISH, and dual-color in situ hybridization. *Cancer Science*, *107*(4), 536–542. http://dx.doi.org/10.1111/cas.12886.

Kreso, A., O'Brien, C. A., van Galen, P., Gan, O. I., Notta, F., Brown, A. M., et al. (2013). Variable clonal repopulation dynamics influence chemotherapy response in colorectal cancer. *Science (New York, N.Y.)*, *339*(6119), 543–548. http://dx.doi.org/10.1126/science.1227670.

Kurczy, M. E., Zhu, Z. J., Ivanisevic, J., Schuyler, A. M., Lalwani, K., Santidrian, A. F., et al. (2015). Comprehensive bioimaging with fluorinated nanoparticles using breathable liquids. *Nature Communications*, *6*, 5998. http://dx.doi.org/10.1038/ncomms6998.

Lanekoff, I., Stevens, S. L., Stenzel-Poore, M. P., & Laskin, J. (2014). Matrix effects in biological mass spectrometry imaging: Identification and compensation. *The Analyst*, *139*(14), 3528–3532. http://dx.doi.org/10.1039/c4an00504j.

Lawrence, M. S., Stojanov, P., Polak, P., Kryukov, G. V., Cibulskis, K., Sivachenko, A., et al. (2013). Mutational heterogeneity in cancer and the search for new cancer-associated genes. *Nature*, *499*(7457), 214–218. http://dx.doi.org/10.1038/nature12213.

Lazova, R., Seeley, E. H., Keenan, M., Gueorguieva, R., & Caprioli, R. M. (2012). Imaging mass spectrometry—A new and promising method to differentiate Spitz nevi from Spitzoid malignant melanomas. *The American Journal of Dermatopathology*, *34*(1), 82–90. http://dx.doi.org/10.1097/DAD.0b013e31823df1e2.

Le Faouder, J., Laouirem, S., Alexandrov, T., Ben-Harzallah, S., Leger, T., Albuquerque, M., et al. (2014). Tumoral heterogeneity of hepatic cholangiocarcinomas revealed by MALDI imaging mass spectrometry. *Proteomics*, *14*(7–8), 965–972. http://dx.doi.org/10.1002/pmic.201300463.

Lipinski, K. A., Barber, L. J., Davies, M. N., Ashenden, M., Sottoriva, A., & Gerlinger, M. (2016). Cancer evolution and the limits of predictability in precision cancer medicine. *Trends in Cancer*, *2*(1), 49–63. http://dx.doi.org/10.1016/j.trecan.2015.11.003.

Lou, S., Balluff, B., de Graaff, M. A., Cleven, A. H., Briaire-de Bruijn, I., Bovee, J. V., et al. (2016). High-grade sarcoma diagnosis and prognosis: Biomarker discovery by mass spectrometry imaging. *Proteomics*, *16*(11–12), 1802–1813. http://dx.doi.org/10.1002/pmic.201500514.

Mainini, V., Bovo, G., Chinello, C., Gianazza, E., Grasso, M., Cattoretti, G., et al. (2013). Detection of high molecular weight proteins by MALDI imaging mass spectrometry. *Molecular BioSystems*, *9*(6), 1101–1107. http://dx.doi.org/10.1039/c2mb25296a.

Maley, C. C., Galipeau, P. C., Finley, J. C., Wongsurawat, V. J., Li, X., Sanchez, C. A., et al. (2006). Genetic clonal diversity predicts progression to esophageal adenocarcinoma. *Nature Genetics*, *38*(4), 468–473. http://dx.doi.org/10.1038/ng1768.

Mao, X., He, J., Li, T., Lu, Z., Sun, J., Meng, Y., et al. (2016). Application of imaging mass spectrometry for the molecular diagnosis of human breast tumors. *Scientific Reports*, *6*, 21043. http://dx.doi.org/10.1038/srep21043.

McDonnell, L. A., Corthals, G. L., Willems, S. M., van Remoortere, A., van Zeijl, R. J., & Deelder, A. M. (2010). Peptide and protein imaging mass spectrometry in cancer research. *Journal of Proteomics*, *73*(10), 1921–1944. http://dx.doi.org/10.1016/j.jprot.2010.05.007.

Mroz, E. A., Tward, A. D., Hammon, R. J., Ren, Y., & Rocco, J. W. (2015). Intra-tumor genetic heterogeneity and mortality in head and neck cancer: Analysis of data from the Cancer Genome Atlas. *PLoS Medicine*, *12*(2), e1001786. http://dx.doi.org/10.1371/journal.pmed.1001786.

Murphy, K. P. (2012). *Machine learning: A probabilistic perspective*. Cambridge, MA: MIT Press.

Nowell, P. C. (1976). The clonal evolution of tumor cell populations. *Science (New York, N.Y.)*, *194*(4260), 23–28.

Palmer, A. D., Bunch, J., & Styles, I. B. (2015). The use of random projections for the analysis of mass spectrometry imaging data. *Journal of the American Society for Mass Spectrometry*, *26*(2), 315–322. http://dx.doi.org/10.1007/s13361-014-1024-7.

Patel, A. P., Tirosh, I., Trombetta, J. J., Shalek, A. K., Gillespie, S. M., Wakimoto, H., et al. (2014). Single-cell RNA-seq highlights intratumoral heterogeneity in primary glioblastoma. *Science (New York, N.Y.)*, *344*(6190), 1396–1401. http://dx.doi.org/10.1126/science.1254257.

Prideaux, B., & Stoeckli, M. (2012). Mass spectrometry imaging for drug distribution studies. *Journal of Proteomics*, *75*(16), 4999–5013. http://dx.doi.org/10.1016/j.jprot.2012.07.028.

Rauser, S., Marquardt, C., Balluff, B., Deininger, S. O., Albers, C., Belau, E., et al. (2010). Classification of HER2 receptor status in breast cancer tissues by MALDI imaging mass spectrometry. *Journal of Proteome Research*, 9(4), 1854–1863. http://dx.doi.org/10.1021/pr901008d.

Rubakhin, S. S., Jurchen, J. C., Monroe, E. B., & Sweedler, J. V. (2005). Imaging mass spectrometry: Fundamentals and applications to drug discovery. *Drug Discovery Today*, 10(12), 823–837. http://dx.doi.org/10.1016/s1359-6446(05)03458-6.

Ruschoff, J., Hanna, W., Bilous, M., Hofmann, M., Osamura, R. Y., Penault-Llorca, F., et al. (2012). HER2 testing in gastric cancer: A practical approach. *Modern Pathology: An Official Journal of the United States and Canadian Academy of Pathology, Inc*, 25(5), 637–650. http://dx.doi.org/10.1038/modpathol.2011.198.

Sarkari, S., Kaddi, C. D., Bennett, R. V., Fernandez, F. M., & Wang, M. D. (2014). Comparison of clustering pipelines for the analysis of mass spectrometry imaging data. *Conference Proceedings: ... Annual International Conference of the IEEE Engineering in Medicine and Biology Society. IEEE Engineering in Medicine and Biology Society. Annual Conference*, 2014, 4771–4774. http://dx.doi.org/10.1109/embc.2014.6944691.

Schone, C., Hofler, H., & Walch, A. (2013). MALDI imaging mass spectrometry in cancer research: Combining proteomic profiling and histological evaluation. *Clinical Biochemistry*, 46(6), 539–545. http://dx.doi.org/10.1016/j.clinbiochem.2013.01.018.

Shibata, D. (2006). Clonal diversity in tumor progression. *Nature Genetics*, 38(4), 402–403. http://dx.doi.org/10.1038/ng0406-402.

Suzuki, H., Aoki, K., Chiba, K., Sato, Y., Shiozawa, Y., Shiraishi, Y., et al. (2015). Mutational landscape and clonal architecture in grade II and III gliomas. *Nature Genetics*, 47(5), 458–468. http://dx.doi.org/10.1038/ng.3273.

Swanton, C. (2012). Intratumor heterogeneity: Evolution through space and time. *Cancer Research*, 72(19), 4875–4882. http://dx.doi.org/10.1158/0008-5472.can-12-2217.

Tata, A., Zheng, J., Ginsberg, H. J., Jaffray, D. A., Ifa, D. R., & Zarrine-Afsar, A. (2015). Contrast agent mass spectrometry imaging reveals tumor heterogeneity. *Analytical Chemistry*, 87(15), 7683–7689. http://dx.doi.org/10.1021/acs.analchem.5b01992.

Turajlic, S., McGranahan, N., & Swanton, C. (2015). Inferring mutational timing and reconstructing tumour evolutionary histories. *Biochimica et Biophysica Acta*, 1855(2), 264–275. http://dx.doi.org/10.1016/j.bbcan.2015.03.005.

Turner, N. C., & Reis-Filho, J. S. (2012). Genetic heterogeneity and cancer drug resistance. *The Lancet Oncology*, 13(4), e178–e185. http://dx.doi.org/10.1016/s1470-2045(11)70335-7.

Vegvari, A., Shavkunov, A. S., Fehniger, T. E., Grabau, D., Nimeus, E., & Marko-Varga, G. (2016). Localization of tamoxifen in human breast cancer tumors by MALDI mass spectrometry imaging. *Clinical and Translational Medicine*, 5(1), 10. http://dx.doi.org/10.1186/s40169-016-0090-9.

Veselkov, K. A., Mirnezami, R., Strittmatter, N., Goldin, R. D., Kinross, J., Speller, A. V., et al. (2014). Chemo-informatic strategy for imaging mass spectrometry-based hyperspectral profiling of lipid signatures in colorectal cancer. *Proceedings of the National Academy of Sciences of the United States of America*, 111(3), 1216–1221. http://dx.doi.org/10.1073/pnas.1310524111.

Vogelstein, B., Papadopoulos, N., Velculescu, V. E., Zhou, S., Diaz, L. A., Jr., & Kinzler, K. W. (2013). Cancer genome landscapes. *Science (New York, N.Y.)*, 339(6127), 1546–1558. http://dx.doi.org/10.1126/science.1235122.

Waclaw, B., Bozic, I., Pittman, M. E., Hruban, R. H., Vogelstein, B., & Nowak, M. A. (2015). A spatial model predicts that dispersal and cell turnover limit intratumour heterogeneity. *Nature*, 525(7568), 261–264. http://dx.doi.org/10.1038/nature14971.

Walch, A. K., Zitzelsberger, H. F., Bink, K., Hutzler, P., Bruch, J., Braselmann, H., et al. (2000). Molecular genetic changes in metastatic primary Barrett's adenocarcinoma and related lymph node metastases: Comparison with nonmetastatic Barrett's adenocarcinoma. *Modern Pathology: An Official Journal of the United States and Canadian Academy of Pathology, Inc*, 13(7), 814–824. http://dx.doi.org/10.1038/modpathol.3880143.

Widlak, P., Mrukwa, G., Kalinowska, M., Pietrowska, M., Chekan, M., Wierzgon, J., et al. (2016). Detection of molecular signatures of oral squamous cell carcinoma and normal epithelium—Application of a novel methodology for unsupervised segmentation of imaging mass spectrometry data. *Proteomics*, *16*(11–12), 1613–1621. http://dx.doi.org/10.1002/pmic.201500458.

Willems, S. M., van Remoortere, A., van Zeijl, R., Deelder, A. M., McDonnell, L. A., & Hogendoorn, P. C. (2010). Imaging mass spectrometry of myxoid sarcomas identifies proteins and lipids specific to tumour type and grade, and reveals biochemical intra-tumour heterogeneity. *The Journal of Pathology*, *222*(4), 400–409. http://dx.doi.org/10.1002/path.2771.

World Health Organization. (1992). *ICD-10: International statistical classification of diseases and related health problems* (10th revision ed.) . Geneva: World Health Organization.

Wu, J. M., Halushka, M. K., & Argani, P. (2010). Intratumoral heterogeneity of HER-2 gene amplification and protein overexpression in breast cancer. *Human Pathology*, *41*(6), 914–917. http://dx.doi.org/10.1016/j.humpath.2009.10.022.

CHAPTER NINE

Ambient Mass Spectrometry in Cancer Research

Z. Takats[*,1], N. Strittmatter[†], J.S. McKenzie[*]

[*]Imperial College London, London, United Kingdom
[†]Drug Safety and Metabolism, AstraZeneca, Cambridge, United Kingdom
[1]Corresponding author: e-mail address: z.takats@imperial.ac.uk

Contents

1. Desorption Electrospray Ionization	232
2. Intraoperative Mass Spectrometry	240
3. REIMS Instrumentation	244
4. DESI-MSI for Drug Imaging in Cancer Research	252
References	253

Abstract

Ambient ionization mass spectrometry was developed as a sample preparation-free alternative to traditional MS-based workflows. Desorption electrospray ionization (DESI)-MS methods were demonstrated to allow the direct analysis of a broad range of samples including unaltered biological tissue specimens. In contrast to this advantageous feature, nowadays DESI-MS is almost exclusively used for sample preparation intensive mass spectrometric imaging (MSI) in the area of cancer research. As an alternative to MALDI, DESI-MSI offers matrix deposition-free experiment with improved signal in the lower (<500 m/z) range. DESI-MSI enables the spatial mapping of tumor metabolism and has been broadly demonstrated to offer an alternative to frozen section histology for intraoperative tissue identification and surgical margin assessment. Rapid evaporative ionization mass spectrometry (REIMS) was developed exclusively for the latter purpose by the direct combination of electrosurgical devices and mass spectrometry. In case of the REIMS technology, aerosol particles produced by electrosurgical dissection are subjected to MS analysis, providing spectral information on the structural lipid composition of tissues. REIMS technology was demonstrated to give real-time information on the histological nature of tissues being dissected, deeming it an ideal tool for intraoperative tissue identification including surgical margin control. More recently, the method has also been used for the rapid lipidomic phenotyping of cancer cell lines as it was demonstrated in case of the NCI-60 cell line collection.

1. DESORPTION ELECTROSPRAY IONIZATION

Ambient ionization mass spectrometry is a collective term describing all mass spectrometric ionization methods, which are able to ionize constituents of natural samples under ambient conditions (Cooks, Ouyang, Takats, & Wiseman, 2006). In this regard, ambient MS methods represent an ideal means to study tissue samples without any chemical modification, ideally in vivo, giving outstanding significance to these methods in the field of cancer research. The first ambient MS method described was desorption electrospray ionization mass spectrometry (DESI-MS), which was implemented by directing a pneumatically assisted solvent electrospray onto the surface of interest (Takats, Wiseman, Gologan, & Cooks, 2004). A scheme of the experimental setup can be seen in Fig. 1. The electrosprayed solvent droplets dissolve certain chemical constituents of the surface investigated and also induce the formation of secondary—still charged—droplets taking off from the surface. As these secondary droplets already contain constituents of the sample, they produce molecular ions of the analytes on their

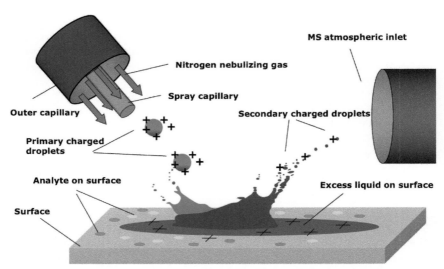

Fig. 1 Schematic of DESI mechanism. Charged primary solvent droplets from a pneumatically assisted electrospray source are directed toward a surface. Primary droplets form a liquid film on sample surface which solvates analyte molecules. The impact of further incoming primary droplets leads to desorption of multiply charged secondary droplets which are subsequently undergoing electrospray-like solvent evaporation and ionization. Produced ions are sampled by an extended mass spectrometer inlet.

evaporation in the atmospheric pressure interface of the mass spectrometer. This way electrospray-like mass spectrometric information is obtained on the sample, featuring multiply charged ions and solvent adducts among other electrospray-specific spectral features.

DESI analysis of tissues was already demonstrated in the first publications, giving detailed information on the small molecular constituents of tissues without sample preparation. Following the initial experiments performed using unmodified tissue samples, the attention was soon turned to the analysis of frozen tissue sections in a spatially resolved experiment (Takats, Wiseman, & Cooks, 2005; Wiseman, Puolitaival, Takáts, Cooks, & Caprioli, 2005). The excitement around the blooming field of MALDI-MS imaging facilitated the development of similar approaches on the DESI field. Although DESI spectra of frozen tissue sections feature only low molecular weight compounds including constituents of core metabolism (TCA cycle, ADP, AMP, acylcarnitines, etc.) and lipids (glycerophospholipids, sphingolipids, sulphatides, bile acids, etc.), the intensity distribution of these features still shows high histological and histopathological specificity (Wiseman et al., 2005). Interestingly, DESI-MS—similarly to ESI—is able to ionize pure protein samples (Takats et al., 2005; Takats, Wiseman, Ifa, & Cooks, 2008); however, the DESI-MSI datasets of tissue specimens rarely contain and ions associated with macromolecular constituents. Furthermore, analytes typically not undergoing ESI usually show poor analytical sensitivity in DESI (Takats et al., 2005). This latter problem was solved by the implementation of the so-called reactive DESI experiment, where compounds not undergoing ionization by ESI are derivatized using a derivatizing agent present in the DESI solvent (Cotte-Rodríguez, Takáts, Talaty, Chen, & Cooks, 2005). Typical examples include the derivatization of carbohydrates by organoboranes (Chen, Cotte-Rodriguez, & Cooks, 2006; Zhang & Chen, 2010) or the derivatization of cholesterol (Wu, Ifa, Manicke, & Cooks, 2009). Another example using a similar strategy used on-tissue derivatization in order to enhance sensitivity for neurotransmitters, their metabolites and neuroactive drugs with primary amine functionality by charge-tagging with 2,4-diphenyl-pyranylium tetrafluoroborate (Shariatgorji et al., 2016).

The general workflow of a DESI imaging experiment is shown in Fig. 2 and shows slight differences in comparison with the analogous MALDI-MSI workflow. The most striking difference is probably the lack of matrix deposition prior to mass spectrometric analysis, which not only simplifies the analytical workflow, but also eliminates the matrix-associated mass spectral

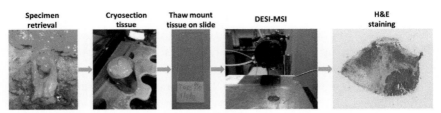

Fig. 2 DESI-MSI workflow. Samples are acquired from various surgical excisions, and typically frozen in liquid nitrogen. Each specimen is then cryosectioned to a thickness of typically 10 μm before being mounted on glass slides. These slides are then analyzed by DESI-MSI. Following imaging, the same sections are stained (typically with hematoxylin and eosin) before undergoing histopathological assessment.

peaks from the lower mass region, making the technique ideal for the metabolic profiling of tissues (Bjarnholt, Li, D'Alvise, & Janfelt, 2014; Vickerman, 2011). DESI-MSI is performed as a series of continuous line scans, where the sample continuously moves under the spray, instead of the stepwise analysis typical for MALDI experiments. While the spatial resolution of MALDI-MSI experiments currently approaches 1 μm (Zavalin, Yang, Hayden, Vestal, & Caprioli, 2015), DESI experiments are typically performed at 35–100 μm feature resolution (lower resolutions of 10–20 μm obtained in unpublished experiments) (Abbassi-Ghadi et al., 2014; Campbell, Ferreira, Eberlin, & Cooks, 2012; Guenther et al., 2015). Similarly to MALDI, the DESI-MSI experiment has also undergone significant developments regarding the speed of analysis. While the initial DESI-MSI studies typically used 1 pixel/s, more recently up to 50 pixels/s imaging speeds have been reported, pushing the time demand for the analysis of a typical 2–3 cm^2 tissue section under an hour (unpublished data by E.A. Jones, Waters Corporation).

DESI-MSI is used for tissue analysis in two markedly different ways, including the conventional imaging experiment where spatially resolved data are collected by the analysis of histological tissue sections and smear analysis where tissue specimens are smeared against a substrate and the material left on the surface is analyzed in a swift manner. In case of imaging experiments the workflow starts with sample collection. Ideally samples are resected from the living organism in form of surgical, core, or pinch biopsies and flash-frozen prior to storage or shipping to the analytical facility. If it is not possible, then the specimens should be flash-frozen as soon as it is possible following their removal from the living organism (Wiseman, Ifa, Venter, & Cooks, 2008). Studying tissue metabolism is possible only in the former case, however, extended storage of the samples at temperatures >4°C induces significant changes in the concentration of structural lipids deeming the samples unsuitable for DESI-MSI analysis, unless the study focuses on the degradation of the samples. Metabolic content of tissues can be considered stable once tissues are stored at −80°C (Goodwin, 2012), although changes in protein and peptide detection are reported after that period (Lemaire et al., 2006). The samples are directly transferred to cryotome without thawing and sectioned at a temperature appropriate for the tissue type. Embedding of samples is possible to simplify the sectioning process; however, embedding media like OCT (optimal cutting temperature) is not ideal as it introduces a strong polymeric signal (polyethylene glycol; PEG) which suppresses tissue signal and may also interfere with data analysis (Nelson, Daniels, Fournie, & Hemmer, 2013). Water ice, carboxymethyl-cellulose, or gelatine are widely reported not to interfere with the mass spectrometric analysis. Section thickness in the range of 4–50 μm has been reported, although most of the published studies use 5–15 μm thick sections for DESI-MSI. The frozen sections are ideally thawed in a low-moisture, oxygen-depleted atmosphere to avoid the precipitation of large amounts of water on the microscope slides and the degradation of oxygen-sensitive metabolites (Wiseman et al., 2008). Following the preparations, the sections are subjected to DESI analysis using a computer-controlled moving stage setup and a pneumatically assisted electrospray emitter. Number of studies focused on optimal geometrical settings for DESI-MSI, briefly the optimal solvent flow rates were reported to be in the 0.2–2 μL/min range with 7–10 bars of nebulizing gas pressure, >60 degree impact angle, and <10 degree of collection angle (further parameters are shown in Fig. 3) (Abbassi-Ghadi et al., 2015; Campbell et al., 2012; Kertesz & Van Berkel, 2008).

There has been a broad disagreement on the optimal solvent composition used for tissue analysis. Initial studies used methanol/water 1:1 mixture (Wiseman, Ifa, Song, & Cooks, 2006; Wiseman et al., 2005), which was changed to solvent systems containing dimethyl-formamide and acetonitrile in order to save tissue structure for subsequent staining and morphological assessment (Eberlin et al., 2011). Other groups reported excellent results by using methanol/water 95:5, which avoids the introduction of high boiling point solvents into the instrument (Abbassi-Ghadi et al., 2014; Guenther et al., 2015; Swales et al., 2014). In general high organic content solvent systems were observed to leave cellular structure intact, likely by denaturing structural proteins. Not only morphological staining

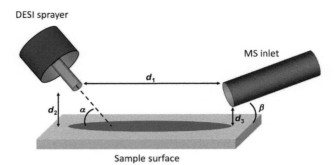

Fig. 3 Schematic of DESI experiment with relevant geometrical parameters. The performance of DESI typically depends on the relative position of the sample, sprayer, and MS inlet to each other. The various parameters are α, incidence angle of sprayer to sample surface; β, collection angle of MS inlet capillary relative to sample surface; d_1, distance of sprayer tip to MS inlet capillary; d_2, distance of sprayer tip to sample surface; d_3, distance of MS inlet capillary to sample surface.

(e.g., hematoxylin–eosin) but also immunohistochemical stains were applied successfully following the MSI experiment. Fig. 4 shows the image coregistration and histology-guided feature extraction for imaging experiments.

A few papers focusing specifically on the analysis of DESI-MSI datasets have been published. The proposed workflows comprise (1) data recalibration and alignment, (2) background subtraction, (3) normalization and log-transformation, (4) coregistration with histological data, (5) pixel-wise multivariate or univariate analysis of the data (Veselkov et al., 2014). In case of the identification of unknown tissue types, precalculated multivariate models are used for the pixel-wise classification (Dill et al., 2011; Guenther et al., 2015; Veselkov et al., 2014).

The principal application area of DESI-MSI has been cancer research with primary focus on histological and drug distribution applications. Deployment of the former in interventional environment has been pursued by a number of research groups, with the rationale of replacing frozen section histology for the margin control in course of cancer resection surgery (Calligaris et al., 2014, 2013; Eberlin et al., 2013, 2014). The histological applications with regard to tumor tissue detection was first demonstrated in case of liver cancer, where—in spite of the lack of protein signal—excellent discrimination performance was achieved using individual tissue markers (Wiseman et al., 2005). This study not only heralded the success of implementing DESI for tissue imaging, but also turned the attention of the field to the less-studied structural lipid composition of tissues. Structural

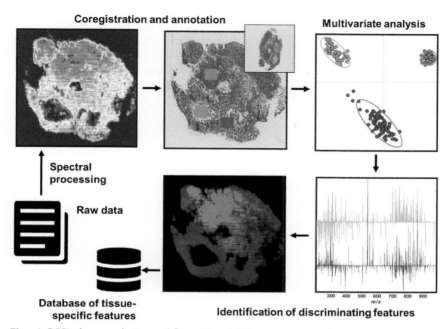

Fig. 4 DESI data analysis workflow. The DESI imaging workflow involves spectral processing from raw data files to produce a mass spectral image. Following coregistration of the optical H&E stained image, specific tissue regions are annotated by a histopathologist. Multivariate analysis can be used to discriminate between such regions and identifies ions that discriminate between the tissue types. These characteristic spectral features can be saved to a database and used to predict the presence of certain tissue types in other tissue sections without the need for histopathological annotation.

lipids had been detected in tissues using both NMR and MS, however, this information was widely regarded as useless until several secondary ion mass spectrometry (SIMS), MALDI, and DESI studies proved that the structural lipid composition of tissues perfectly follows their histological (or histopathological) classification. A typical DESI spectrum of colorectal adenocarcinoma recorded in negative ion mode can be seen in Fig. 5.

Following the initial success with DESI-MSI, a number of studies focusing on different cancer types were published. These studies not only provided further evidence for the histological specificity of the DESI-MSI data, but also led to the discovery of a number of novel tissue cancer biomarkers including ascorbic acid and cholesterol-sulfate in case of prostate cancer (Eberlin et al., 2010). The initial cancer diagnostic studies used tissue samples containing tumor and noncancerous tissue and performed line scans

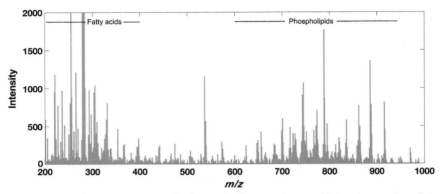

Fig. 5 A typical DESI spectrum of colorectal adenocarcinoma. Major detected analyte groups in negative ion mode are fatty acids typically in the *m/z* range spanning from 200 to 400, while phospholipids appear over the *m/z* range from 600 to 1000.

across the tumor border (Wiseman et al., 2005). As in the early days of the technology development mostly complex lipids were detected from tissues, a number of structural lipid constituents were found to be overproduced in one or another tissue. The initially deployed instrumentation providing low mass spectrometric resolution did not allow the proper identification of these ions; however, later studies confirmed their chemical identities. The first larger-scale real imaging studies were performed using healthy and matched normal tissue specimens, focusing on the spectral differences between the averaged spectra obtained from cancer and healthy tissues. In a study aimed at the DESI-MSI characterization of canine invasive translational cell carcinoma of the urinary bladder, significant differences were found with regard to both complex lipid composition and fatty acid composition of the two different tissue classes (Dill et al., 2009). In positive ion mode phosphatidylcholine (PC) and sphingomyelin species were found to show higher concentration in the tumor, while in negative mode certain fatty acids, phosphatidylinositols (PI), phosthatidylserines (PS), and phosphatidylglycerol species were found to serve as tissue-specific markers. The study also demonstrated the application of unsupervised multivariate statistical methods in the form of principal component analysis (PCA) for the visualization of these spectral differences. In another study 68 human samples with prostate cancer, prostate intraepithelial neoplasia (PIN), and healthy prostate cancer specimens were analyzed and cholesterol-sulfate was found to be an exclusive marker for not only cancerous tissue but also for precancerous lesions represented by PIN (Eberlin et al., 2010).

A subsequent study aimed at the DES-MSI examination of human bladder cancer using 20 healthy-cancerous matched pairs of samples found a number of glycerophospholipids contributing to the spectral differences detected between samples. This study pioneered the application of supervised multivariate analysis-driven pixel-wise classification for DESI-MSI data represented by partial least squares discriminant analysis of spectra (Dill et al., 2011). Fig. 6 shows supervised multivariate statistical representation of a DESI image of an FFPE-fixed colorectal tissue specimen.

A number of following studies applied identical or similar data treatment for studies targeted on the DESI-MS analysis of various human cancer cohorts including seminoma (Masterson et al., 2011), colorectal

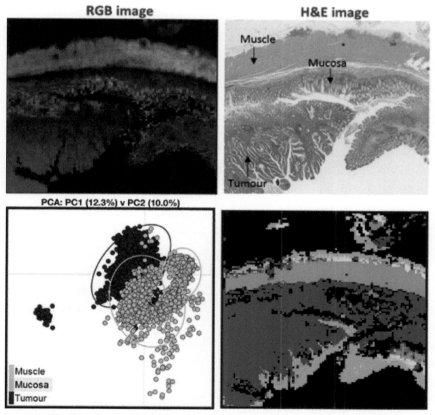

Fig. 6 DESI-MSI analysis of a human colorectal sample in positive ionization mode. RGB image of overlaid *m/z* ions (*m/z* 269.16—*green*, muscle; 665.38—*red*, mucosa; 349.12—*blue*, tumor). H&E stained optical image of tissue sample analyzed. PCA of all pixels and MMC components RGB image also shown.

adenocarcinoma (Gerbig et al., 2012), and gliomas (Eberlin et al., 2012). The common conclusion of these studies included that (1) the structural lipid profile of tissues obtained by DESI-MSI shows histological specificity, (2) the specificity is sufficiently high to allow not only pair-wise differentiation but identification of tissues, (3) there are certain tumor type-specific individual metabolites (typically not structural lipids) which can be used as univariate markers for the detection of a the specific tumor. Further studies revealed that the specificity of the detected lipid profiles goes beyond histopathological specificity and differences pertaining to the molecular mechanism of tumorigenesis can also be identified using DESI-MS of cell cultures or DESI-MSI of human tissues. Characteristic spectral features differentiating KRAS-mutant and wild-type colorectal adenocarcinomas were found in a human cohort, indicating a closer relationship between the signaling networks driving tumorigenesis and the structural lipid composition of tumors (Gerbig et al., 2012). Further developments of the technology were focused on the development of DESI-MSI for intraoperative tissue diagnostics replacing frozen section histology (discussed later) and the implementation of DESI-MSI-based methodology for patient stratification and personalized medicine. The difference between these two research directions is clearly demonstrated by two publications on the DESI-MSI of breast cancer. While one focuses on the intraoperative rapid imaging of samples (Calligaris et al., 2014), the other explores the diagnostic and prognostic information content of the DESI-MSI data (Guenther et al., 2015).

2. INTRAOPERATIVE MASS SPECTROMETRY

Originally liver adenocarcinoma liver metastasis specimens were the first human samples analyzed by DESI-MS in the form of line scans and the study demonstrated unique spectral differences between tumor region, transition regions between tumor and normal tissue, and the normal region (Wiseman et al., 2005). While the study was carried out on frozen tissue sections deposited on microscope slides, the spectra were obtained within a minute. This rapid assessment showed the potential of DESI-MS for use in intraoperative applications compatible with surgical decision making.

DESI-MSI provides a significantly higher degree of spatial resolution than a single line scan allowing two and three dimensional images to be reconstructed from the pixel-by-pixel spectral data. The ability to record chemical composition directly from tissue samples has been applied in several studies on different cancer types including brain tumors

(Agar et al., 2011; Calligaris et al., 2013; Eberlin et al., 2012, 2013; Santagata et al., 2014), breast tumors (Calligaris et al., 2014; Guenther et al., 2015), and gastric tumors (Eberlin et al., 2014) as it is discussed earlier. In one of these experiments, DESI-MS was used to measure differences in levels of 2-hydroxyglutarate (Santagata et al., 2014). The metabolite, which is derived from α-ketoglutarate, is known to accumulate in leukemia and specifically in most grade two and three gliomas in adult patients due to a mutation in isocitrate dehydrogenase 1 and 2 and could be used as a biomarker. DESI-MS was able to identify an increase in 2-HG that was independent of the exact mutation of IDH1. This compares well with immunohistology which is only able to recognize certain types of the many IDH1 mutants using a specific antibody (anti-IDH1 R132H). An intraoperative model for the detection of the biomarker was constructed by Santagata et al. and led to detection of increased levels from intraoperative samples of glioma containing a mutation in IDH1 in the operating room (Santagata et al., 2014). Using DESI-MS installed within the operating room, the group was able to detect the tumor metabolite 2-HG within minutes from intraoperative glioma specimens containing mutant variants of IDH1. This is just one example demonstrating how chemical profiles obtained from tissue have potential to offer additional information compared to traditional histology.

In another experiment, statistical analysis of DESI-MS data was able to differentiate between gliomas (oligodendrogliomas, astrocytomas, and oligoastrocytomas) representing a variety of grades using lipid species such as fatty acids, glycerophosphoinositols, glycerophoserines, plasmenyl phosphoethanolamines, and sulfatides (Eberlin et al., 2012). Classifiers constructed from multivariate statistical analysis of spectral signatures and correlated with histopathology revealed a high cross-validated predictive accuracy of 97% for tumor type, grade, and cellular concentration. Models based on lipid profiles have provided similar results when distinguishing gliomas from meningiomas (Eberlin et al., 2013), necrotic tumor tissue from viable glioblastoma multiforme (Calligaris et al., 2013), and in breast cancer margin assessment (Calligaris et al., 2014). The authors of the studies concluded analysis the negative ion mode revealed greater sensitivity for the detection of lipids and metabolite species.

Some experiments have been conducted on animal tissue to modify the DESI-MS sampling probe in order to take intraoperative measurements using the technique. Specifically, Chen et al. tested the possibility of inserting the sampling probe via an endoscope for direct tissue analysis

(Chen et al., 2013). Some of the issues with this type of analysis are the length required for the ion transfer tubing and the inability to use organic solvents for electrospray ionization in the surgical setting. By the removal of high voltage from the application and using pure water for electrospray ionization, the authors achieved good intensity measurements with 0.5 s time delay from positioning the probe against tissue surface. Evaluation of the techniques usefulness for intraoperative use requires measurements from human ex vivo samples and validation for use in the operating theater.

MSI techniques have provided large quantities of data that is highly relevant to tissue identification and understanding biological processes in cancer-host tissue systems. However, all of the current MSI techniques require preparation of frozen tissue sections. This process can lead to spatial dislocation of chemical species or modification of the chemical profile on the tissue and adds to the time it takes to analyze a sample. This inability to produce results within the timeframe of the operation as well as the requirements arising from sample preparation limit their usefulness for intraoperative margin assessment.

Techniques such as MALDI imaging and DESI-MS demonstrate the potential of mass spectrometry for rapid measurement of chemical data directly from tissue, which would be required for intraoperative applications. On the basis of this potential, a range of ambient ionization techniques have been developed toward use in the operating theater. These techniques take into account the requirements for rapid measurements directly from tissue and have been designed based on the current standard practices in the surgical setting. Efficiencies in operating time and economy of movements are important to ensure an optimized surgical workflow and help to reduce operative error. The ideal intraoperative ion source therefore should have minimal impact on surgical workflow. The benefit of using surgical tools already in use in standard surgical practice is that additional steps are not warranted. The discovery that various surgical tools incidentally cause ionization of tissues during routine surgical use has been of great significance and these techniques are described later.

Rapid evaporative ionization mass spectrometry (REIMS) was first described in 2009 as the direct utilization of surgical diathermy for the ionization of biological tissue constituents in vivo (Schäfer et al., 2009). Surgical diathermy is a universally used dissection and hemostasis tool, utilizing radio-frequency alternating current for heating and thermally ablating tissues in course of surgical dissections. In course of diathermal (also known as electrosurgical) manipulations, the patient becomes part of the electric circuit

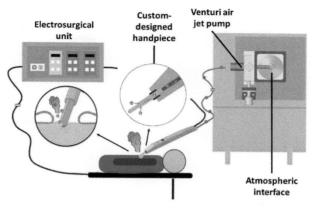

Fig. 7 Monopolar REIMS setup for intraoperative applications. The customized monopolar surgical diathermy device with integrated aerosol aspiration line generates an aerosol which is directed into the atmospheric interface of a mass spectrometer for profiling analysis.

and the electric current flowing through ("dia-" = through) the tissue induces heat ("therme" = heat) dissipation due to the high impedance of biological tissues. Surgical diathermy is implemented in two markedly different ways. In case of the so-called monopolar electrosurgery the patient lies on a large surface area electrode, while the surgeon has a handheld, sharp electrode for dissection. As the circuit closes, the current density reaches its maximal value at the contact point of the electrode and becomes sufficiently high to thermally ablate or evaporate the tissue material. This setup is shown in Fig. 7.

In case of the "bipolar" setup both electrodes are handheld in the form of a pair of forceps. Tissues (or blood vessels) held between the tips of the forceps are heated up by the electric current, avoiding stray currents causing undesired side effects, e.g., functional neurological damage. This setup is shown in Fig. 8. While monopolar electrosurgery is used for macroscopic dissection and hemostasis of minor bleedings, bipolar electrosurgery is used only for fine tissue manipulations, usually stopping capillary bleeds on the surgical area. The latter technology is primarily used in neurosurgery, while the former is used generally in multiple fields of surgery. Electrosurgery shows particular advantages with regard to cancer surgery as all the cells involved in the surgical manipulation are exposed to sufficient heat to induce necrosis, hence minimizing the chance of surgically induced metastasis formation. As surgical diathermy shows physicochemical analogy with certain mass spectrometric ionization methods (with special emphasis on

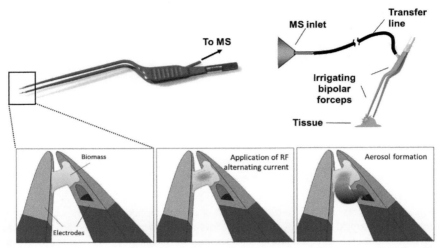

Fig. 8 REIMS setup for analysis of unicellular organisms and tissues using a bipolar handpiece. The irrigating bipolar forceps contains an aspiration channel down one of the prongs which is directly connected to the mass spectrometer inlet using a polymer transfer line. RF alternating electric current is applied between the electrodes of the forceps which results in the formation of an aerosol similarly to the monopolar handpiece.

thermospray and laser desorption ionization), it was tested as a potential ion source for mass spectrometry and found to produce organic ion population associated with the structural lipid content of tissues. As the spectral appearance of REIMS signal shows high similarity to DESI-MS data, it was also expected to elicit good histological specificity (Golf, Muirhead, et al., 2015).

3. REIMS INSTRUMENTATION

The initial REIMS setups featured commercially available electrosurgical handpieces equipped with an additional fluid line, which was used for the aspiration of nascent aerosol formed during tissue manipulation. The aerosol was directly introduced into the mass spectrometer without any means of postionization (Schäfer et al., 2009). Due to contamination effects, this setup was replaced by next-generation system featuring a Venturi air jet pump to facilitate the efficient transfer of aerosol from the surgical site to the mass spectrometer (Balog et al., 2013). The sampling of the aerosol was implemented using an orthogonal MS inlet tube to exert some level of momentum threshold for the sampled particles and to keep soot and lipid droplets outside of the vacuum regime of the instrument. It was discovered that the actual ion formation, i.e., the formation of ions detected by the mass

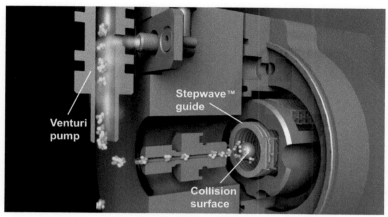

Fig. 9 A cut-away schematic of the REIMS atmospheric interface. The transfer line into the Venturi pump introduces the aerosol into the mass spectrometer. The collision surface separates the incoming ion clusters before entry into the Stepwave™ ion guide and subsequent mass analysis.

spectrometer takes place in the atmospheric interface of instruments via droplet-surface collision phenomenon, initiating a further refinement of the setup (Balog et al., 2016). This refinement led to the introduction of a heated collision surface into the setup, which is integral part of the commercialized solution marketed by Waters Corporation (Milford, USA). A scheme of such a modified atmospheric pressure interface is shown in Fig. 9. The in vivo human applications require a modified mass spectrometer with all functional elements (instrument, vacuum pump, and computer) embedded into a single, mobile housing.

The surgical application of the REIMS technology is often termed "iKnife" or intelligent surgical device in the literature. The surgical applications are predominantly focused on cancer surgery, in a narrower sense, the assessment of surgical margin clearance and identification of unknown tissue features for diagnostic purposes. In course of the in vivo applications the surgical professionals use an electrosurgical handpiece (also known as "pencil," shown in Fig. 10) equipped with aerosol evacuation line, which is connected to the mass spectrometer standing in 2–3 m distance from the surgical table. On tissue dissection the aerosol is aspirated by the instrument and subsequently analyzed.

The resulting mass spectra are recalibrated, background subtracted, normalized, and subjected to multivariate statistical analysis. The latter step comprises the localization of the data point in a precalculated statistical

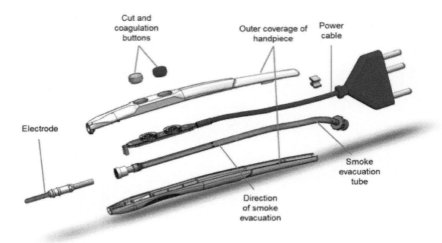

Fig. 10 Schematic of monopolar handpiece is used for REIMS analysis. The custom-build monopolar handpiece is used for in vivo and ex vivo tissue analysis. It is a modified version of a typical surgical diathermy handpiece. The smoke evacuation line is connected to the mass spectrometer via polymer transfer tubing.

model, which was obtained by the PCA and linear discriminant analysis of the histologically annotated reference spectra (Balog et al., 2013). Since the overall time demand for aerosol transfer, mass analysis, and data classification is 0.5–2 s, the instrument is used for the continuous histological tracking of the surgical procedure. In case of margin clearance application the output data of the setup is limited to healthy vs cancerous, while in case of unknown tissue identification the output is a histological (or histopathological) classification result accompanied by the probability of correct classification given in percentage.

Although the REIMS technology was developed exclusively for intraoperative tissue identification, more recently laboratory applications were also developed. These applications include the analysis of ex vivo surgical tissue specimens and the analysis of in vitro cancer models including cell lines and organoids. The primary rationale of the former application is to create reference data for the in vivo identification of tissues; however, the same approach can also be used for the analysis of biopsy specimens for rapid (<30 s) tissue diagnostics. In case of the acquisition of reference data, the surgical specimens are delivered to the histopathology facility, where parts of the specimen not influencing the histological diagnosis are separated and sent to the mass spectrometry laboratory. At the MS lab the specimens are cut to macroscopic slices and sampled using either a handheld monopolar surgical tool or an automated REIMS imaging system capable of spatially resolved data collection (Golf, Strittmatter, et al., 2015). Following the acquisition of MS data the specimens are formalin-fixed, paraffin-embedded, sectioned, and stained. The sections are examined by a histopathology professional and the histological environment of burn marks caused by REIMS sampling are used for the classification of REIMS data.

Analysis of cultured cell lines by REIMS has been demonstrated recently using bipolar diathermy (Strittmatter et al., 2016). This setup as shown in Fig. 8 has also been used to analyze microorganisms (Strittmatter et al., 2013, 2016). The cell lines were grown as a confluent layer, harvested via trypsinization, and mobilized cells were centrifuged to form a pellet. 0.2–1.5 mg of biomass from the pellets were taken up using the bipolar forceps and subjected to REIMS analysis. The resulting spectra—similarly to the REIMS spectra of tissues—feature predominantly phospholipids; however, they are markedly different from the analogous tumor tissue they were isolated from. Nevertheless, the structural lipid profiles were shown to be characteristic to the cell lines, which allowed the identification of the members of the NCI-60 cell line collection. Fig. 11 shows a classification model for three individual cell lines obtained using REIMS which resulted in 100% correct cross-validation using leave-one-out cross-validation.

As in earlier metabolic studies of cell lines, only a weak correlation with tissue type of origin was found within the NCI-60 cell line panel; however,

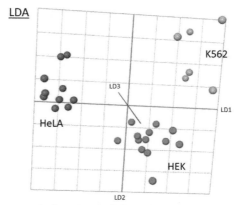

Fig. 11 Multivariate statistical analysis of REIMS cell line data. Supervised analysis of three distinct cell lines by REIMS by linear discriminants analysis. Using leave-one-sample-out cross-validation, the model's classification performance is 100%. The three cell lines shown are HEK (*red*), HeLa (*green*), and K562 (*orange*).

metabolic changes observed were found to be in good agreement with LC-MS data of the same cell lines. The data was also used to assess the detection of mycoplasma infection and could in future find application to determine the viability of cells with potential applications in high-throughput drug screening, especially in combination with recently developed automated REIMS platforms (Bolt et al., 2016).

Laser desorption ionization-mass spectrometry (LDI-MS) is an ambient ionization technique that can utilize surgical lasers in the ultraviolet and far-infrared wavelengths (Schäfer, Szaniszló, et al., 2011). CO_2 lasers are one of the most common lasers used in a variety of surgical procedures. They offer precise cutting and coagulation with minimal thermal spread. Recent advancements have heralded the flexible CO_2 laser that enables use in a wider range of applications including brain, and endoscopic surgeries (Au et al., 2012; Ryan et al., 2010). Aerosol produced as a direct result of laser application on tissue and can be aspirated continuously into a mass spectrometer (Schäfer, Szaniszló, et al., 2011). The ionization takes place as the laser causes boiling of the intracellular water, which results in the aerosol containing water vapor and thermal degradation of cellular products. Analysis of ex vivo samples of human liver colorectal metastasis and human renal cell carcinoma using LDI-MS showed good separation with multivariate statistics. CO_2 lasers produced better classification results compared to Nd:YAG laser with better reproducibility and less signal-to-noise ratio. A schematic of both setups is shown in Fig. 12. The laser can be used both in continuous

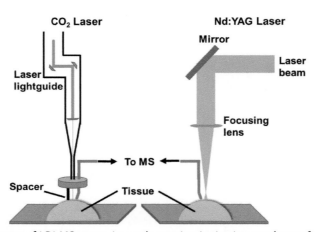

Fig. 12 Scheme of LDI-MS setup. Laser desorption ionization can be performed using either a CO_2 laser (shown *left*) or Nd:YAG laser (shown *right*). Focusing lenses are used for both setups in order to collimate the beam onto the tissue surface. The desorbed ions are aspirated into the MS interface via PTFE tubing connected to a Venturi air jet pump.

wave mode and pulsed mode. LDI-MS almost exclusively produces ions representative of various lipid compounds such as phospholipids and free fatty acids. In the analysis of liver tissues deprotonated molecular ions of bile acids were also detected. Notably photochemical effects were not observed with LDI-MS in contrast to other mass spectrometry techniques such as MALDI. Laser energy was proportional to the recorded spectral intensity of phospholipid ions (Schäfer, Szaniszló, et al., 2011).

Cavitron ultrasonic surgical aspirator (CUSA) uses ultrasound waves to cause mechanical fragmentation of tissues with suction to clear away the resultant tissue debris. Primarily used in neurosurgery, it works as a precise technique for selectively excising parenchymal tissue while sparing blood vessels, it can also be used in operations targeting the liver and pancreas. Venturi easy ambient sonic-spray ionization (V-EASI) was developed by directly linking the liquefied tissue extraction tubing to a Venturi air jet pump at the mass spectrometer interface (Schäfer, Balog, et al., 2011). The liquid readily converts into gas phase by nebulization and ions presumed to be predominantly lipid species are directed into the mass spectrometer. This setup is shown in Fig. 13. Other small molecules, not readily detected by other ambient ionization mass spectrometry techniques using a thermal process, were also detected. The technique has been used to analyzes ex vivo brain and liver samples (Schäfer, Balog, et al., 2011). A clear difference was

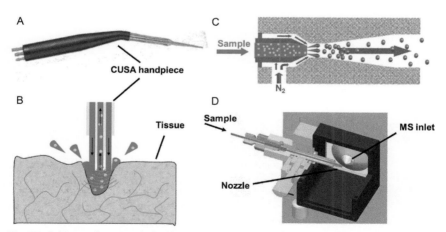

Fig. 13 Scheme showing CUSA setup. (A) Cavitron ultrasonic surgical aspirator (CUSA) handpiece. (B) CUSA handpiece is applied to tissue specimen where ultrasonic vibrations dislodge surface molecules. (C) Schematic view of Venturi air jet pump nebulizing aspirated fluids. (D) Schematic of mass spectrometer inlet with orthogonally oriented Venturi setup.

seen between spectra corresponding to gray and white brain matter as well as different types of brain tumors. Liver tumors also separated well from healthy liver parenchyma. The analysis was performed in 2–3 s from CUSA application, leading to near instant measurements that are significantly faster than the sample preparation for DESI-MS described in a previous paper (Wiseman et al., 2005).

A third ambient ionization method that can be coupled to a surgical instrument for intraoperative measurements is REIMS (Balog et al., 2013). This technique utilizes an electrosurgical tool often known as "diathermy" that is used in many types of surgeries for the cutting and coagulation of tissues. An electrosurgical generator delivers alternating electrical current at radiofrequencies through the tip of the tool which acts as an active electrode. When touched on tissue, it causes impedance to the current resulting in high heat produced at the tip. The rapid vaporization of cells results in aerosol, which is transferred to the mass spectrometer (Schäfer et al., 2009).

REIMS has been extensively tested to date by analyzing 2933 ex vivo specimens consisting of normal and cancerous tissue from 302 individual patients. The samples contained stomach, colon, liver, lung, and breast samples, as well as samples from other tissue types. Spectra from these samples were detected in the mass range of m/z 600–900 representing predominantly lipids and the data collected was correlated with histopathology before applying multivariant statistics (PCA and linear discriminant analysis) (Balog et al., 2013). A typical REIMS workflow is shown in Fig. 14.

Cross validation revealed the sensitivies and specificities of the correct tissue classification ranged from 69% for mixed adenosquamous cell

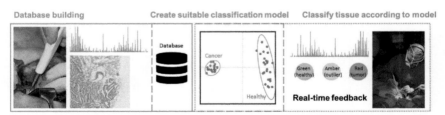

Fig. 14 REIMS data analysis workflow. A database of REIMS spectra covering a range of tissue types and histological types is first created. For classification in real time, the suitable spectra are combined in a multivariate model. Spectra acquired in real time are processed and classified using this model, and a traffic-light system is used to provide feedback regarding the tissue classification.

carcinoma to 100% for healthy tissue from lung and liver. Further validation was obtained by measuring in vivo data in the operating theater for 81 patients. The sensitivities and specificities from this analysis ranged between 95–100% and 92–100%, respectively. The measurement data was obtained within 2.5 s, resulting in near real-time measurements during operation (Balog et al., 2013). The rapid data acquisition and the ease of incorporation into the clinical environment are the main advantages of REIMS. A typical classification model obtained using REIMS is shown in Fig. 15 for healthy and two different types of cancerous breast tissue.

Another class of rapid mass spectrometry techniques is called probe electrospray ionization. These include paper spray mass spectrometry (PS-MS) and probe electrospray ionization mass spectrometry (PESI-MS), which both use ambient ionization techniques to create ions directly from the material for analysis. PS-MS involves placing the sample onto a triangular strip of porous paper and applying a solvent and a high voltage across the paper inducing a spray of charged ions that is aspirated into a mass spectrometer for analysis (Wang et al., 2011). PESI-MS uses a small probe or solid needle for sampling and subsequent electrospray ionization (Hiraoka, Nishidate, Mori, Asakawa, & Suzuki, 2007). While capable of rapid identification of sufficient lipid profiles to classify normal and tumor tissue

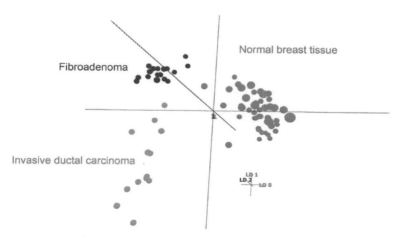

Fig. 15 Multivariate statistical models of REIMS breast cancer profiles. A linear discriminants analysis between three different tissue types commonly encountered during breast cancer surgery. The largest differences across the first latent variable (LD 0, *red*) differentiate between normal and nonnormal tissue. Fibroadenoma and invasive ducal carcinoma separate clearly along the second latent variable (LD 1, *green*).

(Mandal et al., 2013) and even analysis of metabolites in single cells (Gong et al., 2014), these techniques are more suited for analysis of biopsy samples. Their advantages include short processing times of under a minute and higher signal intensity in comparison with DESI-MS from the same tissue for PS-MS (Wang et al., 2011) and the potential to directly measure intracellular lipids without homogenization for PESI-MS (Yoshimura et al., 2012).

4. DESI-MSI FOR DRUG IMAGING IN CANCER RESEARCH

While several studies have been performed using ambient mass spectrometry for drug imaging applications, the majority of these were focused on nontumor tissues such as kidney, brain, or lung. Applications of ambient mass spectrometry in general pharmaceutical research are exemplified by the use of DESI-MSI to visualize the molecular content of crystal-like structures in frozen and paraffin-embedded formalin-fixed rabbit kidney to explain the severe renal toxicity observed in dosed animals (Bruinen et al., 2016). Swales et al. published a study using DESI-MSI (and other MSI modalities) on cassette-dosed rat tissues (brain, kidney), where DESI showed good sensitivity for a wide range of different compounds including antitumor drug erlotinib (Swales et al., 2014). DESI-MSI was also used to analyze propranolol in whole-body tissue sections of mice to complement whole-body autoradiography (WBA) studies (Kertesz et al., 2008). The advantage of MSI over WBA is the label-free nature of MSI allowing earlier application in the drug development pipeline and the ability to distinguish drug from possible drug metabolites carrying the label. However, in this study no images of drug metabolites could be obtained. Using on-tissue derivatization, Shariatgorji et al. reported the visualization of neuroactive drugs amphetamine, sibutramine, and fluvoxamine and their effects on endogenous small molecule neurotransmitters using DESI-MSI (Shariatgorji et al., 2016). Quantitative drug imaging was shown for clozapine and fluvoxamine, respectively, in brain tissues using on-tissue calibration curves (Shariatgorji et al., 2016; Vismeh, Waldon, Teffera, & Zhao, 2012).

To date, only a single study was published by AstraZeneca applying ambient mass spectrometry to imaging of antitumor agents in cancer tissues (Ashton et al., 2016). In this study, Aurora B-kinase inhibitor AZD2811 was dosed directly and in Accurin nanoparticle formulation in tumor tissue using MALDI and DESI-MSI. While both modalities were able to visualize AZD2811 and longer exposure of tumor tissue to drug for the nanoparticle formulation, higher sensitivity was obtained using DESI-MSI. Additionally, the latter allowed visualization of polymeric polylactic acid and thus could be used to trace the drug delivery system itself. This makes DESI-MSI particularly useful for analysis of complex drug formulations and possible distinction of bound and released drug. Although these and other studies exemplified the use of DESI-MSI for pharmaceutical applications, uptake by pharmaceutical companies is still in progress and studies are often published together with academic collaborators. However, several DESI-MSI instruments can now be found in industrial mass spectrometry imaging groups such as in AstraZeneca, United Kingdom (two instruments), and Servier, France (one instrument).

REFERENCES

Abbassi-Ghadi, N., Jones, E. A., Veselkov, K. A., Huang, J., Kumar, S., Strittmatter, N., et al. (2015). Repeatability and reproducibility of desorption electrospray ionization-mass spectrometry (DESI-MS) for the imaging analysis of human cancer tissue: A gateway for clinical applications. *Analytical Methods, 7*(1), 71–80.

Abbassi-Ghadi, N., Veselkov, K., Kumar, S., Huang, J., Jones, E., Strittmatter, N., et al. (2014). Discrimination of lymph node metastases using desorption electrospray ionisation-mass spectrometry imaging. *Chemical Communications, 50*(28), 3661–3664.

Agar, N. Y., Golby, A. J., Ligon, K. L., Norton, I., Mohan, V., Wiseman, J. M., et al. (2011). Development of stereotactic mass spectrometry for brain tumor surgery. *Neurosurgery, 68*(2), 280–289. discussion 290.

Ashton, S., Song, Y. H., Nolan, J., Cadogan, E., Murray, J., Odedra, R., et al. (2016). Aurora kinase inhibitor nanoparticles target tumors with favorable therapeutic index in vivo. *Science Translational Medicine, 8*(325), 325ra317.

Au, J. T., Mittra, A., Wong, J., Carpenter, S., Carson, J., Haddad, D., et al. (2012). Flexible CO_2 laser and submucosal gel injection for safe endoluminal resection in the intestines. *Surgical Endoscopy, 26*(1), 47–52.

Balog, J., Perenyi, D., Guallar-Hoyas, C., Egri, A., Pringle, S. D., Stead, S., et al. (2016). Identification of the species of origin for meat products by rapid evaporative ionization mass spectrometry. *Journal of Agricultural and Food Chemistry, 64*(23), 4793–4800.

Balog, J., Sasi-Szabó, L., Kinross, J., Lewis, M. R., Muirhead, L. J., Veselkov, K., et al. (2013). Intraoperative tissue identification using rapid evaporative ionization mass spectrometry. *Science Translational Medicine, 5*(194), 194ra193.

Bjarnholt, N., Li, B., D'Alvise, J., & Janfelt, C. (2014). Mass spectrometry imaging of plant metabolites—Principles and possibilities. *Natural Product Reports, 31*(6), 818–837.

Bolt, F., Cameron, S. J. S., Karancsi, T., Simon, D., Schaffer, R., Rickards, T., et al. (2016). Automated high-throughput identification and characterization of clinically important bacteria and fungi using rapid evaporative ionization mass spectrometry. *Analytical Chemistry, 88*(19), 9419–9426.

Bruinen, A. L., van Oevelen, C., Eijkel, G. B., Van Heerden, M., Cuyckens, F., & Heeren, R. M. A. (2016). Mass spectrometry imaging of drug related crystal-like structures in formalin-fixed frozen and paraffin-embedded rabbit kidney tissue sections. *Journal of the American Society for Mass Spectrometry, 27*(1), 117–123.

Calligaris, D., Caragacianu, D., Liu, X., Norton, I., Thompson, C. J., Richardson, A. L., et al. (2014). Application of desorption electrospray ionization mass spectrometry imaging in breast cancer margin analysis. *Proceedings of the National Academy of Sciences*, *111*(42), 15184–15189.

Calligaris, D., Norton, I., Feldman, D. R., Ide, J. L., Dunn, I. F., Eberlin, L. S., et al. (2013). Mass spectrometry imaging as a tool for surgical decision-making. *Journal of Mass Spectrometry*, *48*(11), 1178–1187.

Campbell, D. I., Ferreira, C. R., Eberlin, L. S., & Cooks, R. G. (2012). Improved spatial resolution in the imaging of biological tissue using desorption electrospray ionization. *Analytical and Bioanalytical Chemistry*, *404*(2), 389–398.

Chen, H., Cotte-Rodriguez, I., & Cooks, R. G. (2006). cis-Diol functional group recognition by reactive desorption electrospray ionization (DESI). *Chemical Communications*, *6*, 597–599.

Chen, C.-H., Lin, Z., Garimella, S., Zheng, L., Shi, R., Cooks, R. G., et al. (2013). Development of a mass spectrometry sampling probe for chemical analysis in surgical and endoscopic procedures. *Analytical Chemistry*, *85*(24), 11843–11850.

Cooks, R. G., Ouyang, Z., Takats, Z., & Wiseman, J. M. (2006). Ambient mass spectrometry. *Science*, *311*(5767), 1566–1570.

Cotte-Rodríguez, I., Takáts, Z., Talaty, N., Chen, H., & Cooks, R. G. (2005). Desorption electrospray ionization of explosives on surfaces: Sensitivity and selectivity enhancement by reactive desorption electrospray ionization. *Analytical Chemistry*, *77*(21), 6755–6764.

Dill, A. L., Eberlin, L. S., Costa, A. B., Zheng, C., Ifa, D. R., Cheng, L., et al. (2011). Multivariate statistical identification of human bladder carcinomas using ambient ionization imaging mass spectrometry. *Chemistry—A European Journal*, *17*(10), 2897–2902.

Dill, A. L., Ifa, D. R., Manicke, N. E., Costa, A. B., Ramos-Vara, J. A., Knapp, D. W., et al. (2009). Lipid profiles of canine invasive transitional cell carcinoma of the urinary bladder and adjacent normal tissue by desorption electrospray ionization imaging mass spectrometry. *Analytical Chemistry*, *81*(21), 8758–8764.

Eberlin, L. S., Dill, A. L., Costa, A. B., Ifa, D. R., Cheng, L., Masterson, T., et al. (2010). Cholesterol sulfate imaging in human prostate cancer tissue by desorption electrospray ionization mass spectrometry. *Analytical Chemistry*, *82*(9), 3430–3434.

Eberlin, L. S., Ferreira, C. R., Dill, A. L., Ifa, D. R., Cheng, L., & Cooks, R. G. (2011). Non-destructive, histologically compatible tissue imaging by desorption electrospray ionization mass spectrometry. *Chembiochem: A European Journal of Chemical Biology*, *12*(14), 2129–2132.

Eberlin, L. S., Norton, I., Dill, A. L., Golby, A. J., Ligon, K. L., Santagata, S., et al. (2012). Classifying human brain tumors by lipid imaging with mass spectrometry. *Cancer Research*, *72*(3), 645–654.

Eberlin, L. S., Norton, I., Orringer, D., Dunn, I. F., Liu, X., Ide, J. L., et al. (2013). Ambient mass spectrometry for the intraoperative molecular diagnosis of human brain tumors. *Proceedings of the National Academy of Sciences*, *110*(5), 1611–1616.

Eberlin, L. S., Tibshirani, R. J., Zhang, J., Longacre, T. A., Berry, G. J., Bingham, D. B., et al. (2014). Molecular assessment of surgical-resection margins of gastric cancer by mass-spectrometric imaging. *Proceedings of the National Academy of Sciences*, *111*(7), 2436–2441.

Gerbig, S., Golf, O., Balog, J., Denes, J., Baranyai, Z., Zarand, A., et al. (2012). Analysis of colorectal adenocarcinoma tissue by desorption electrospray ionization mass spectrometric imaging. *Analytical and Bioanalytical Chemistry*, *403*(8), 2315–2325.

Golf, O., Muirhead, L. J., Speller, A., Balog, J., Abbassi-Ghadi, N., Kumar, S., et al. (2015). XMS: Cross-platform normalization method for multimodal mass spectrometric tissue profiling. *Journal of the American Society for Mass Spectrometry*, *26*(1), 44–54.

Golf, O., Strittmatter, N., Karancsi, T., Pringle, S. D., Speller, A. V. M., Mroz, A., et al. (2015). Rapid evaporative ionization mass spectrometry imaging platform for direct mapping from bulk tissue and bacterial growth media. *Analytical Chemistry, 87*(5), 2527–2534.

Gong, X., Zhao, Y., Cai, S., Fu, S., Yang, C., Zhang, S., et al. (2014). Single cell analysis with probe ESI-mass spectrometry: Detection of metabolites at cellular and subcellular levels. *Analytical Chemistry, 86*(8), 3809–3816.

Goodwin, R. J. A. (2012). Sample preparation for mass spectrometry imaging: Small mistakes can lead to big consequences. *Journal of Proteomics, 75*(16), 4893–4911.

Guenther, S., Muirhead, L. J., Speller, A. V., Golf, O., Strittmatter, N., Ramakrishnan, R., et al. (2015). Spatially resolved metabolic phenotyping of breast cancer by desorption electrospray ionization mass spectrometry. *Cancer Research, 75*(9), 1828–1837.

Hiraoka, K., Nishidate, K., Mori, K., Asakawa, D., & Suzuki, S. (2007). Development of probe electrospray using a solid needle. *Rapid Communications in Mass Spectrometry, 21*(18), 3139–3144.

Kertesz, V., & Van Berkel, G. J. (2008). Improved imaging resolution in desorption electrospray ionization mass spectrometry. *Rapid Communications in Mass Spectrometry, 22*(17), 2639–2644.

Kertesz, V., Van Berkel, G. J., Vavrek, M., Koeplinger, K. A., Schneider, B. B., & Covey, T. R. (2008). Comparison of drug distribution images from whole-body thin tissue sections obtained using desorption electrospray ionization tandem mass spectrometry and autoradiography. *Analytical Chemistry, 80*(13), 5168–5177.

Lemaire, R., Wisztorski, M., Desmons, A., Tabet, J. C., Day, R., Salzet, M., et al. (2006). MALDI-MS direct tissue analysis of proteins: Improving signal sensitivity using organic treatments. *Analytical Chemistry, 78*(20), 7145–7153.

Mandal, M. K., Saha, S., Yoshimura, K., Shida, Y., Takeda, S., Nonami, H., et al. (2013). Biomolecular analysis and cancer diagnostics by negative mode probe electrospray ionization. *Analyst, 138*(6), 1682–1688.

Masterson, T. A., Dill, A. L., Eberlin, L. S., Mattarozzi, M., Cheng, L., Beck, S. D. W., et al. (2011). Distinctive glycerophospholipid profiles of human seminoma and adjacent normal tissues by desorption electrospray ionization imaging mass spectrometry. *Journal of the American Society for Mass Spectrometry, 22*(8), 1326–1333.

Nelson, K. A., Daniels, G. J., Fournie, J. W., & Hemmer, M. J. (2013). Optimization of whole-body zebrafish sectioning methods for mass spectrometry imaging. *Journal of Biomolecular Techniques, 24*(3), 119–127.

Ryan, R. W., Wolf, T., Spetzler, R. F., Coons, S. W., Fink, Y., & Preul, M. C. (2010). Application of a flexible CO_2 laser fiber for neurosurgery: Laser-tissue interactions. *Journal of Neurosurgery, 112*(2), 434–443.

Santagata, S., Eberlin, L. S., Norton, I., Calligaris, D., Feldman, D. R., Ide, J. L., et al. (2014). Intraoperative mass spectrometry mapping of an onco-metabolite to guide brain tumor surgery. *Proceedings of the National Academy of Sciences, 111*(30), 11121–11126.

Schäfer, K.-C., Balog, J., Szaniszló, T., Szalay, D., Mezey, G., Dénes, J., et al. (2011). Real time analysis of brain tissue by direct combination of ultrasonic surgical aspiration and sonic spray mass spectrometry. *Analytical Chemistry, 83*(20), 7729–7735.

Schäfer, K.-C., Dénes, J., Albrecht, K., Szaniszló, T., Balog, J., Skoumal, R., et al. (2009). In vivo, in situ tissue analysis using rapid evaporative ionization mass spectrometry. *Angewandte Chemie, International Edition, 48*(44), 8240–8242.

Schäfer, K.-C., Szaniszló, T., Günther, S., Balog, J., Dénes, J., Keserű, M., et al. (2011). In situ, real-time identification of biological tissues by ultraviolet and infrared laser desorption ionization mass spectrometry. *Analytical Chemistry, 83*(5), 1632–1640.

Shariatgorji, M., Strittmatter, N., Nilsson, A., Källback, P., Alvarsson, A., Zhang, X., et al. (2016). Simultaneous imaging of multiple neurotransmitters and neuroactive substances in the brain by desorption electrospray ionization mass spectrometry. *NeuroImage, 136*, 129–138.

Strittmatter, N., Jones, E. A., Veselkov, K. A., Rebec, M., Bundy, J. G., & Takats, Z. (2013). Analysis of intact bacteria using rapid evaporative ionisation mass spectrometry. *Chemical Communications, 49*(55), 6188–6190.

Strittmatter, N., Lovrics, A., Sessler, J., McKenzie, J. S., Bodai, Z., Doria, M. L., et al. (2016). Shotgun lipidomic profiling of the NCI60 cell line panel using rapid evaporative ionization mass spectrometry. *Analytical Chemistry, 88*(15), 7507–7514.

Swales, J. G., Tucker, J. W., Strittmatter, N., Nilsson, A., Cobice, D., Clench, M. R., et al. (2014). Mass spectrometry imaging of cassette-dosed drugs for higher throughput pharmacokinetic and biodistribution analysis. *Analytical Chemistry, 86*(16), 8473–8480.

Takats, Z., Wiseman, J. M., & Cooks, R. G. (2005). Ambient mass spectrometry using desorption electrospray ionization (DESI): Instrumentation, mechanisms and applications in forensics, chemistry, and biology. *Journal of Mass Spectrometry, 40*(10), 1261–1275.

Takats, Z., Wiseman, J. M., Gologan, B., & Cooks, R. G. (2004). Mass spectrometry sampling under ambient conditions with desorption electrospray ionization. *Science, 306*(5695), 471–473.

Takats, Z., Wiseman, J. M., Ifa, D. R., & Cooks, R. G. (2008). Desorption electrospray ionization (DESI) analysis of intact proteins/oligopeptides. *Cold Spring Harbor Protocols. 2008*(4). pdb.prot4992. http://dx.doi.org/10.1101/pdb.prot4992.

Veselkov, K. A., Mirnezami, R., Strittmatter, N., Goldin, R. D., Kinross, J., Speller, A. V. M., et al. (2014). Chemo-informatic strategy for imaging mass spectrometry-based hyperspectral profiling of lipid signatures in colorectal cancer. *Proceedings of the National Academy of Sciences, 111*(3), 1216–1221.

Vickerman, J. C. (2011). Molecular imaging and depth profiling by mass spectrometry-SIMS, MALDI or DESI? *Analyst, 136*(11), 2199–2217.

Vismeh, R., Waldon, D. J., Teffera, Y., & Zhao, Z. (2012). Localization and quantification of drugs in animal tissues by use of desorption electrospray ionization mass spectrometry imaging. *Analytical Chemistry, 84*(12), 5439–5445.

Wang, H., Manicke, N. E., Yang, Q., Zheng, L., Shi, R., Cooks, R. G., et al. (2011). Direct analysis of biological tissue by paper spray mass spectrometry. *Analytical Chemistry, 83*(4), 1197–1201.

Wiseman, J. M., Ifa, D. R., Song, Q., & Cooks, R. G. (2006). Tissue imaging at atmospheric pressure using desorption electrospray ionization (DESI) mass spectrometry. *Angewandte Chemie, International Edition, 45*(43), 7188–7192.

Wiseman, J. M., Ifa, D. R., Venter, A., & Cooks, R. G. (2008). Ambient molecular imaging by desorption electrospray ionization mass spectrometry. *Nature Protocols, 3*(3), 517–524.

Wiseman, J. M., Puolitaival, S. M., Takáts, Z., Cooks, R. G., & Caprioli, R. M. (2005). Mass spectrometric profiling of intact biological tissue by using desorption electrospray ionization. *Angewandte Chemie, International Edition, 44*(43), 7094–7097.

Wu, C., Ifa, D. R., Manicke, N. E., & Cooks, R. G. (2009). Rapid, direct analysis of cholesterol by charge labeling in reactive desorption electrospray ionization. *Analytical Chemistry, 81*(18), 7618–7624.

Yoshimura, K., Chen, L. C., Mandal, M. K., Nakazawa, T., Yu, Z., Uchiyama, T., et al. (2012). Analysis of renal cell carcinoma as a first step for developing mass spectrometry-based diagnostics. *Journal of the American Society for Mass Spectrometry, 23*(10), 1741–1749.

Zavalin, A., Yang, J., Hayden, K., Vestal, M., & Caprioli, R. M. (2015). Tissue protein imaging at 1 μm laser spot diameter for high spatial resolution and high imaging speed using transmission geometry MALDI TOF MS. *Analytical and Bioanalytical Chemistry, 407*(8), 2337–2342.

Zhang, Y., & Chen, H. (2010). Detection of saccharides by reactive desorption electrospray ionization (DESI) using modified phenylboronic acids. *International Journal of Mass Spectrometry, 289*(2–3), 98–107.

CHAPTER TEN

Rapid Mass Spectrometry Imaging to Assess the Biochemical Profile of Pituitary Tissue for Potential Intraoperative Usage

K.T. Huang*, S. Ludy*, D. Calligaris*, I.F. Dunn*, E. Laws*, S. Santagata*,†, N.Y.R. Agar*,†,1

*Brigham and Women's Hospital, Harvard Medical School, Boston, MA, United States
†Dana-Farber Cancer Institute, Harvard Medical School, Boston, MA, United States
1Corresponding author: e-mail address: nathalie_agar@dfci.harvard.edu

Contents

1. Introduction	257
2. Current Imaging and Visualization Techniques	259
3. Mass Spectrometry in Clinical Usage	261
3.1 Atmospheric Pressure Ionization Mass Spectrometry	261
3.2 Matrix-Assisted Laser Desorption/Ionization Mass Spectrometry	271
4. Future Directions	275
References	276

Abstract

Pituitary adenomas are relatively common intracranial neoplasms that are frequently treated with surgical resection. Rapid visualization of pituitary tissue remains a challenge as current techniques either produce little to no information on hormone-secreting function or are too slow to practically aid in intraoperative or even perioperative decision-making. Matrix-assisted laser desorption/ionization mass spectrometry imaging (MALDI MSI) represents a powerful method by which molecular maps of tissue samples can be created, yielding a two-dimensional representation of the expression patterns of small molecules and proteins from biologic samples. In this chapter, we review the use of MALDI MSI, its application to the characterization of the pituitary gland, and its potential applications for guiding the management of pituitary adenomas.

1. INTRODUCTION

Pituitary adenomas are common intracranial neoplasms that can cause significant morbidity through either the hypersecretion of pituitary hormones or the compression of neighboring neural or endocrine tissues. The tumors can cause a wide spectrum of morbidity, including signs and symptoms as disparate as gigantism, acromegaly, Cushing's disease, impaired libido, amenorrhea, headaches, cranial nerve palsies, and bitemporal vision loss (Arafah et al., 2000). Pituitary adenomas are classified as either micro- or macroadenomas based on their size, less than or greater than 1 cm, respectively. These tumors are also classified according to the hormones which they secrete (Kovacs, Horvath, & Vidal, 2001). The prevalence of these tumors has been estimated at approximately 68–94 per 100,000 in cross-sectional registry studies and as high as 16.7% in postmortem and radiographic screening studies (Daly et al., 2006; Ezzat et al., 2004; Fernandez, Karavitaki, & Wass, 2010; Raappana et al., 2010).

The current standard of treatment for many of these lesions is surgical resection, either through a transnasal, transphenoidal route, or in select cases, via a craniotomy (Cappabianca et al., 2002; Komotar et al., 2012; Mortini et al., 2005). Unlike in many other types of tumor surgery, however, intraoperative examination of surgical pathology specimens can often be of limited utility, particularly in helping to assess the extent of tumor resection. Gross pathology examination is limited by a number of factors, including the inability to grossly discriminate pituitary tumors from normal pituitary tissue and generally small size of many symptomatic tumors. Moreover, differentiating pathologic from normal tissue can be challenging by microscopic analysis, the only routinely available intraoperative method for tissue evaluation. This approach is also limited by processing times that only allow pathologists to evaluate a small number of specimens in any given case. Thus, residual tumor can be left after surgery, leading to persistent symptoms, tumor regrowth, and even to possibly a second operation (Pouratian et al., 2007). Moreover, very small pituitary adenomas can still produce symptoms due to hormone hypersecretion. As a result, a subset of pituitary adenomas is particularly hard to identify during surgery, and their successful resection can often be difficult.

A number of novel techniques have recently been developed to aid in the characterization of intraoperative surgical pathologic specimens. The goal of these techniques is the rapid processing of intraoperative surgical pathology specimens to allow topographic mapping of hormone expression within resection samples with the aim of providing surgeons intraoperative feedback.

Later, we review both the current state of imaging and visualization techniques of pituitary adenomas, with special emphasis on the development of mass spectrometry methods for rapid characterization of biomolecular expression in pathologic tissue.

2. CURRENT IMAGING AND VISUALIZATION TECHNIQUES

Currently, the diagnosis of pituitary adenomas is made by considering a combination of factors including patient history, physical examination, tests that measure serum hormone levels, and noninvasive brain imaging such as magnetic resonance imaging. After diagnosis, attempts are made to characterize the hormone-secreting profile of the tumor. Though prolactin-secreting tumors and clinically asymptomatic tumors may reasonably be treated with chemotherapy and/or observation, most other symptomatic tumors benefit from surgical resection.

Currently, options for visualizing pituitary tissue samples are similar to those used in other areas of surgical oncology. The only widely available option for intraoperative feedback regarding the nature of resected tissue is the examination of frozen sections of samples by surgical pathologists. First described over 100 years ago by Louis B. Wilson, frozen section preparation features the immersion of tissue samples in a gel-like medium and then rapidly cooling to subzero temperatures (Wilson, 1905). These samples are subsequently sectioned, stained with standard dyes such as hematoxylin and eosin, and then examined under the light microscope. Frozen section pathology benefits from rapid turn-around time, and as such it has become the standard of care in many types of surgical oncology procedures. In many different tissue types, ranging from ovarian to head and neck malignancies, it has proven accurate and reliable (Ferreiro, Myers, & Bostwick, 1995; Obiakor et al., 1991; Remsen, Lucente, & Biller, 1984; Sparkman, 1962).

On H&E-stained sections, adenomas are discriminated from normal pituitary based on several factors. While normal pituitary cells are clustered in discrete nests surrounded by reticulin, this architecture is effaced with marked distortion or abrogation of the nesting pattern on H&E sections. In addition, these normal nests of pituitary cells comprise a broad range of cell types which stain differently using H&E, leading to a heterogeneous appearance of the cells within these nests. The staining of adenoma cells is more uniform, leading to a monomorphic appearance. These features are not well captured in frozen sections. Freezing tissue often introduces artifacts that can distort tissue architecture, alter nuclear morphology and change staining characteristics (Burger, 1985). Moreover, discriminating adenomas from normal pituitary can be challenging even under the more optimal conditions proved by the evaluation of permanent sections that have undergone formalin fixation and paraffin embedding (Powell, 2005). Due to their friability and the relative ease with which adenomatous cells detach from a tumor, many have described the usefulness of touch, imprint, and smear preparations in the pathologic analysis pituitary tumors in particular (Burger, 1985; Powell, 2005). These techniques do not preserve underlying spatial architecture of cells but can provide important information in certain circumstances. Of utmost importance, no established technique allows for intraoperative characterization of the functional secretory status of pituitary tissue.

Permanent sections of tissue embedded in paraffin blocks, provide superior tissue preservation compared to frozen section processing, and can significantly aid in the diagnosis of pituitary lesions. In such sections, pituitary adenomas can be distinguished based on the larger size of secretory acini in adenomatous tissue, the increase in nuclear to cytoplasmic ratios, and the increased homogeneity of cell type staining in pituitary adenomas (Burger, 1985; Scheithauer et al., 1986). Standard staining techniques, however, offer little insight into the underlying function of the adenoma. Indeed, though pituitary tumors were originally classified based on their staining affinity as either acidophilic, basophilic, or chromophobic, that classification has since fallen out of favor as it is limited in providing information about the biologic behavior and pathophysiology of the underlying tumor (Kovacs, Horvath, & Ezrin, 1977; Scheithauer et al., 1986). Instead, currently adenomas are often classified as either typical, atypical (MIB-1 index $\geq 3\%$ and extensive p53 immunoreactivity), or carcinomatous and they are also classified by hormone-secreting function (somatotroph, lactotroph, thyrotroph, corticotroph, gonadotroph, null cell, or plurihormonal) (Al-Shraim & Asa, 2004). Even this classification scheme has been critiqued by some, however, as the atypical designation does not correlate strongly with clinical aggressiveness (Di Ieva et al., 2014; Saeger et al., 2016a). Moreover, current classification guidelines do not account for the differential expression of metalloproteinases or molecular profiles in tumors seen in familial syndromes (e.g., multiple endocrine neoplasia or McCune-Albright syndrome), which some have argued are also important in predicting clinical course (Saeger et al., 2016b). It is clear, though, that as diagnostic schemes have evolved, immunohistochemistry has become fundamental to the modern classification of pituitary adenomas.

Originally developed and described by Albert Coons in 1941, immunohistochemistry allows for the selective labeling of individual polypeptides in tissue by direct antibody labeling (Coons, Creech, & Jones, 1941; Hsu, Raine, & Fanger, 1981; Taylor, 1978). This method has grown to become the primary method by which the hormonal function of pituitary tissue is characterized, and can even characterize the expression of hormones may not be reflected in a patient's serum hormone levels (Black et al., 1987; Randall et al., 1985). Though accurate and useful for understanding a given piece of pituitary tissue, immunohistochemistry often takes several days to perform, and as such does not contribute to decision-making at the time of surgery.

3. MASS SPECTROMETRY IN CLINICAL USAGE

Though techniques such as magnetic resonance imaging, immunohistochemistry, and permanent tissue section inspection are helpful in identifying and classifying pituitary tumors, there is a need for faster, more specific, intraoperative techniques to improve tumor resection. Such needs have led researchers to explore the use of mass spectrometry in vivo. Mass spectrometry is a widely employed analytical chemistry technique used to detect, identify, and quantitate molecules such as proteins, peptides, lipids, metabolites, and drugs based on their mass-to-charge ratio (m/z) in a given sample. There are a variety of mass spectrometry related methods that have been applied to clinical research more recently such as DESI (Eberlin et al., 2010a), REIMS (Balog et al., 2013), LDI-MS (Sachfer et al., 2011), CUSA-MS (Schafer et al., 2011), PESI-MS (Mandal et al., 2012a, 2013), and SPA-ESI (Mandal et al., 2012b) methods. Some of these methods will briefly be explained and their clinical relevance discussed.

3.1 Atmospheric Pressure Ionization Mass Spectrometry

Electrospray (ESI) techniques were developed, and over the past few decades, clinical usage of mass spectrometry has expanded considerably to cover a broad spectrum of biomolecular analyses (Fenn et al., 1989). Mass spectrometry has been widely used in newborn screenings, forensic analysis, as well as drug detection (Takats et al., 2004). Newborn screenings in particular offer a key example of the expanding role of mass spectrometry in clinical usage. Though neonatal screenings began in the 1960s with testing for phenylalanine deficiencies, as more metabolic disorders were added to the typical PKU screening, clinicians needed a faster and more reliable way to analyze samples (Garg & Dasouki, 2006). Therefore, in the 1990s, tandem mass spectrometry was introduced as a diagnostic tool for newborn screenings to evaluate in a single assay multiple analytes such as amino acid, organic acid, and fatty acid involved in disorders (Table 1) (Garg & Dasouki, 2006). As such, it has become a faster and more cost efficient technique for screenings in newborns compared to older methods and has enjoyed extreme popularity in modern clinical laboratories.

Ambient ionization mass spectrometry techniques are increasingly important methods in tissue analysis due to the minimal need for sample preparation (Takats et al., 2004). Desorption electrospray ionization (DESI) and rapid evaporative ionization mass spectrometry (REIMS) are two specific variants of ambient ionization mass spectrometry that are employed (Ifa et al., 2010). DESI is a combination of electrospray and desorption, where the electrospray droplets are directed to the sample surface, the sample is desorbed and ionized, and those ions travel through air into the atmospheric pressure interface connected to the mass spectrometer (Ifa et al., 2010). Therefore, DESI-MS can ionize gases, liquids, and solids in open air under ambient conditions, allowing for a form of "soft" ionization that enables the study of small biomolecules that would fracture or breakdown under harsher ionization techniques (Takats et al., 2004). The lack of need for tissue sample preparation has prompted interest in its use as a rapid turn-around analysis tool for clinical pathology specimens (Calligaris et al., 2013). For instance, DESI-MS has been used to define tumor margins in breast cancer surgery, discriminating invasive ductal carcinomas from normal breast tissue based on metabolite profiles (Calligaris et al., 2014) (Fig. 1). Oleic acid, a lipid involved in a range of processes that are central to invasive breast cancer cells, as well as another biomarker, were discovered to be highly expressed in tumor tissue supporting the use of DESI-MS for the intraoperative identification of breast cancer cells. Other studies using DESI-MSI have shown high sensitivity and specificity for the rapid intraoperative detection and diagnosis of certain glioma tumors. Santagata et al. used the specific oncometabolite 2-HG (2-hydroxyglutarate) found in gliomas with IDH1 and IDH2 mutations to identify and differentiate pathologic tissue from normal brain tissue (Fig. 2) (Santagata et al., 2014). The researchers identified the specific metabolite with its characteristic peak, found the limit of detection in frozen samples, and defined the tumor margins. After these experiments proved successful, they performed DESI-MS intraoperatively to assist in the surgical resection of the tumors.

The clinical use of DESI-MS continues to expand. Many oncological studies in particular, spanning a wide variety of tumor types, use DESI techniques. In just the past decade, DESI has been used to study other brain

Table 1 MS/MS Detectable Disorders

Targeted Condition	Metabolic Defect(s)	Primary Marker Metabolite on MS/MS	Confirmatory/Follow-Up Studies	Finding(s) on Confirmatory Follow-Up Studies	Comments
Disorders of amino acids/urea cycle disorders					
Argininemia	Arginase 1 (liver)	↑ Arginine	Plasma NH3, PAA, enzyme assay	↑ NH3, ↑ arginine on PAA, ↓ hepatic arginase activity	Low arginine occurs in CPS, NAGS and OTC deficiencies. Low levels are difficult to detect by MS/MS
Argininosuccinic aciduria (ASA) or ASA lyase deficiency	Argininosuccinate lyase (ASL)	↑ Citrulline	Plasma NH3, UAA, PAA, enzyme assay	↑ NH3, ↑ argininosuccinic acid on UAA and PAA, ↓ fibroblast/liver ASL activity	As citrulline is elevated in both ASA and classical citrullinemia, MS/MS cannot distinguish these two disorders. ASL is expressed in the liver, kidney, fibroblasts, and intestines
Citrullinemia Type 1 "Neonatal" citrullinemia	Argininosuccinate synthase (ASS)	↑ Citrulline	Plasma NH3, PAA	↑ NH3, ↑ citrulline on PAA, ↓ fibroblast/liver ASS activity	Low citrulline (difficult to detect by MS/MS) occurs in CPS, NAGS, and OTC deficiencies. ASS expression is similar to ASL
Homocystinuria	Cystathionine β-synthase (CBS)	↑ Methionine	PAA, UAA, UOA	↑ Blood and urine homocyst(e)ine on PAA and UAA; ↑ urine methylmalonic acid on UOA in cobalamin C, D, F synthesis defects	Currently, homocysteine is not measured by MS/MS. Homocystinuria due to MTHFR deficiency and cobalamin (A–G) synthesis defects are not detected by MS/MS as plasma methionine levels are not increased

Continued

Table 1 MS/MS Detectable Disorders—cont'd

Targeted Condition	Metabolic Defect(s)	Primary Marker Metabolite on MS/MS	Confirmatory/Follow-Up Studies	Finding(s) on Confirmatory Follow-Up Studies	Comments
Maple syrup urine disease (MSUD)	Branched chain α-ketoacid dehydrogenase	↑ Total "leucine, isoleucine, alloisoleucine" and ↑ valine	PAA, urine DNPH, UOA	↑ Leucine, isoleucine, alloisoleucine and valine on PAA; positive DNPH; ↑ branched chain α-keto and hydroxyl acids on UOA	MS/MS does not distinguish between leucine, isoleucine, and alloisoleucine. Also, on MS/MS hydroxyproline appears in the same peak as "leucines"
Phenylketonuria	Phenylalanine hydroxylase (>98%), BH4 synthesis/regeneration defects (<2%)	↑ Phenylalanine, ↑ phenylalanine/tyrosine ratio	PAA, urine and/or blood or CSF neopterin and biopterin studies	↑ Phenylalanine on PAA, ↑ phenylalanine/tyrosine ratio; abnormal urinary and/or blood or CSF pterins in BH4 synthesis defects	↑ Phenylalanine/tyrosine ratio has better predictive value than phenylalanine level alone (Kovacs et al., 1977)
Tyrosinemia type 1	Fumarylacetoacetate hydrolase (FAH)	↑ Tyrosine	PAA, UOA	↑ Tyrosine and methionine on PAA; ↑ succinylacetone and tyrosine metabolites on UOA	As succinylacetone is generally not detected by MS/MS, therefore different types of tyrosinemia are indistinguishable
Tyrosinemia type 2/oculocutaneous tyrosinemia	Tyrosine aminotransferase	↑ Tyrosine	PAA, UOA	↑ Tyrosine on PAA; ↑ tyrosine metabolites without increased succinylacetone on UOA	Plasma tyrosine level is generally higher in type 2 than in type 1
Organic acidemias/acidurias					
Glutaric aciduria 1 (GA I)	Glutaryl-CoA dehydrogenase	↑ Glutarylcarnitine (C5-dicarboxylic)	UOA, PACP	↑ Glutaric acid, 3-hydroxyglutaric acid, glutaconic acid on UOA; ↑ glutarylcarnitine (C5-dicarboxylic) on PACP	3-Hydroxyglutaric acid may be difficult to detect on UOA

HMG-CoA lyase deficiency	HMG-CoA lyase	↑ 3-Hydroxyisovalerylcarnitine (C5-OH)	UOA, PACP	↑ 3-Hydroxyisovaleric, 3-methylglutaconic, 3-methylglutaric, 3-hydroxy-3-methylglutaric acids on UOA; ↑ C5-hydroxyisovalerylcarnitine (C5-OH), 3 methylglutarylcarnitine (C6DC) on PACP	3-Hydroxyisovalerylcarnitine is also high in 3-MCC, 3-methylglutaconyl-CoA hydratase, holocarboxylase and biotinidase deficiencies. 3-Hydroxyisovalerylcarnitine is indistinguishable from 2-methyl-3-hydroxybutyrylcarnitine
Isovaleric acidemia	Isovaleryl-CoA dehydrogenase	↑ Isovalerylcarnitine (C5)	UOA, PACP	↑ Isovalerylglycine, 3-hydroxyisovaleric acid on UOA; ↑ isovalerylcarnitine (C5) on PACP	On MS/MS isovalerylcarnitine is indistinguishable from 2-methylbutyrylcarnitine and pivaloylcarnitine derived from antibiotics containing pivoxilsulbactam
3-Keto(oxo) thiolase deficiency	3-Keto(oxo)thiolase	↑ Tiglylcarnitine (C5:1), ↑ 3-hydroxy-2-methylbutyryl carnitine (C5-OH)	UOA, PACP	↑ 2-Methyl-3-hydroxybutyrate, 2-methylacetoacetic, tiglylglycine on UOA; ↑ tiglylcarnitine (C5:1). ↑ 3-hydroxy-2-methyl butyrylcarnitine (C5-OH) on PACP	On MS/MS 3-hydroxy-2-methylbutyrylcarnitine is indistinguishable from 3-hydroxyisovalerylcarnitine
3-MCC deficiency	3-MCC	↑ 3-Hydroxyisovalerylcarnitine (C5-OH)	UOA, PACP	↑ 3-Hydroxyisovaleric, 3-methylcrotonylglycine on UOA; ↑ 3-hydroxyisovaleryl carnitine (C5-OH) on PACP	See comment in HMG-CoA lyase deficiency. 3-Methylcrotonylcarnitine may also be high

Continued

Table 1 MS/MS Detectable Disorders—cont'd

Targeted Condition	Metabolic Defect(s)	Primary Marker Metabolite on MS/MS	Confirmatory/Follow-Up Studies	Finding(s) on Confirmatory Follow-Up Studies	Comments
2-Methylbutyryl CoA dehydrogenase deficiency	2-Methylbutyryl CoA dehydrogenase	↑ 2-Methylbutyrylcarnitine (C5)	UOA	↑ 2-Methylbutyrylglycine on PACP	On MS/MS 2-methylbutyrylcarnitine is indistinguishable from isovalerylcarnitine
3-Methylglutanoyl CoA hydratase deficiency	3-Methylglutanoyl CoA hydratase	↑ 3-Hydroxyisovalerylcarnitine (C5-OH)	UOA, PACP	↑ 3-Hydroxyisovaleric, 3-methylglutaconic, 3-methylglutaric on UOA; ↑ 3-hydroxyisovaleryl carnitine (C5-OH) on PACP	See comment in HMG-CoA lyase deficiency
Methylmalonic acidemia	Methylmalonyl CoA mutase, Cobalamin (C, D, F) deficiency	↑ Propionylcarnitine (C3)	UOA, PACP	↑ Methylmalonic, 3-hydroxypropionate, methylcitrate, propionylglycine on UOA; ↑ propionylcarnitine (C3) on PACP	Elevated C3 acylcarnitine also occurs in propionic academia, holocarboxylase, biotinidase, and vitamin B12 deficiencies
Multiple CoA carboxylase deficiency	Holocarboxylase synthetase	↑ Propionylcarnitine (C3), ↑ 3-hydroxyisovalerylcarnitine (C5-OH)	UOA, PACP	↑ 3-OH-isovaleric, 3-methylcrotonylglycine, methylcitrate, 3-OH-propionic, lactate, pyruvate, acetoacetate, 3-OH-butyrate on UOA; ↑ propionylcarnitine (C3), ↑ 3-hydroxyisovaleryl carnitine (C5-OH) on PACP	Deficiency of holocarboxylase synthetase results in functional deficiency of all carboxylases: "pyruvate carboxylase, propionyl CoA carboxylase, MCC, acetyl CoA carboxylase"

Propionic acidemia	Propionyl-CoA carboxylase	↑ Propionylcarnitine (C3)	UOA, PACP	↑ 3-Hydroxypropionate, methylcitrate, propionylglycine; ↑ propionylcarnitine (C3) on PACP	C3 is also high in methylmalonic acidemia, holocarboxylase and biotinidase and vitamin B12 deficiencies

Mitochondrial fatty acid oxidation disorders

Carnitine acylcarnitine translocase (CACT) deficiency	Carnitine acylcarnitine translocase	↑ C16–C18 acylcarnitines, ↓ free carnitine	PACP, CK, glucose, NH3	↑ C16–C18 acylcarnitines on PACP; ↓ free carnitine, ↑ CK, ↓ glucose, ↑ NH3 on plasma	
Carnitine palmitoyl transferase type 1 (CPT-1) deficiency	CPT-1	↓ C16–C18 acylcarnitines, ↑–↑ free carnitine	PACP, CK, glucose, NH3	↓ C16–C18 acylcarnitines on PACP; ↑–↑ free carnitine, ↑ CK, ↓ glucose, ↑ NH3 on plasma	Not reliably detected by MS/MS as instruments may not be optimized to measure low acylcarnitine levels
Carnitine palmitoyl transferase type 2 (CPT-2) deficiency	CPT-2	↑ C16–C18 acylcarnitines, ↓ free carnitine	PACP, CK, glucose, NH3	↑ C16–C18 acylcarnitines on PACP; ↓ free carnitine, ↑ CK, ↓ glucose, ↑ NH3 on plasma	
Carnitine uptake/transporter defect	SLC22A5 (encodes the sodium ion-dependent carnitine transporter OCTN2)	↓ C16–C18 acylcarnitines, ↓ free carnitine	PACP, urine carnitine, CK, glucose, NH3	↓ C16–C18 acylcarnitines on PACP; ↑ urine carnitine; ↓ free carnitine, ↑ CK, ↓ glucose, ↑ NH3 on plasma	Not reliably detected by MS/MS as the instruments are not optimized to measure low concentrations
3-Hydroxy long chain acyl-CoA dehydrogenase deficiency (LCHAD/MTP)	LCHAD/MTP	↑ Long chain 3-hydroxy acylcarnitines	PACP, UOA, CK, glucose, NH3	↑ Long chain 3-hydroxy acylcarnitines on PACP; ↑ 3-OH dicarboxylic acids on UOA; ↑ CK, ↓ glucose, ↑ NH3 on plasma	Mitochondrial trifunctional protein (MTP) is a heterooctamer of four α subunits (dehydrogenase and hydratase activities) and four β subunits (thiolase activity). The α-subunit "G1528C" mutation is the most common

Continued

Table 1 MS/MS Detectable Disorders—cont'd

Targeted Condition	Metabolic Defect(s)	Primary Marker Metabolite on MS/MS	Confirmatory/ Follow-Up Studies	Finding(s) on Confirmatory Follow-Up Studies	Comments
Medium chain acyl-CoA dehydrogenase (MCAD) deficiency	Medium chain acyl-CoA dehydrogenase	↑ C8–C10 acylcarnitines	PACP, UOA, CK, glucose, NH3	↑ C8–C10 acylcarnitines on PACP; ↑ dicarboxylic acids, hexanoylglycine, phenylpropionylglycine and suberylglycine on UOA; ↑ CK, ↓ glucose, ↑ NH3 on plasma	MCAD deficiency is the most common disorder of fatty acid oxidation. 50% of affected individuals are homozygous while 40% are compound heterozygous for the common A985G (K304E) mutation
Multiple acyl-CoA dehydrogenase deficiency (MADD) or glutaric acidemia-type 2	Electron transfer flavoprotein (ETF) α/β subunits, ETF dehydrogenase, Riboflavin responsive MADD	↑ Multiple acylcarnitines	PACP, UOA, CK, glucose, NH3	↑ Multiple acylcarnitines on PACP; ↑ glutaric, ethylmalonic, dicarboxylic acids, hexanoylglycine, phenylpropionylglycine and suberylglycine on UOA; ↑ CK, ↓ glucose, ↑ NH3 on plasma	
Short chain acyl-CoA dehydrogenase deficiency (SCAD)	Short chain acyl-CoA dehydrogenase	↑ Butyrylcarnitine (C4)	PACP, UOA	↑ Butyrylcarnitine on PACP; ↑ ethylmalonic, methylsuccinic, butyrylglycine on UOA	Elevated C4 acylcarnitine is also seen in isobutyryl CoA dehydrogenase, MAD deficiencies and ethylmalonic encephalopathy (caused by ETH1 mutations)
Very long chain acyl-CoA dehydrogenase deficiency (VLCAD)	Very long chain acyl-CoA dehydrogenase	↑ C14, C14:1, C14:2 acylcarnitines, ↓ Free carnitine	PACP, CK, glucose, NH3	↑ Long chain acylcarnitines on PACP; ↑ CK, ↓ glucose, ↑ NH3 on plasma	Liver enzymes may also be high

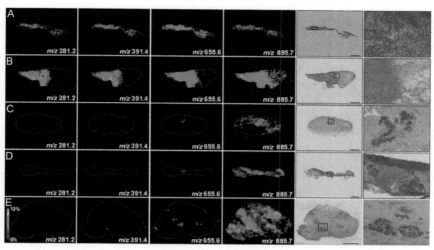

Fig. 1 DESI-MS images from the tumor center (A), the tumor edge (B), 2 cm away from the tumor (C), 5 cm away from the tumor (D), and contralateral side (E) tissue sections from research subject 9 showing the distributions of ions at m/z 281.2, m/z 391.4, m/z 655.6, and m/z 885.7. (*Right*) Light microscopy images of the H&E-stained sections are shown. Scale bars = 2 mm.

tumors including astrocytomas, oligodendrogliomas, glioblastomas, and meningiomas (Calligaris et al., 2015a; Eberlin et al., 2010b, 2012, 2013) as well as adenocarcinomas of the human liver, breast, colorectal, and prostate (Dill et al., 2009; Eberlin et al., 2010a; Gerbig et al., 2012; Kerian, Jarmusch, & Cooks, 2014; Kerian et al., 2015; Wiseman et al., 2005), breast cancer (Guenther et al., 2015; Tata et al., 2015), seminomas (Masterson et al., 2011), gastrointestinal cancers (Eberlin et al., 2014b), papillary, renal cell, and transitional cell carcinomas of the bladder (Dill et al., 2010, 2011; Mandal et al., 2012b), and lymphomas and lymph node metastases (Abbassi-Ghadi et al., 2014; Eberlin et al., 2014a). These techniques allowed researchers to discriminate various cancers from normal tissues, to detect numerous compounds with altered expression during tumorigenesis, and to use statistical data evaluation and machine learning approaches to enhance diagnostic accuracy (Ifa & Eberlin, 2016).

REIMS is a newer technique that utilizes aerosol vapor released during electrosurgical dissection to analyze tissue samples in vivo. Balog et al. compared REIMS methods to other oncosurgical and histological techniques to determine its efficacy and success as a suitable intraoperative replacement (Balog et al., 2013). With tissue samples from brain, liver, lung, breast,

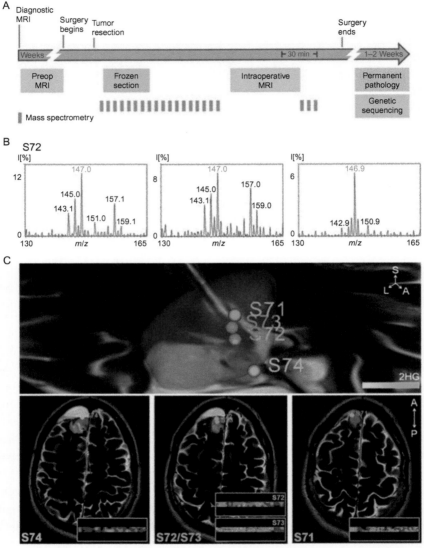

Fig. 2 (A) Time course and work flow of patient care associated with a typical 5-h neurosurgery in the AMIGO, MRI-equipped, operative suite at Brigham, and Women's Hospital. See Supporting Information of the cited article for additional description. (B) Negative ion mode DESI mass spectra obtained using an amaZon Speed ion trap from m/z 130–165 (Bruker Daltonics) from a swab (*left*), a smear (*center*), and a section (*right*) for sample S72. (C) Normalized 2-HG signal is represented with a warm color scale as indicated by the scale bar, set from the lowest (*yellow*) to highest (*orange*) levels detected from this individual case. Stereotactic positions were digitally registered to the preoperative MRI using neuronavigation (BrainLab system) in a standard operating room. The 3D tumor volume is shown (*upper*). Classification results of samples S74, S72, S73, and S71 are further visualized on axial sections (*lower*).

and colorectal tumors, they created a lipid profile and used it to differentiate between healthy tissue and primary or metastatic tumor tissue. Their results suggested that REIMS is able to characterize distinct tissue types that matched the histopathological tissue analyses exactly. Most notably, the speed of analysis of REIMS is nearly unmatched, with a reported feedback time of 0.7–2.5 s (Eberlin et al., 2010a). In another study by Golf et al., REIMS technique was also used to detect and distinguish various bacterial strains from colorectal adenocarcinoma metastases in human liver tissue (Golf et al., 2015). As the above examples demonstrate, the lipid profiles used in both DESI and REIMS techniques are particularly significant in identifying negative margins in tumor resection. The ability to resect clean margins without compromising healthy tissue has numerous clinical advantages, including decreasing recurrence and increasing survival rates (Balog et al., 2013; Calligaris et al., 2014; Santagata et al., 2014; Thomas et al., 2013).

Performing mass spectrometry in ambient conditions without the need for sample preparation is a key advantage when used in in vivo tissue sampling. These techniques have been shown to be just as efficient as previous oncosurgical techniques in analyzing and distinguishing various lipid profiles with the added benefit of providing that analysis in a shorter time frame (Balog et al., 2013). However, there are some limitations with these techniques when considering usage in pituitary tumor resection in particular. The primary molecules of analysis in pituitary tumors are peptide hormones, which are differentially over- or underexpressed in pituitary adenomas depending on their functional status, and which have substantially higher molecular weights than the lipids, fatty acids, and small metabolites previously mentioned. As such, DESI and REIMS, though showing significant interest in the application of tumor surgery elsewhere in the body, are poorly suited to imaging of the pituitary gland and of pituitary tumors. In their stead, researchers with a pituitary focus have turned toward alternative techniques, such as matrix-assisted laser ionization, that can be geared specifically toward protein and peptide characterization (Trauger, Webb, & Siuzdak, 2002).

3.2 Matrix-Assisted Laser Desorption/Ionization Mass Spectrometry

Matrix-assisted laser desorption/ionization mass spectrometry (MALDI MS) has been used successfully in a large variety of clinical and investigational endeavors. One area of extensive application of MALDI MS has been in clinical microbiology laboratories. As traditional bacterial strain identification is time-consuming and often requires a combination of Gram staining, culture results, and biochemical characterization, MALDI MS has been proposed as a quicker, and potentially less-expensive method of bacterial species identification (Anhalt & Fenselau, 1975; Seng et al., 2009). Recent investigations have demonstrated the ability of mass spectrometry to accurately identify as high as 93% of bacterial and yeast isolates up to the species level, and up to an additional 5% at the genus level (Bizzini et al., 2010; Dhiman et al., 2011; La Scola & Raoult, 2009; van Veen, Claas, & Kuijper, 2010). Though a wide range of clinical species were able to be identified, lower accuracy rates have been noted with identification of *Streptococcus*, *Shigella*, HACEK organisms (Blondiaux, Gaillot, & Courcol, 2010; Carbonnelle et al., 2011). Not only has MALDI MS been proven useful in the speciation of samples, but it has also shown utility in phenotypic characterization of antimicrobial resistance patterns (Fournier et al., 2013). The technology has been demonstrated to be able to identify beta-lactamase activity and carbapenemase activity in bacterial isolates with high sensitivity and specificity (Hooff et al., 2012; Hrabak et al., 2011). Thus, though still not common place in clinical microbiologic laboratories, MALDI MS may also soon yield significant improvements in the speed and cost of microbiologic analysis.

Another area of significant research has been the application of MALDI MS to better characterize human neoplasms. It has been used to subclassify human gliomas based on their proteomic expression pattern, potentially allowing for improved classification schemes of human gliomas (Schwartz et al., 2005). In addition, it has been used to classify nonsmall cell lung cancers according to the histologic type and to identify nodal involvement in nonsmall cell lung cancer (Lee et al., 2012; Yanagisawa et al., 2003). Similar studies have validated the use of MALDI MS to effectively differentiate cancerous for normal tissue in breast, prostate, thyroid, and colon samples (Cazares et al., 2009; Chaurand et al., 2001; Cornett et al., 2006; Ryu et al., 2013).

MALDI mass spectrometry imaging (MALDI MSI) has also proven to be extremely powerful to study the distribution of molecules within tissue sections (Cornett et al., 2007). The pituitary gland has been a key focus of this field from its very first description by Caprioli, Farmer, and Gile (1997). Due to the known anatomic compartmentalization between anterior and posterior pituitary glands of small hormone peptides amenable to MALDI MS analysis, the pituitary gland was an ideal candidate to test the ability of the technology to produce potential imaging data. Numerous human cancers, such as COX7A2, TAGLN2, and S100 expression in Barrett's adenocarcinoma (Elsner et al., 2012), CRIP1, HNP-1, and S100 presence in gastric cancer (Balluff et al., 2011), and HER2 receptor status in breast cancer tissue (Fig. 3) (Rauser et al., 2010), were also analyzed by MALDI MSI. It has also been used to study the angiotensin metabolism in samples of mouse kidneys (Grobe et al., 2012). Neural tissue in particular has been an area of significant study as differing functional regions often have similar microanatomic structure but potential different proteomic expression. MALDI MSI has successfully been used to image the molecular expression patterns in glioblastoma (Stoeckli et al., 2001), lipid changes in stroke

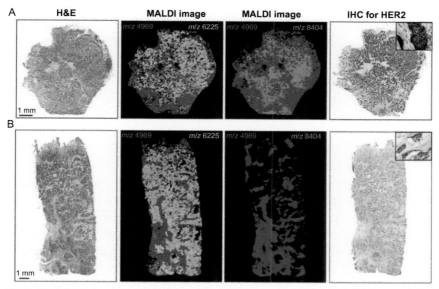

Fig. 3 Visualization of *m/z* 8404 (shown in *red*) as an example of a highly upregulated spectral feature in HER2-positive breast cancer tissues: optical microscopic images of the MALDI-IMS measured and subsequently H&E-stained original tissue sections of a HER2-positive (A) and a HER2-negative (B) case. Immunohistochemistry for HER2 of the respective case on a consecutive serial section is shown on the very *right side*. Note the strong immunoreactivity (score 3+) of the positive case (A), while the reaction in the negative case indicates a score 0 (B). The visualization of the *m/z* 8404 (shown in *red*) found by MALDI-IMS as highly upregulated in HER2-positive tissues (A) clearly shows that this spectral feature is specific for cancer cells in this HER2-positive case but absent in cancer cells of the HER2-negative case (B). The spectral feature of *m/z* 6225 (shown in *yellow*) is an example specific for cancer cells but not distinguishing between HER2-positive and HER2-negative tissues and is therefore present in both cases (A and B); *m/z* 4969 (shown in *blue*) is a *m/z* species specific for tumor stroma and thus also present in both cases (A and B). Scale bars = 1 mm.

(Wang et al., 2012), and differential neuropeptide signaling in supraoptic nucleus and caudate-putamen-hypothalamus circuits (Fournier, Day, & Salzet, 2003). As these and many more examples demonstrate, the ability of this technology to produce visual representation of the molecular expression in tissue samples represents a powerful tool in studying a wide range of pathologies in a wide variety of different tissues.

MALDI MS technology is particularly useful in the imaging of pituitary tissue as the molecules of interest are often hormone peptides, which have significantly higher molecular weights than the lipids and thus are poor targets for ambient ionization techniques. These have paved the way for use of the technology in actual clinical use, and key studies involving MALDI MS in pituitary imaging are reviewed later.

In 1997, Caprioli et al. described the use of MALDI MS to isolate polypeptides in samples of rat pituitary (Caprioli et al., 1997). This report was notable for using the technology to determine not just the presence of individual substances within a tissue, but specifically determining the spatial distribution of those substances within a given piece of mammalian tissue. Altelaar et al. improved upon the originally described technique through use a stigmatic mass microscope (Altelaar et al., 2006, 2007; Luxembourg et al., 2004). Through use of this instrument, which preserved the spatial configuration of molecules as they were liberalized from the original tissue, the authors were able to improve their resolution down to $\sim$4 μm with a higher speed of analysis. As a result, rather than the linear scans described by Caprioli et al., the authors were able to produce full two-dimensional representations of rat, mouse, and human pituitary gland with spatial representations of vasopressin, oxytocin, and alpha-melanocyte-stimulating hormone.

The throughput and time-to-image techniques have improved further in recent years. Recently, members of our group have described the further refining of the above techniques to allow for possibility of intraoperative feedback of pituitary tissue resection (Calligaris et al., 2015b). Through improvements in automation of both data acquisition and statistical processing, the time from tissue acquisition to completed analysis has been reduced to less than 30 min, well within the time of a typical pituitary surgery (Calligaris et al., 2015b; Kertesz et al., 2015). Moreover, the ability to confirm the identity of the hormone using in-source decay directly from the tissue section surface provides unprecedented specificity in the MALDI MSI characterization of the tissue. Though MALDI MS has been used to study proteomic difference in normal vs adenomatous pituitary tissue previously

(Desiderio & Zhan, 2003), this represents the first time this technology has advanced to the point where it useful in a clinical scenario.

By giving intraoperative feedback on the functional qualities of resected pituitary tissue, a number of important possibilities are opened up. The preservation of normal pituitary gland can be enhanced as surgeons can have real-time feedback on it, and how much, normal pituitary tissue has been resected. Correspondingly, degree of adenoma resection can also be enhanced in scenarios where surgeons learn that no normal pituitary has yet been resected, implying that residual tumor may yet still remain. Moreover, this technology represents particular promise in the resection of small corticotropic adenomas. Though still large enough to be symptomatic, some of these tumors are small enough to where they are undetectable even on high-resolution magnetic resolution imaging (Salenave et al., 2004; Semple et al., 2000; Tyrrell et al., 1978). Significant lengths are often undertaken to correctly identify and to help localize such small corticotropic adenomas (Oldfield et al., 1985, 1991). As such, the ability to determine if an intraoperative sample contains the suspected adenoma would prove very useful, and aid in the maximal preservation of normal pituitary function.

MALDI MSI represents a powerful tool that allows near real-time spatial resolution of the hormone-secreting profile of pituitary tissue. Future applications should allow for effective intraoperative differentiation between pathologic and normal pituitary tissue during pituitary tumor surgery, thereby reducing complication rates and the needs for reoperation.

4. FUTURE DIRECTIONS

Though this technology has many possible applications to current clinical practice, most of those possibilities have yet to be realized. To date, no clinical studies have been conducted on the use of intraoperative MALDI MS. Thus, it has not yet been established whether it is feasible to incorporate MALDI MS analysis within the surgical workflow on a large volume of patients, let alone if it is comparable or superior to currently available surgical pathology techniques. These represent substantial questions that will need to be addressed before MALDI MS can be considered for more widespread use.

From a technical standpoint, the next 5 years also presents the challenge of improving the speed, resolution, and cost of MALDI MS technology. Though 30 min turn-around times are in principle feasible for providing intraoperative feedback, any delay during surgery represents a suboptimal situation with potentially increased morbidity and wasted operating room resources. Moreover, if there does prove to be a use for real-time mass spectrometry in intraoperative guidance, the cost of adopting the technology, both in terms of acquiring new equipment and training/retraining personnel may prove daunting or even prohibitive for implementation.

Current efforts to further improve the speed and turn-around times have focused on liquid microjunction surface sampling, which combined with liquid chromatography and an ambient ionization mass spectrometry have led to promising improvements in processing speed (Kertesz et al., 2015). This approach relies on an automated, continuous-flow system of liquid extraction that may have superior throughput in processing samples as compared to laser-assisted ionization (Van Berkel & Kertesz, 2013). This liquid microjunction sampling has the advantage of in situ ambient condition ionization of tissue, obviating the need for sample preparation in a fixed matrix, and even allowing for analysis of samples directly taken from living tissue (Hsu et al., 2013). Thus, while MALDI MSI has shown great promise for the molecular imaging of pituitary tissues, in the next 5 years other types of imaging mass spectrometry may also provide significant advantages.

In addition to potential advances in imaging techniques, there are still many other unanswered questions left regarding how best to utilize imaging mass spectrometry from a surgical perspective. How much tissue is required and how best to prepare tissue for analysis has yet to be determined. Moreover, how surgery should be modified to incorporate the information yielded by new imaging technologies is still completely unexplored—how many samples to send, and what, if any, attempts should be made to preserve the integrity of pathology specimens will only be determined once there is more experience with intraoperative use of mass spectrometric imaging.

Thus, while important steps have been made in illuminating the possibility for MALDI MSI for use in analyzing intraoperative pituitary tissue samples, significant challenges remain before its use becomes commonplace in routine clinical practice.

REFERENCES

Abbassi-Ghadi, N., et al. (2014). Discrimination of lymph node metastases using desorption electrospray ionisation-mass spectrometry imaging. *Chemical Communications (Cambridge, England), 50*(28), 3661–3664.

Al-Shraim, M., & Asa, S. L. (2004). The 2004 World Health Organization classification of pituitary tumors: What is new? *Acta Neuropathologica, 111*(1), 1–7.

Altelaar, A. F., et al. (2006). Gold-enhanced biomolecular surface imaging of cells and tissue by SIMS and MALDI mass spectrometry. *Analytical Chemistry, 78*(3), 734–742.

Altelaar, A. M., et al. (2007). High-resolution MALDI imaging mass spectrometry allows localization of peptide distributions at cellular length scales in pituitary tissue sections. *International Journal of Mass Spectrometry, 260*(2), 203–211.

Anhalt, J. P., & Fenselau, C. (1975). Identification of bacteria using mass spectrometry. *Analytical Chemistry, 47*(2), 219–225.

Arafah, B. M., et al. (2000). The dominant role of increased intrasellar pressure in the pathogenesis of hypopituitarism hyperprolactinemia, and headaches in patients with pituitary adenomas. *The Journal of Clinical Endocrinology and Metabolism, 85*(5), 1789–1793.

Balluff, B., et al. (2011). MALDI imaging identifies prognostic seven-protein signature of novel tissue markers in intestinal-type gastric cancer. *The American Journal of Pathology, 179*(6), 2720–2729.

Balog, J., et al. (2013). Intraoperative tissue identification using rapid evaporative ionization mass spectrometry. *Science Translational Medicine, 5*(194), 194ra93.

Bizzini, A., et al. (2010). Performance of matrix-assisted laser desorption ionization-time of flight mass spectrometry for identification of bacterial strains routinely isolated in a clinical microbiology laboratory. *Journal of Clinical Microbiology, 48*(5), 1549–1554.

Black, P. M., et al. (1987). Hormone production in clinically nonfunctioning pituitary adenomas. *Journal of Neurosurgery, 66*(2), 244–250.

Blondiaux, N., Gaillot, O., & Courcol, R. J. (2010). MALDI-TOF mass spectrometry to identify clinical bacterial isolates: Evaluation in a teaching hospital in Lille. *Pathologie-Biologie, 58*(1), 55–57.

Burger, P. C. (1985). Use of cytological preparations in the frozen section diagnosis of central nervous system neoplasia. *The American Journal of Surgical Pathology, 9*(5), 344–354.

Calligaris, D., et al. (2013). Mass spectrometry imaging as a tool for surgical decision-making. *Journal of Mass Spectrometry, 48*(11), 1178–1187.

Calligaris, D., et al. (2014). Application of desorption electrospray ionization mass spectrometry imaging in breast cancer margin analysis. *Proceedings of the National Academy of Sciences of the United States of America, 111*(42), 15184–15189.

Calligaris, D., et al. (2015a). Molecular typing of meningiomas by desorption electrospray ionization mass spectrometry imaging for surgical decision-making. *International Journal of Mass Spectrometry, 377*, 690–698.

Calligaris, D., et al. (2015b). MALDI mass spectrometry imaging analysis of pituitary adenomas for near-real-time tumor delineation. *Proceedings of the National Academy of Sciences of the United States of America, 112*(32), 9978–9983.

Cappabianca, P., et al. (2002). Surgical complications associated with the endoscopic endonasal transsphenoidal approach for pituitary adenomas. *Journal of Neurosurgery, 97*(2), 293–298.

Caprioli, R. M., Farmer, T. B., & Gile, J. (1997). Molecular imaging of biological samples: Localization of peptides and proteins using MALDI-TOF MS. *Analytical Chemistry, 69*(23), 4751–4760.

Carbonnelle, E., et al. (2011). MALDI-TOF mass spectrometry tools for bacterial identification in clinical microbiology laboratory. *Clinical Biochemistry, 44*(1), 104–109.

Cazares, L. H., et al. (2009). Imaging mass spectrometry of a specific fragment of mitogen-activated protein kinase/extracellular signal-regulated kinase kinase kinase 2 discriminates cancer from uninvolved prostate tissue. *Clinical Cancer Research, 15*(17), 5541–5551.

Chaurand, P., et al. (2001). Profiling proteins from azoxymethane-induced colon tumors at the molecular level by matrix-assisted laser desorption/ionization mass spectrometry. *Proteomics, 1*(10), 1320–1326.

Coons, A. H., Creech, H. J., & Jones, R. N. (1941). Immunological properties of an antibody containing a fluorescent group. *Experimental Biology and Medicine, 47*(2), 200–202.

Cornett, D. S., et al. (2006). A novel histology-directed strategy for MALDI-MS tissue profiling that improves throughput and cellular specificity in human breast cancer. *Molecular & Cellular Proteomics, 5*(10), 1975–1983.

Cornett, D. S., et al. (2007). MALDI imaging mass spectrometry: Molecular snapshots of biochemical systems. *Nature Methods, 4*(10), 828–833.

Daly, A. F., et al. (2006). High prevalence of pituitary adenomas: A cross-sectional study in the province of Liege, Belgium. *The Journal of Clinical Endocrinology and Metabolism, 91*(12), 4769–4775.

Desiderio, D. M., & Zhan, X. (2003). The human pituitary proteome: The characterization of differentially expressed proteins in an adenoma compared to a control. *Cellular and Molecular Biology (Noisy-le-Grand, France), 49*(5), 689–712.

Dhiman, N., et al. (2011). Performance and cost analysis of matrix-assisted laser desorption ionization-time of flight mass spectrometry for routine identification of yeast. *Journal of Clinical Microbiology, 49*(4), 1614–1616.

Di Ieva, A., et al. (2014). Aggressive pituitary adenomas—Diagnosis and emerging treatments. *Nature Reviews. Endocrinology, 10*(7), 423–435.

Dill, A. L., et al. (2009). Mass spectrometric imaging of lipids using desorption electrospray ionization. *Journal of Chromatography. B, Analytical Technologies in the Biomedical and Life Sciences, 877*(26), 2883–2889.

Dill, A. L., et al. (2010). Multivariate statistical differentiation of renal cell carcinomas based on lipidomic analysis by ambient ionization imaging mass spectrometry. *Analytical and Bioanalytical Chemistry, 398*(7–8), 2969–2978.

Dill, A. L., et al. (2011). Multivariate statistical identification of human bladder carcinomas using ambient ionization imaging mass spectrometry. *Chemistry, 17*(10), 2897–2902.

Eberlin, L. S., et al. (2010a). Cholesterol sulfate imaging in human prostate cancer tissue by desorption electrospray ionization mass spectrometry. *Analytical Chemistry, 82*(9), 3430–3434.

Eberlin, L. S., et al. (2010b). Discrimination of human astrocytoma subtypes by lipid analysis using desorption electrospray ionization imaging mass spectrometry. *Angewandte Chemie (International Ed. in English), 49*(34), 5953–5956.

Eberlin, L. S., et al. (2012). Classifying human brain tumors by lipid imaging with mass spectrometry. *Cancer Research, 72*(3), 645–654.

Eberlin, L. S., et al. (2013). Ambient mass spectrometry for the intraoperative molecular diagnosis of human brain tumors. *Proceedings of the National Academy of Sciences of the United States of America, 110*(5), 1611–1616.

Eberlin, L. S., et al. (2014a). Alteration of the lipid profile in lymphomas induced by MYC overexpression. *Proceedings of the National Academy of Sciences of the United States of America, 111*(29), 10450–10455.

Eberlin, L. S., et al. (2014b). Molecular assessment of surgical-resection margins of gastric cancer by mass-spectrometric imaging. *Proceedings of the National Academy of Sciences of the United States of America, 111*(7), 2436–2441.

Elsner, M., et al. (2012). MALDI imaging mass spectrometry reveals COX7A2, TAGLN2 and S100–A10 as novel prognostic markers in Barrett's adenocarcinoma. *Journal of Proteomics, 75*(15), 4693–4704.

Ezzat, S., et al. (2004). The prevalence of pituitary adenomas: A systematic review. *Cancer, 101*(3), 613–619.

Fenn, J. B., et al. (1989). Electrospray ionization for mass spectrometry of large biomolecules. *Science, 246*(4926), 64–71.

Fernandez, A., Karavitaki, N., & Wass, J. A. (2010). Prevalence of pituitary adenomas: A community-based, cross-sectional study in Banbury (Oxfordshire, UK). *Clinical Endocrinology, 72*(3), 377–382.

Ferreiro, J. A., Myers, J. L., & Bostwick, D. G. (1995). Accuracy of frozen section diagnosis in surgical pathology: Review of a 1-year experience with 24,880 cases at Mayo clinic Rochester. *Mayo Clinic Proceedings, 70*(12), 1137–1141.

Fournier, I., Day, R., & Salzet, M. (2003). Direct analysis of neuropeptides by in situ MALDI-TOF mass spectrometry in the rat brain. *Neuro Endocrinology Letters, 24*(1–2), 9–14.

Fournier, P. E., et al. (2013). Modern clinical microbiology: New challenges and solutions. *Nature Reviews. Microbiology, 11*(8), 574–585.

Garg, U., & Dasouki, M. (2006). Expanded newborn screening of inherited metabolic disorders by tandem mass spectrometry: Clinical and laboratory aspects. *Clinical Biochemistry, 39*(4), 315–332.

Gerbig, S., et al. (2012). Analysis of colorectal adenocarcinoma tissue by desorption electrospray ionization mass spectrometric imaging. *Analytical and Bioanalytical Chemistry, 403*(8), 2315–2325.

Golf, O., et al. (2015). Rapid evaporative ionization mass spectrometry imaging platform for direct mapping from bulk tissue and bacterial growth media. *Analytical Chemistry, 87*(5), 2527–2534.

Grobe, N., et al. (2012). Mass spectrometry for the molecular imaging of angiotensin metabolism in kidney. *American Journal of Physiology. Endocrinology and Metabolism, 302*(8), E1016–E1024.

Guenther, S., et al. (2015). Spatially resolved metabolic phenotyping of breast cancer by desorption electrospray ionization mass spectrometry. *Cancer Research, 75*(9), 1828–1837.

Hooff, G. P., et al. (2012). Characterization of beta-lactamase enzyme activity in bacterial lysates using MALDI-mass spectrometry. *Journal of Proteome Research, 11*(1), 79–84.

Hrabak, J., et al. (2011). Carbapenemase activity detection by matrix-assisted laser desorption ionization-time of flight mass spectrometry. *Journal of Clinical Microbiology, 49*(9), 3222–3227.

Hsu, S. M., Raine, L., & Fanger, H. (1981). A comparative study of the peroxidase-antiperoxidase method and an avidin-biotin complex method for studying polypeptide hormones with radioimmunoassay antibodies. *American Journal of Clinical Pathology, 75*(5), 734–738.

Hsu, C. C., et al. (2013). Real-time metabolomics on living microorganisms using ambient electrospray ionization flow-probe. *Analytical Chemistry, 85*(15), 7014–7018.

Ifa, D. R., & Eberlin, L. S. (2016). Ambient ionization mass spectrometry for cancer diagnosis and surgical margin evaluation. *Clinical Chemistry, 62*(1), 111–123.

Ifa, D. R., et al. (2010). Desorption electrospray ionization and other ambient ionization methods: Current progress and preview. *Analyst, 135*(4), 669–681.

Kerian, K. S., Jarmusch, A. K., & Cooks, R. G. (2014). Touch spray mass spectrometry for in situ analysis of complex samples. *Analyst, 139*(11), 2714–2720.

Kerian, K. S., et al. (2015). Differentiation of prostate cancer from normal tissue in radical prostatectomy specimens by desorption electrospray ionization and touch spray ionization mass spectrometry. *Analyst, 140*(4), 1090–1098.

Kertesz, V., et al. (2015). Profiling of adrenocorticotropic hormone and arginine vasopressin in human pituitary gland and tumor thin tissue sections using droplet-based liquid-microjunction surface-sampling-HPLC-ESI-MS-MS. *Analytical and Bioanalytical Chemistry, 407*(20), 5989–5998.

Komotar, R. J., et al. (2012). Endoscopic endonasal compared with microscopic transsphenoidal and open transcranial resection of giant pituitary adenomas. *Pituitary, 15*(2), 150–159.

Kovacs, K., Horvath, E., & Ezrin, C. (1977). Pituitary adenomas. *Pathology Annual, 12*(Pt. 2), 341–382.

Kovacs, K., Horvath, E., & Vidal, S. (2001). Classification of pituitary adenomas. *Journal of Neuro-Oncology, 54*(2), 121–127.

La Scola, B., & Raoult, D. (2009). Direct identification of bacteria in positive blood culture bottles by matrix-assisted laser desorption ionisation time-of-flight mass spectrometry. *PLoS One, 4*(11). e8041.

Lee, G. K., et al. (2012). Lipid MALDI profile classifies non-small cell lung cancers according to the histologic type. *Lung Cancer, 76*(2), 197–203.

Luxembourg, S. L., et al. (2004). High-spatial resolution mass spectrometric imaging of peptide and protein distributions on a surface. *Analytical Chemistry, 76*(18), 5339–5344.

Mandal, M. K., et al. (2012a). Application of probe electrospray ionization mass spectrometry (PESI-MS) to clinical diagnosis: Solvent effect on lipid analysis. *Journal of the American Society for Mass Spectrometry, 23*(11), 2043–2047.

Mandal, M. K., et al. (2012b). Solid probe assisted nanoelectrospray ionization mass spectrometry for biological tissue diagnostics. *Analyst, 137*(20), 4658–4661.

Mandal, M. K., et al. (2013). Biomolecular analysis and cancer diagnostics by negative mode probe electrospray ionization. *Analyst, 138*(6), 1682–1688.

Masterson, T. A., et al. (2011). Distinctive glycerophospholipid profiles of human seminoma and adjacent normal tissues by desorption electrospray ionization imaging mass spectrometry. *Journal of the American Society for Mass Spectrometry, 22*(8), 1326–1333.

Mortini, P., et al. (2005). Results of transsphenoidal surgery in a large series of patients with pituitary adenoma. *Neurosurgery, 56*(6), 1222–1233. discussion 1233.

Obiakor, I., et al. (1991). The accuracy of frozen section in the diagnosis of ovarian neoplasms. *Gynecologic Oncology, 43*(1), 61–63.

Oldfield, E. H., et al. (1985). Preoperative lateralization of ACTH-secreting pituitary microadenomas by bilateral and simultaneous inferior petrosal venous sinus sampling. *The New England Journal of Medicine, 312*(2), 100–103.

Oldfield, E. H., et al. (1991). Petrosal sinus sampling with and without corticotropin-releasing hormone for the differential diagnosis of Cushing's syndrome. *New England Journal of Medicine, 325*(13), 897–905.

Pouratian, N., et al. (2007). Outcomes and management of patients with Cushing's disease without pathological confirmation of tumor resection after transsphenoidal surgery. *The Journal of Clinical Endocrinology and Metabolism, 92*(9), 3383–3388.

Powell, S. Z. (2005). Intraoperative consultation, cytologic preparations, and frozen section in the central nervous system. *Archives of Pathology & Laboratory Medicine, 129*(12), 1635–1652.

Raappana, A., et al. (2010). Incidence of pituitary adenomas in Northern Finland in 1992–2007. *The Journal of Clinical Endocrinology and Metabolism, 95*(9), 4268–4275.

Randall, R. V., et al. (1985). Pituitary adenomas associated with hyperprolactinemia: A clinical and immunohistochemical study of 97 patients operated on transsphenoidally. *Mayo Clinic Proceedings, 60*(11), 753–762.

Rauser, S., et al. (2010). Classification of HER2 receptor status in breast cancer tissues by MALDI imaging mass spectrometry. *Journal of Proteome Research, 9*(4), 1854–1863.

Remsen, K. A., Lucente, F. E., & Biller, H. F. (1984). Reliability of frozen section diagnosis in head and neck neoplasms. *Laryngoscope, 94*(4), 519–524.

Ryu, J., et al. (2013). Lipid MALDI MS profiling accurately distinguishes papillary thyroid carcinoma from normal tissue. *Journal of Proteomics & Bioinformatics, 6*, 065–071.

Sachfer, K. C., et al. (2011). In situ, real-time identification of biological tissues by ultraviolet and infrared laser desorption ionization mass spectrometry. *Analytical Chemistry, 83*(5), 1632–1640.

Saeger, W., et al. (2016a). Clinical impact of the current WHO classification of pituitary adenomas. *Endocrine Pathology, 27*(2), 104–114.

Saeger, W., et al. (2016b). Emerging histopathological genetic parameters of pituitary adenomas: Clinical impact and recommendation for future WHO classification. *Endocrine Pathology*, *27*(2), 115–122.

Salenave, S., et al. (2004). Pituitary magnetic resonance imaging findings do not influence surgical outcome in adrenocorticotropin-secreting microadenomas. *The Journal of Clinical Endocrinology and Metabolism*, *89*(7), 3371–3376.

Santagata, S., et al. (2014). Intraoperative mass spectrometry mapping of an onco-metabolite to guide brain tumor surgery. *Proceedings of the National Academy of Sciences of the United States of America*, *111*(30), 11121–11126.

Schafer, K. C., et al. (2011). Real time analysis of brain tissue by direct combination of ultrasonic surgical aspiration and sonic spray mass spectrometry. *Analytical Chemistry*, *83*(20), 7729–7735.

Scheithauer, B. W., et al. (1986). Pathology of invasive pituitary tumors with special reference to functional classification. *Journal of Neurosurgery*, *65*(6), 733–744.

Schwartz, S. A., et al. (2005). Proteomic-based prognosis of brain tumor patients using direct-tissue matrix-assisted laser desorption ionization mass spectrometry. *Cancer Research*, *65*(17), 7674–7681.

Semple, P. L., et al. (2000). Transsphenoidal surgery for Cushing's disease: Outcome in patients with a normal magnetic resonance imaging scan. *Neurosurgery*, *46*(3), 553–558. discussion 558–559.

Seng, P., et al. (2009). Ongoing revolution in bacteriology: Routine identification of bacteria by matrix-assisted laser desorption ionization time-of-flight mass spectrometry. *Clinical Infectious Diseases*, *49*(4), 543–551.

Sparkman, R. S. (1962). Reliability of frozen sections in the diagnosis of breast lesions. *Annals of Surgery*, *155*, 924–934.

Stoeckli, M., et al. (2001). Imaging mass spectrometry: A new technology for the analysis of protein expression in mammalian tissues. *Nature Medicine*, *7*(4), 493–496.

Takats, Z., et al. (2004). Mass spectrometry sampling under ambient conditions with desorption electrospray ionization. *Science*, *306*(5695), 471–473.

Tata, A., et al. (2015). Contrast agent mass spectrometry imaging reveals tumor heterogeneity. *Analytical Chemistry*, *87*(15), 7683–7689.

Taylor, C. R. (1978). Immunoperoxidase techniques: Practical and theoretical aspects. *Archives of Pathology & Laboratory Medicine*, *102*(3), 113–121.

Thomas, A., et al. (2013). Histology-driven data mining of lipid signatures from multiple imaging mass spectrometry analyses: Application to human colorectal cancer liver metastasis biopsies. *Analytical Chemistry*, *85*(5), 2860–2866.

Trauger, S. A., Webb, W., & Siuzdak, G. (2002). Peptide and protein analysis with mass spectrometry. *Journal of Spectroscopy*, *16*(1), 15–28.

Tyrrell, J. B., et al. (1978). Cushing's disease—Selective trans-sphenoidal resection of pituitary microadenomas. *The New England Journal of Medicine*, *298*(14), 753–758.

Van Berkel, G. J., & Kertesz, V. (2013). Continuous-flow liquid microjunction surface sampling probe connected on-line with high-performance liquid chromatography/mass spectrometry for spatially resolved analysis of small molecules and proteins. *Rapid Communications in Mass Spectrometry*, *27*(12), 1329–1334.

van Veen, S. Q., Claas, E. C., & Kuijper, E. J. (2010). High-throughput identification of bacteria and yeast by matrix-assisted laser desorption ionization-time of flight mass spectrometry in conventional medical microbiology laboratories. *Journal of Clinical Microbiology*, *48*(3), 900–907.

Wang, H. Y., et al. (2012). MALDI-mass spectrometry imaging of desalted rat brain sections reveals ischemia-mediated changes of lipids. *Analytical and Bioanalytical Chemistry*, *404*(1), 113–124.

Wilson, L. B. (1905). A method for the rapid preparation of fresh tissues for the microscope. *Journal of the American Medical Association, 45*(23), 1737.

Wiseman, J. M., et al. (2005). Mass spectrometric profiling of intact biological tissue by using desorption electrospray ionization. *Angewandte Chemie (International Ed. in English), 44*(43), 7094–7097.

Yanagisawa, K., et al. (2003). Proteomic patterns of tumour subsets in non-small-cell lung cancer. *Lancet, 362*(9382), 433–439.

CHAPTER ELEVEN

Mass Spectrometry Imaging in Cancer Research: Future Perspectives

L.A. McDonnell[*,†,1], **P.M. Angel**[‡], **S. Lou**[†], **R.R. Drake**[‡]
[*]Fondazione Pisana per la Scienza ONLUS, Pisa, Italy
[†]Leiden University Medical Center, Leiden, The Netherlands
[‡]Medical University of South Carolina, Charleston, SC, United States
[1]Corresponding author: e-mail address: l.a.mcdonnell@oulook.com

Contents

1. MSI-Based Diagnostics	286
2. Biological Insights	287
3. Multimodal MSI	287
4. Targeted MSI	288
5. Summary	288
References	289

Abstract

In the last decade mass spectrometry imaging has developed rapidly, in terms of multiple new instrumentation innovations, expansion of target molecules, and areas of application. Mass spectrometry imaging has already had a substantial impact in cancer research, uncovering biomolecular changes associated with disease progression, diagnosis, and prognosis. Many new approaches are incorporating the use of readily available formalin-fixed paraffin-embedded cancer tissues from pathology centers, including tissue blocks, biopsy specimens, and tumor microarrays. It is also increasingly used in drug formulation development as an inexpensive method to determine the distributions of drugs and their metabolites. In this chapter, we offer a perspective in the current and future methodological developments and how these may open up new vistas for cancer research.

Mass spectrometry imaging (MSI) can provide the spatial distribution of hundreds of biomolecules directly from tissue (McDonnell & Heeren, 2007). Its ability to discover and localize biomolecules in tissues, without prior knowledge of their presence and in a label-free manner, makes it a powerful discovery tool (Caprioli, 2015; Walch, Rauser, Deininger, &

Hofler, 2008). In 2017, the 20th year anniversary of the first publication for mass spectrometry tissue imaging will occur, and it will be the 16th year since application of the method to tumor tissues (Caprioli, Farmer, & Gile, 1997; Stoeckli, Chaurand, Hallahan, & Caprioli, 2001). Especially in the last decade, the field has continued to grow rapidly, encompassing multiple new instrumentation innovations and expansion of target molecules beyond proteins and peptides to include lipids, metabolites, glycans, drugs, and their metabolites. MALDI MSI has already had a substantial impact in clinical research, uncovering biomolecular changes associated with disease progression, diagnosis, and prognosis (Schwamborn & Caprioli, 2010). It is also used in drug formulation development as an inexpensive method to determine the distributions of drugs and their metabolites (Prideaux & Stoeckli, 2012; Rauser, Deininger, Suckau, Hofler, & Walch, 2010).

These advancements have facilitated a move toward clinically directed studies. Clinical specimens may be investigated by MALDI IMS as tissue sections, biopsies, or tissue microarrays. Histology-guided MSI data analysis is now an established approach for biomarker discovery, and MSI-based molecular pathology has been shown to complement established histopathological methods (Casadonte & Caprioli, 2011). The integration of MSI with histology allows cell-specific molecular profiles to be extracted from the often histologically heterogeneous tumor tissues routinely encountered in the clinic.

The current state of MALDI IMS is due to significant advancements in both sample preparation and instrumentation over the last 10 years. Sample preparation techniques are now reproducible and robust, utilizing well-characterized robotics for high-throughput spray application of thin layers of chemical, enzyme, or matrix (Bouschen, Schulz, Eikel, & Spengler, 2010), or matrix sublimation (Hankin, Barkley, & Murphy, 2007). Current MALDI MSI instrumentation is predominantly microprobe analysis systems, in which a focused laser beam is used to analyze a localized area of the sample. The resulting mass spectrum is then stored along with the coordinates of the irradiated area. A number of developments have combined to enable higher speed, higher resolution, and higher sensitivity MALDI MSI, and many of these advances can be attributed to laser improvements combined with significantly improved matrix application techniques.

(i) laser spot size—determines the achievable spatial resolution of the microprobe experiment. Commercial instrumentation now routinely achieves 10 μm or better spatial resolution. Note oversampling has been used to increase the spatial resolution of microprobe experiments beyond the limits imposed by the size of the laser spot (Jurchen, Rubakhin, & Sweedler, 2005).

(ii) laser speed—laser speed has increased from 200 Hz of 8–10 years ago to commercially available lasers on MSI instruments of 2–10 kHz.

(iii) laser shots—due to improvements in sample preparation that achieve fine submicrometer crystal sizes, laser fluence no longer needs to be increased inordinately to achieve sensitive detection of analytes. The number of laser shots is minimal, moving from routinely collected 1500 laser shots a pixel of years ago to most often 100–500 shots per pixel. Current techniques use a limited number of laser shots, to deplete all signal from a single pixel. This has had a huge impact in being able to remove matrix and perform histology staining on the exact same tissue section, allowing molecular signatures to be mapped to their histopathological origin.

These technological developments now allow fast, routine imaging at spatial resolutions that now enable routine MALDI MSI ≤ 10 µm pixel size. The latest commercial instrumentation combines continuous sample stage movement with a synchronized laser raster to provide 10 kHz MS scan rate and 50 pixels per second. Tissue images can thus be produced in acquisition times that allow exploration of molecular data from the tissue near the single-cell level.

Currently, there is discussion within the field (both academic and by MSI instrument suppliers) of exploiting the rapid, unlabelled, and spatially resolved molecular analysis provided by MALDI MSI to develop a multiplexed molecular addition to conventional histopathological practice (Aichler & Walch, 2015). In this approach, MALDI MSI results need not be validated by immunohistochemistry or extraction-based methods, but the individual biomarkers or multivariate signatures are validated by independent multicentre studies using standardized methods (Dekker et al., 2014).

In this scenario, a key element is the reproducibility of the results in other clinical centres. Gurdak et al. reported a desorption electrospray ionization (DESI) interlaboratory study focused on intensity repeatability and constancy (Gurdak et al., 2014). In a similar manner, Erich et al. have recently reported a statistical analysis of on-tissue digestion-based MALDI MSI, to determine which of the many reported methods were most reproducible (Erich, Sammour, Marx, & Hopf, 2016). Dekker et al. reported the first multicentre clinical study and demonstrated that protein biomarkers of stromal activation in breast cancer could be reproducibly detected in two centres, in which different patient collections were analyzed by different analysts, using slightly different methods and instruments (Dekker et al., 2014). The MALDI MSI average mass spectra recorded in the two centres appeared very different, but the *univariate* biomarkers of stromal activation in breast cancer reported by the discovery dataset (recorded at the Helmoltz Zentrum Munich) were also statistically significant in the validation dataset (recorded at LUMC). Or in other words, while >95% of the mass spectra exhibited differences in intensity, the clinical biomarkers of interest, if detected, were statistically significant in both centers. The importance of the reproducibility of the entire MSI dataset depends on how much of the dataset is used in the clinical assay, and which metric is used.

It may be argued that for biomarker discovery-type experiments further improvements are needed to better utilize the higher spatial resolution measurements now achievable: an MSI datasets measured with a pixel size of 10 μm will generate a total data load 100× greater than if it were measured at 100 μm. Once the biomarker ions/signatures are defined, however, the high speed and high spatial resolution capabilities can be fully utilized, e.g., for tumor tissues associated with skin cancers and colon polyps that represent high volume analysis needs in pathology facilities. For example, emerging approaches for evaluating panels of metabolites (Ly et al., 2016) or *N*-glycans (Heijs et al., 2016; Powers et al., 2014) directly in FFPE tissues offer new cancer biomarker profiling capabilities using rapid MALDI. Combined with reproducible and automated sample preparation protocols, the rapid rate of progress in the field is expected to continue, enabling more patient samples to be analyzed and with higher spatial detail. There are several distinct areas in which MSI is expected to grow in the near future.

1. MSI-BASED DIAGNOSTICS

MSI-based diagnostics require that the assays are reproducible in different clinical centers (as discussed earlier). Once successful, this would enable the full capabilities to be leveraged for improved patient and disease stratification as diagnosis would not be based solely on detailed histopathological analysis, but combined with complementary MSI data, immunohistochemical, and molecular assays. This would enable biomarkers from proteins, metabolites, lipids, and glycans to be used for improved personalization of healthcare.

The Biotyper, a MALDI-based identification of microorganisms based on standardized MALDI MS analysis of colonies followed by classification, is now an FDA-approved technology and has had a large clinical impact (Faron et al., 2015), primarily through the reduction in health care costs associated with faster and less expensive methods for microorganism identification. The transformative success of the Biotyper in the clinical microbiology laboratory testifies to the promise of a MALDI-based Tissuetyper for clinical pathology applications. Arguably the biggest challenge for the clinical translation of MSI is that it has been developed mainly by analytical chemists, not clinicians, and so the focus has been on the development of innovative methods rather than standardization. Significant input from pathologists is needed in regards to defining where a Tissuetyper approach could tangibly improve diagnosis or analysis workflows for routine tissue sample analysis performed daily in pathology laboratories. This is beginning to change (Crutchfield, Thomas, Sokoll, & Chan, 2016; Kriegsmann, Kriegsmann, & Casadonte, 2015), and positive demonstrations are expected within the next 5 years.

2. BIOLOGICAL INSIGHTS

The molecular histology capabilities of MALDI MSI, especially the identification of tumor subpopulation associated with patient outcome, may be used to direct microdissection followed by either quantitative proteomics, metabolomics, or transcriptomics analysis. The development of liquid extraction surface analysis methods (Wisztorski et al., 2013) as well as MSI-directed laser capture dissection (Longuespee et al., 2014), both of which provide protein samples from localized regions for LC-MS/MS characterization, begin to provide the capabilities necessary to bridge MSI and established proteomics/metabolomics methods. An anticipated advancement here is the ability to perform targeted experiments on multiple species, gaining further specificity of tandem mass spectrometry data. The ability to perform multiple reaction monitoring experiments by MSI would provide improved determinations of the global changes in molecular interactions across tissue, including the monitoring of targeted interaction networks, metabolic, and signal transduction pathways.

3. MULTIMODAL MSI

MALDI MSI is routinely used to analyze metabolite, lipids, peptides, proteins, and N-linked glycans. An emerging paradigm is to combine MSI datasets from different molecular classes for a more complete molecular description. The combination of biomarkers from different molecular classes should lead to improved classifier performance and could reveal, e.g., characteristic dysregulations in metabolism involving specific changes to key metabolic enzymes. In particular for peptides and glycans (Heijs et al., 2016), identification of specific tumor targets could be adapted to developing antibody or lectin-based reagents that could be applied to in vivo imaging detection by MRI or related methods. Studies that link clinical imaging modalities used routinely in radiology with molecular analysis of the same tissues using MSI approaches are not common, but warranted for future emphasis.

4. TARGETED MSI

The development of imaging mass cytometry, in which antibodies are functionalized with polymers containing isotopically purified lanthanides and then analyzed by laser ablation inductively coupled plasma mass spectrometry (LA-ICP-MS), enables submicron spatial resolution MSI of targeted antigens (Giesen et al., 2014). Imaging mass cytometry can be considered highly multiplexed immunohistochemistry, in which the use of different lanthanides enables >30 antigens to be simultaneously analyzed. Imaging mass cytometry could have a significant impact in cancer research, because it is based on the immunohistochemistry protocols widely established in pathology. The biggest challenge, arguably, pertains to the high cost and potential reproducibility (Baker, 2015) of such highly multiplexed assays based on antibodies.

5. SUMMARY

MSI methodology has significantly advanced over the last 10 years to be reproducible and robust with faster acquisition and higher spatial resolutions. Especially for cancer, this has facilitated an upsurge in the number of large clinical studies examining prognosis, diagnosis, and disease status. In order to continue to move toward being placed in a clinical setting, further improvements need to be made toward decreasing sample preparation time and expertise, instrumentation that is easy to use and maintain, and standardized production of libraries of tissue signatures for routine matching to patient samples. With current advancing technologies in the era of big data and "omic" level studies, new avenues of research using MALDI MSI will continue to emerge.

REFERENCES

Aichler, M., & Walch, A. (2015). MALDI imaging mass spectrometry: Current frontiers and perspectives in pathology research and practice. *Laboratory Investigation, 95*, 422–431.

Baker, M. (2015). Reproducibility crisis: Blame it on the antibodies. *Nature, 521*, 274–276.

Bouschen, W., Schulz, O., Eikel, D., & Spengler, B. (2010). Matrix vapor deposition/recrystallization and dedicated spray preparation for high-resolution scanning microprobe matrix-assisted laser desorption/ionization imaging mass spectrometry (SMALDI-MS) of tissue and single cells. *Rapid Communications in Mass Spectrometry, 24*, 355–364.

Caprioli, R. M. (2015). Imaging mass spectrometry: Enabling a new age of discovery in biology and medicine through molecular microscopy. *Journal of the American Society for Mass Spectrometry, 26*, 850–852.

Caprioli, R. M., Farmer, T. B., & Gile, J. (1997). Molecular imaging of biological samples: Localization of peptides and proteins using MALDI-TOF MS. *Analytical Chemistry, 69*, 4751–4760.

Casadonte, R., & Caprioli, R. M. (2011). Proteomic analysis of formalin-fixed paraffin embedded tissue by MALDI imaging mass spectrometry. *Nature Protocols, 6*, 1695–1709.

Crutchfield, C. A., Thomas, S. N., Sokoll, L. J., & Chan, D. W. (2016). Advances in mass spectrometry-based clinical biomarker discovery. *Clinical Proteomics, 13*, 1.

Dekker, T. J., Balluff, B. D., Jones, E. A., Schone, C. D., Schmitt, M., Aubele, M., et al. (2014). Multicenter matrix-assisted laser desorption/ionization mass spectrometry imaging (MALDI MSI) identifies proteomic differences in breast-cancer-associated stroma. *Journal of Proteome Research, 13*, 4730–4738.

Erich, K., Sammour, D. A., Marx, A., & Hopf, C. (2016). Scores for standardization of on-tissue digestion of formalin-fixed paraffin-embedded tissue in MALDI-MS imaging. *Biochimica et Biophysica Acta.* http://dx.doi.org/10.1016/j.bbapap.2016.08.020 (e-pub ahead of print).

Faron, M. L., Buchan, B. W., Hyke, J., Madisen, N., Lillie, J. L., Granato, P. A., et al. (2015). Multicenter evaluation of the bruker MALDI biotyper CA system for the identification of clinical aerobic Gram-negative bacterial isolates. *PloS One, 10*e0141350.

Giesen, C., Wang, H. A. O., Schapiro, D., Zivanovic, N., Jacobs, A., Hattendorf, B., et al. (2014). Highly multiplexed imaging of tumor tissues with subcellular resolution by mass cytometry. *Nature Methods, 11*, 417–422.

Gurdak, E., Green, F. M., Rakowska, P. D., Seah, M. P., Salter, T. L., & Gilmore, I. S. (2014). VAMAS interlaboratory study for desorption electrospray ionization mass spectrometry (DESI MS) intensity repeatability and constancy. *Analytical Chemistry, 86*, 9603–9611.

Hankin, J. A., Barkley, R. M., & Murphy, R. C. (2007). Sublimation as a method of matrix application for mass spectrometric imaging. *Journal of the American Society for Mass Spectrometry, 18*, 1646–1652.

Heijs, B., Holst, S., Briaire-de Bruijn, I. H., van Pelt, G. W., de Ru, A. H., van Veelen, P. A., et al. (2016). Multimodal mass spectrometry imaging of N-glycans and proteins from the same tissue section. *Analytical Chemistry, 88*, 7745–7753.

Jurchen, J. C., Rubakhin, S. S., & Sweedler, J. V. (2005). MALDI-MS imaging of features smaller than the size of the laser beam. *Journal of the American Society for Mass Spectrometry, 16*, 1654–1659.

Kriegsmann, J., Kriegsmann, M., & Casadonte, R. (2015). MALDI TOF imaging mass spectrometry in clinical pathology: A valuable tool for cancer diagnostics. *International Journal of Oncology, 46*, 893–906.

Longuespee, R., Fleron, M., Pottier, C., Quesada-Calvo, F., Meuwis, M. A., Baiwir, D., et al. (2014). Tissue proteomics for the next decade? Towards a molecular dimension in histology. *OMICS, 18*, 539–552.

Ly, A., Buck, A., Balluff, B., Sun, N., Gorzolka, K., Feuchtinger, A., et al. (2016). High-mass-resolution MALDI mass spectrometry imaging of metabolites from formalin-fixed paraffin-embedded tissue. *Nature Protocols, 11*, 1428–1443.

McDonnell, L. A., & Heeren, R. M. (2007). Imaging mass spectrometry. *Mass Spectrometry Reviews, 26*, 606–643.

Powers, T. W., Neely, B. A., Shao, Y., Tang, H., Troyer, D. A., Mehta, A. S., et al. (2014). MALDI imaging mass spectrometry profiling of N-glycans in formalin-fixed paraffin embedded clinical tissue blocks and tissue microarrays. *PloS One, 9*e106255.

Prideaux, B., & Stoeckli, M. (2012). Mass spectrometry imaging for drug distribution studies. *Journal of Proteomics, 75*, 4999–5013.

Rauser, S., Deininger, S. O., Suckau, D., Hofler, H., & Walch, A. (2010). Approaching MALDI molecular imaging for clinical proteomic research: Current state and fields of application. *Expert Review of Proteomics, 7*, 927–941.

Schwamborn, K., & Caprioli, R. M. (2010). Molecular imaging by mass spectrometry— Looking beyond classical histology. *Nature Reviews Cancer, 10*, 639–646.

Stoeckli, M., Chaurand, P., Hallahan, D. E., & Caprioli, R. M. (2001). Imaging mass spectrometry: A new technology for the analysis of protein expression in mammalian tissues. *Nature Medicine, 7*, 493–496.

Walch, A., Rauser, S., Deininger, S. O., & Hofler, H. (2008). MALDI imaging mass spectrometry for direct tissue analysis: A new frontier for molecular histology. *Histochemistry and Cell Biology, 130*, 421–434.

Wisztorski, M., Fatou, B., Franck, J., Desmons, A., Farre, I., Leblanc, E., et al. (2013). Microproteomics by liquid extraction surface analysis: Application to FFPE tissue to study the fimbria region of tubo-ovarian cancer. *Proteomics. Clinical Applications, 7*, 234–240.

INDEX

Note: Page numbers followed by "*f*" indicate figures, and "*t*" indicate tables.

A

Active pharmaceutical ingredient (API), 143–144, 146–147
ADCs. *See* Antibody–drug conjugates (ADCs)
Adenocarcinoma, colorectal, 87
Adenomas, pituitary, 257–259
Aerosol, 247–248
Alcian blue (AB), 96–97
Ambient ionization mass spectrometry, 232–233, 262
Ambient ionization method, 250
Ambient mass spectrometry, 252–253
Angiogenesis, 7–8
Antibody, anticarbohydrate, 95–96
Antibody–drug conjugates (ADCs), 155–156
Anticancer
  drugs, 7–8
  nanoparticles, 146*t*
Anticarbohydrate antibodies, 95–96
API. *See* Active pharmaceutical ingredient (API)
Atmospheric pressure ionization mass spectrometry, 261–271

B

BBB. *See* Blood–brain barrier (BBB)
Bcrp1. *See* Breast cancer resistance protein (Bcrp1)
Biomarker, 86–87, 111, 284–288
  for efficacy, 144–145
  prednisone-responsive, 125–126
Biopsies
  pretherapeutic, 9–12
  tumor, 9–10
Biosynthesis, *N*-linked glycan, 89
Biotyper, 286–287
Bipolar electrosurgery, 243–244
Blood–brain barrier (BBB), 143–144
  penetration, 150–153

Breast cancer, 9–10
  margin assessment, 241
  metastasis in, 218*f*
  tissues, 273*f*
Breast cancer resistance protein (Bcrp1), 152–153

C

Cancer metabolism, 138–139
Cancer research
  applications to, 78–82
  drug imaging in, 50–53
    DESI-MSI for, 252–253
  glycan MSI in, 47–50
  lipid MSI in, 45–46
  protein MSI in, 33–45
Cancer stem cells (CSCs), 46
Candidate drug, 135–137
Carrier-mediated drugs, 146–147
Cavitron ultrasonic surgical aspirator (CUSA), 249–250, 249*f*
Central nervous system (CNS), 143–144
Chemoresponse, 39
Chemotherapy, 134
Cholangiocarcinoma, 69, 80–82, 81*f*
Cholesterol, 69, 73–74, 78–83, 79*f*
Cholesterol ester, 69, 75–77, 78*f*
CID. *See* Collision-induced dissociation (CID)
Clonal evolution of cancer, 205*f*
Coagulation artifacts, of transurethral resections specimen, 15–16, 15*f*
Collision-induced dissociation (CID), 86–87
Colorectal adenocarcinoma, 87
CSCs. *See* Cancer stem cells (CSCs)
CUSA. *See* Cavitron ultrasonic surgical aspirator (CUSA)
α-Cyano-4-hydroxycinnamic acid (CHCA), 43

D

DCIS. *See* Ductal carcinoma in situ (DCIS)
DESI. *See* Desorption electrospray ionization (DESI)
Design-make-test-analyze (DMTA) cycle, 144–145
Desorption electrospray ionization (DESI), 33, 232–240, 262, 271, 285–286
    data analysis workflow, 237*f*
    DESI-MS, 262–269, 269–270*f*
    experiment, 236*f*
        imaging, 233–234, 234*f*
    mechanism, 232*f*
    spectrum of colorectal adenocarcinoma, 238*f*
    tissue analysis, 233
Desorption electrospray ionization mass spectrometry (DESI-MS), 232–233, 239–240, 242
Desorption electrospray ionization mass spectrometry imaging (DESI-MSI), 207–208, 219, 222, 233–235, 237–241
    for drug imaging in cancer research, 252–253
    human colorectal sample analysis, 239*f*
    lipids in cancer by, 45–46
Detectable disorders, MS/MS, 263–268*t*
Diathermy, surgical, 242–244
2,5-Dihydroxybenzoic acid (DHB), 43
DNA-damaging radiotherapy, 134
Drug
    attrition, 134
    distribution, MSI analysis, 138–139
    quantitation, 141–142
    radiolabeled, 146–147
Drug delivery, 145–147
    pharmacokinetic principles in, 135–137
    technologies, 134
Drug discovery and development, 134–135
Drug imaging, in cancer research, 50
    MALDI-MSI on tissue sections, 51–52
Ductal carcinoma in situ (DCIS), 9

E

Electrospray ionization (ESI), 261–262
    desorption, 232–240

Electrosurgery, 243–244
Endocervical mucinous borderline tumors, 17–19
Endogenous masses, 144–145
Endogenous molecular targets, 139–140
Endogenous molecules, 138
Endoplasmic reticulum (ER), 89
Enhanced permeability retention (EPR) effect, 145–147
Enzymatic digestion, in situ, 179–180
Enzymes, intracellular, 139
EPR effect. *See* Enhanced permeability retention (EPR) effect
ESI. *See* Electrospray ionization (ESI)

F

Fatty acids (FAs), 69
Fatty acid synthase (FASN) inhibitors, 138–139
FFPE. *See* Formalin-fixation and paraffin-embedding (FFPE)
Fingerprints, metabolic, 125–126
Formalin-fixed paraffin-embedded (FFPE) tissue, 86–87, 120–124, 175
Formalin-fixed paraffin-embedded TMA
    classification, 56–57
    construction, 176–178, 176*f*
    MALDI IMS on, application of, 188–192
Fourier transform ion cyclotron resonance (FT-ICR) mass spectrometers, 181–182
Fresh-frozen-*vs.* formalin-fixed paraffin-embedded tissue samples, 120–124
Fucosylation, 105–106

G

Gastric cancer, MSI spectra of, 214–215*f*
Girard T reagent, 139
Glycans, 88*f*, 93*f*, 106–108.
    See also N-glycans; N-linked glycans
    biosynthesis, N-linked, 89
    isomer problem, 105–106
    mammalian, 88
    MSI in cancer research, 47–50, 49*f*
    N-linked, and cancer, 89–92
    normal tissue, 103–104
    visualization in tissues, 95–96
Glycobiology, 86–87

Index

Glycomics, 107–108
Glycoproteins, 86–87, 89–92
  carrier, 86–87, 106–107
  functions and regulation of, 87–88
  mucin, 97
  N-linked, 95–96
  O-linked, 95–96
Glycosylation, 47
  function and types, 87–88
Gold-assisted IMS
  gold-coated glass slides preparation, 75
  gold deposition on tissue section, 77
  LDI data acquisition of a CBS-Au-coated tissue section, 77–78, 78f
  sodium salt deposition, 75–76
  tissue deposition on ITO/gold-coated slides, 75
Gold-coated glass slides preparation, 75

H

HER2 expression, heterogeneous, 203f
Heterogeneity. *See also* Tumor heterogeneity
  interpatient, 29
  intertumoral, 126–127
  intratumor, 29, 126–127, 203–204, 203f
  phenotypic, 5–7
Heterogeneous HER2 expression, 203f
High-grade (HG) cancer, 16–17
High-mannose N-glycans, 102–103
Histochemical techniques, 119–120
Histochemistry stains, 96–97
Human gastric mucosa, 154, 155f
Hypoxia, 8, 150

I

IHC. *See* Immunohistochemistry (IHC)
Imaging mass spectrometry (IMS), 68–69, 175. *See also* Gold-assisted IMS
  MALDI, 175, 178f, 181–183
  specificity in, 82
Immune cells, innate, 7
Immunohistochemical techniques, 119–120
Immunohistochemistry (IHC), 39, 206
IMS. *See* Imaging mass spectrometry (IMS)
Inductively coupled plasma mass spectrometry (ICP-MS), 154
Inhibitors, fatty acid synthase, 138–139
Innate immune cells, 7
In situ enzymatic digestion, 179–180
In situ metabolomics, by MALDI imaging, 119–120
Intertumoral heterogeneity, 126–127
Intracellular enzymes, 139
Intracranial neoplasms, 257–258
Intraoperative mass spectrometry, 240–244
Intratumor heterogeneity (ITH), 6f, 7, 11f, 29, 126–127, 203–206, 203f
  clinical relevance of, 206
  on different molecular levels, investigation, 222
  genetic diversity, 221f
  investigating the degree of, 219–222
  phenotypic, 5–7
  revealing, by clustering, 213–217
  supervised classification of, 218–219, 220f
  techniques to study spatial organization of, 206–207
In vivo efficacy models, 148–149t
Ionization, desorption electrospray, 232–240
Ion mobility mass spectrometry, 99–100
ITH. *See* Intratumor heterogeneity (ITH)
ITO-coated slides, tissue deposition on, 70

K

Kidney cancer, 17–19

L

Laser ablation inductively coupled plasma mass spectrometry (LA-ICP-MS), 154, 288
Laser desorption ionization-mass spectrometry (LDI-MS), 247–248, 248f
LDA. *See* Linear discriminant analysis (LDA)
LDI-MS. *See* Laser desorption ionization-mass spectrometry (LDI-MS)
Lectins, 95–96
LESA. *See* Liquid extraction surface analysis (LESA)
Linear discriminant analysis (LDA), 184–185
Lipid, 68–69, 82
  in cancer by
    DESI-MSI, 45–46
    SIMS-MSI, 46
  MSI in cancer research, 45–46

Liquid extraction surface analysis (LESA), 150–151, 155–156
Low-grade (LG) cancer, 16–17
Lymph node metastasis (LNM), 38, 217
Lymphoma, non-Hodgkin, 13–14f

M

MALDI. *See* Matrix-assisted laser desorption/ionization (MALDI)
Malignant melanoma (MM), 51–52
Malignant tumor, TNM classification of, 17–19
Mammalian glycans, 88
Mass spectrometry (MS)
  ambient ionization, 232–233, 262
  atmospheric pressure ionization, 261–271
  in clinical usage, 261–275
  detectable disorders, 263–268t
  Fourier transform ion cyclotron resonance, 181–182
  intraoperative, 240–244
  ion mobility, 99–100
Mass spectrometry imaging (MSI), 28–29, 135, 283–284. *See also* Matrix-assisted laser desorption/ionization mass spectrometry imaging (MALDI MSI)
  advantages, 29–30, 207–208
  applications, 223–224
    investigating the degree of ITH, 219–222
    investigation of ITH on different molecular levels, 222
    revealing ITH by clustering, 213–217, 214–215f
    supervised classification of ITH, 218–219
  based diagnostics, 286–287
  biological tissue, 208f
  in cancer research, 33–45
  data analysis
    dimension reduction, 56–57
    FFPE-TMAs classification, 56–57
    peak detection, 55–56
    spatial information, 53–55, 54f
  DESI, 33, 207–208
  drug safety study, 143–144
  errors cause by
    ill-defined sample groups, 16–19
    sample inherent factors, 9–12
    sample preparations, 12–16
    tissue inherent factors, 4–9
  future of, 57
  importance of pathology in, 2–4
  integration of, 284
  MALDI, 12, 119–120
  MALDI-FT-ICR, 32
  MALDI-MSI, 34–36, 41–42, 50
    advantages, 36
    on 3D tissue cultures, 53
    on tissue sections, 51–52
    on whole body sections, 53
    work flow, 35f
  MALDI-TOF, 31–32
  metabolite distributions by, 210f
  molecular pathology, 57
  multimodal, 287–288
  multivariate data analysis strategies in, 209–212
    projection methods, 212
    supervised classification, 212
    unsupervised analysis, 211
  in oncology drug discovery
    BBB penetration, 150–153
    biodistribution, 137–138
    biomarkers for efficacy, 144–145
    clinical translation, 156–157
    drug delivery, 145–147
    drug distribution, 138–139
    drug quantitation, 141–142
    metrology, 160–162
    pharmacokinetic-pharmacodynamic relationship, 135–137
    sample preparation, 139–141
    small molecules, 153–156
    spatial resolution, 159–160
    spheroids, applications, 157–159, 159f
    toxicity and safety assessment, 142–144
    tumor metabolism, 138–139
    tumor microenvironment, 147–150
    in vivo efficacy models, 148–149t
peptide
  analysing chemoresponse by, 39
  characterisation of intra-and intertumor variability by, 41–42

identification, 44–45
identification of diagnostic and
 prognostic markers by, 39–41, 40f
mass analysers, 44
practical considerations for proteolytic,
 42–43
prediction of metastasis by,
 37–39
spatial resolution, 43–44
tissue types by, 36–37
tumor margins determination by, 37
principles, 30–33
quantitative, 137
SIMS-TOF, 32–33
targeted, 288
tumor heterogeneity study, 207–209
MATH. *See* Mutant-allele tumor
 heterogeneity (MATH)
Matrix-assisted laser desorption/ionization
 (MALDI), 31, 68–69
 imaging, 3–4, 7–8, 12, 16–17,
 124–126, 242
 in situ metabolomics by, 119–120
 MALDI-FT-ICR, 32, 105–106
 N-glycans and peptides detection by,
 94f
Matrix-assisted laser desorption/ionization
 imaging mass spectrometry
 (MALDI-IMS). *See* Matrix-assisted
 laser desorption/ionization mass
 spectrometry imaging (MALDI
 MSI)
Matrix-assisted laser desorption/ionization
 mass spectrometry (MALDI MS),
 271–276
Matrix-assisted laser desorption/ionization
 mass spectrometry imaging (MALDI
 MSI), 50, 86–87, 119–124,
 126–127, 141–142, 150–151, 157,
 175, 207–208, 272–276, 283–288
 on FFPE TMAs, application, 188–192
 on tissue sections, 51–52
Matrix-assisted laser desorption/ionization
 (MALDI) TOF, 31–32
 MSI, 152f
Metabolic changes, 118
Metabolic fingerprints, 125–126
Metabolism, tumor, 138–139

Metabolite, 117–118, 120–124, 135–137,
 240–241
 tissue section imaging, 123f
Metabolome, 117–119
Metabolomics
 in cancer, 117–119
 in situ, 119–120
 untargeted, 119
Metastasis
 in breast cancer, 218f
 lymph nodes, 217
Metrology, for MS imaging, 160–162
Microprobe mode MSI, 30
MM. *See* Malignant melanoma (MM)
Monopolar electrosurgery, 243–244
Monopolar REIMS, 243f
MS. *See* Mass spectrometry (MS)
MSI. *See* Mass spectrometry imaging (MSI)
Mucin glycoprotein, 97
Multimodal MSI, 287–288
Multitumor TMA, 108–110
Multivariate data analysis, strategies in MSI,
 209–212
 projection methods, 212
 supervised classification, 212
 unsupervised analysis, 211
Mutant-allele tumor heterogeneity
 (MATH), 220
MVA methods, 209–212, 215–216
Myxofibrosarcoma, 7
Myxoid liposarcoma, 7

N

Nanomedicines, 134
Necrosis, 8
Neoplasms, intracranial, 257–258
Neuroendocrine tumor, 69
Neurosurgery, 270f
Next-generation sequencing (NGS)
 technologies, 204
N-linked glycans
 biosynthesis, 89
 branching and sialylation, 104–105
 and cancer, 89–92
 colocalization of, 94f
 core fucosylated, biantennary, 91f
 detection, 90f
 by MALDI-FT-ICR, 94f

N-linked glycans (*Continued*)
  detection by MALDI imaging
    glycan visualization in tissues, 95–96
    histochemistry stains, 96–97
    matrix and instrumentation choices for N-glycans, 98–100
    peptide N-glycosidase F, 97–98
    structural confirmation, 100–101
    tissue sources, 92–95
  distribution linked with histopathology
    fucosylation and the glycan isomer problem, 105–106
    high-mannose *N*-glycans, 102–103
    N-glycan branching and sialylation, 104–105
    nontumor stroma and normal tissue glycans, 103–104
    structural classes, 101–102
  high-mannose, 90*f*, 102–103
  imaging, 86–87
  linkage to genomic studies, 110
  MALDI-MSI data for a tissue microarray, 109*f*
  matrix and instrumentation choices for, 98–100
  MSI data, potential clinical diagnostic applications of, 110–111
  nonfucosylated, biantennary, 92*f*
  paucimannose, 90*f*
  structures, 86–87
Non-Hodgkin lymphoma, 13–14*f*
Noninvasive papillary urothelial carcinoma, 17–19
Nonsmall-cell lung carcinoma (NSCLC), 181–182, 190
Nonstem cancer cells (NSCCs), 46
Nontumor stroma, 103–104
Normal tissue glycans, 103–104
n-PALDI (nanoparticle-assisted laser desorption ionization) protocol, 52
NSCLC. *See* Nonsmall-cell lung carcinoma (NSCLC)

O

OCT. *See* Optimal cutting temperature (OCT)
Oleic acid, 262
O-linked glycoproteins, 95–96

Oncology, 134
  MSI drug safety study, 143–144
Oncology bioscience and drug discovery, 160–161
Oncology drug discovery, MSI in
  BBB penetration, 150–153
  biodistribution, 137–138
  biomarkers for efficacy, 144–145
  clinical translation, 156–157
  drug delivery, 145–147
  drug distribution, 138–139
  drug quantitation, 141–142
  metrology, 160–162
  pharmacokinetic-pharmacodynamic relationship, 135–137
  sample preparation, 139–141
  small molecules, 153–156
  spatial resolution, 159–160
  spheroids, applications, 157–159, 159*f*
  toxicity and safety assessment, 142–144
  tumor metabolism, 138–139
  tumor microenvironment, 147–150
  in vivo efficacy models, 148–149*t*
Oncology-targeted approach, 145–146
Optimal cutting temperature (OCT), 234–235
Ovarian cancer, glycan analysis of, 49, 49*f*

P

Papillary thyroid carcinoma (PTC), 38
PAS stain. *See* Periodic acid schiff (PAS) stain
PCA. *See* Principal component analysis (PCA)
Penning trap, 32
Peptide
  detection by MALDI-FT-ICR, 94*f*
  identification, 186–188
Peptide mass fingerprinting (PMF), 44–45
Peptide MSI, 106–108
  analysing chemoresponse by, 39
  characterisation of intra-and intertumor variability by, 41–42
  identification of diagnostic and prognostic markers by, 39–41, 40*f*
  practical considerations
    identification, 44–45
    mass analysers, 44
    spatial resolution, 43–44

prediction of metastasis by, 37–39
proteolytic, practical considerations for, 42–43
tissue types by, 36–37
tumor margins determination by, 37
Peptide N-glycosidase F (PNGaseF), 97–98
digestion, 31
Periodic acid schiff (PAS) stain, 96–97
PESI. See Probe electrospray ionization (PESI)
PESI-MS. See Probe electrospray ionization mass spectrometry (PESI-MS)
Pharmacodynamics (PD), 135–137
Pharmacokinetic–pharmacodynamic–efficacy (PK/PD/E) relationship, 135–137
Pharmacokinetic–pharmacodynamic (PKPD) relationship, 135–137
Pharmacokinetics (PK), 135–137
Phenotypic intratumor heterogeneity, 5–7, 6f, 11f
PIN. See Prostate intraepithelial neoplasia (PIN)
Pituitary adenomas, 257–259
Pituitary gland, 272–274
Pituitary tumors, 260
pLSA. See Probabilistic latent semantic analysis (pLSA)
PMF. See Peptide mass fingerprinting (PMF)
Prednisone-responsive biomarkers, 125–126
Pretherapeutic biopsies, 9–12
Principal component analysis (PCA), 53–57, 184–185, 212, 237–239
Principal component analysis discriminant analysis (PCA-DA) approach, 184–185
Probabilistic latent semantic analysis (pLSA), 212, 215
Probe electrospray ionization (PESI), 251–252
Probe electrospray ionization mass spectrometry (PESI-MS), 251–252
Prostate cancer, 10–12, 237–239
Prostate intraepithelial neoplasia (PIN), 237–239
Protein glycosylation, 47
Protein MSI, in cancer research, 33–45
Proteomic analysis, 29

PTC. See Papillary thyroid carcinoma (PTC)
Pulmonary adenocarcinoma, 3

Q

Quantitative MSI, 138
Quantitative whole body autoradiography (QWBA), 137

R

Radiolabeled drug, 146–147
Radiotherapy, DNA-damaging, 134
Rapid evaporative ionization mass spectrometry (REIMS), 242–243, 243–244f, 262, 269–271
data analysis workflow, 250f
instrumentation, 244–252, 245–251f
Receiver operating characteristic (ROC) curves, 185
REIMS. See Rapid evaporative ionization mass spectrometry (REIMS)

S

Secondary ion mass spectrometry (SIMS), 159–160, 207–208
SIMS-MSI, lipids in cancer by, 46
SIMS-TOF, 32–33
Sensitivity, 68–69, 82–83
Silver-assisted IMS
fast optimization of new tissue sections, 74–75
LDI data acquisition of a silver-coated tissue section, 73–74
silver-coated glass slides preparation, 70
silver deposition on tissue section, 70–73
tissue deposition on silver/ITO-coated slides, 70
Silver-coated glass slides preparation, 70
Silver deposition, on tissue section, 70–73
SIMS. See Secondary ion mass spectrometry (SIMS)
Sodium salt deposition, 75–76
Spheroids, 157–159, 159f
Sputtering system, 69–70, 72, 74–75, 77
Surgical diathermy, 242–244

T

Targeted MSI, 288
Therapeutic biopsies, 9–12

Therapeutic targets, 138–139
Therapy response prediction and prognosis, 125–126
3D tissue culture, MALDI-MSI on, 53
Tissue-based disease classification, 124–125
Tissue biopsy, 192–193
Tissue deposition
  gold deposition on, 77
  on ITO/gold-coated slides, 75
  on silver/ITO-coated slides, 70
Tissue inherent factors, 4–9
Tissue microarrays (TMAs), 35–36, 86–87, 175
  FFPE TMA construction, 176–178, 176f
  MALDI IMS analysis of, 178f, 181–183
    cutting, deparaffinization, and AR, 178–179
    data analysis, 183–185, 183–184f
    matrix deposition, 180–181
    in situ enzymatic digestion, 179–180
  multitumor, 108–110
  preparation of the donor block, 175–176
Tissues, glycan visualization in, 95–96
Tissuetyper approach, 286–287
TMAs. See Tissue microarrays (TMAs)
Triacylglycerols (TAGs), 69
Trypsin, 179

Tumor
  biopsies, 9–10
  metabolism, 138–139
  microenvironment, 7–8, 147–150
  pituitary, 260
Tumor heterogeneity, 5–7, 6f, 11f, 202–207, 203f
  advantages of MSI to study, 207–209
Tumorigenesis, 89–92

U

Untargeted metabolomics, 119
Urothelial carcinoma, noninvasive papillary, 17–19

V

Venturi easy ambient sonic-spray ionization (V-EASI), 249–250
Visualization technique, imaging and, 259–261

W

Warburg effect, 118, 138–139
Whole body autoradiography (WBA), 53, 252
Wilson's disease, 158f

CPI Antony Rowe
Chippenham, UK
2017-02-10 21:03